ASSEMBLY LINE – THEORY AND PRACTICE

Edited by **Waldemar Grzechca**

Assembly Line – Theory and Practice
Edited by Waldemar Grzechca

Published by InTech
Janeza Trdine 9, 51000 Rijeka, Croatia

Publishing Process Manager Romina Krebel

Technical Editor Teodora Smiljanic

Cover Designer InTech Design Team

Additional hard copies can be obtained from orders@intechopen.com

Assembly Line – Theory and Practice, Edited by Waldemar Grzechca
p. cm.
ISBN 978-953-307-995-0

For EU product safety concerns:
IN TECH d.o.o., Prolaz Marije Krucifikse Kozulić 3, 51000 Rijeka, Croatia,
info@intechopen.com or visit our website at intechopen.com.

Meet the editor

Waldemar Grzechca was born in Gliwice in Poland. He graduated at the Silesian University of Technology in 1989 and at the same time he started working at the Department of Automation Control, Electronics and Computer Science. His research was focused on accuracy of robot trajectories and robot applications in manufacturing companies. Next, he has moved his research and interest to manufacturing systems. He studied different structures of machines (sequencing and scheduling in single machine and parallel machines configurations, job shop system, flow shop system, etc.). He is especially interested in assembly lines balancing problem. He investigates different types of lines and focuses on estimation of final results of balance of single and two-sided lines. He is author of more than 60 conference and journal papers which deal with assembly lines problems.

Contents

Preface

An assembly line is a manufacturing process in which parts are added to a product in a sequential manner using optimally planned logistics to create a finished product in the fastest possible way. It is a flow-oriented production system where the productive units performing the operations, referred to as stations, are aligned in a serial manner. The work pieces visit stations successively as they are moved along the line usually by some kind of transportation system, e.g. a conveyor belt. Assembly lines are mostly designed for a sequential organization of workers, tools or machines, and parts. The motion of workers is minimized to the extent possible. Each worker typically performs one simple operation. All parts or assemblies are handled either by conveyors or motorized vehicles such as forklifts, or gravity, with no manual trucking. Henry Ford was the first who introduced the assembly line and was able to improve other aspects of production. An assembly line is designed to be highly efficient, and very cost effective. The workers focus on a small part of the overall whole, meaning that they do not require extensive training. Parts are fed along a conveyor belt or series of belts for workers to handle, creating a continuous flow of the desired product. At the peak of production, Ford's assembly line turned out a new automobile every three minutes, and modern assembly lines can be even more rapid, especially when they combine automated machinery with human handlers.

The present edited book is a collection of 12 chapters written by experts and well-known professionals of the field. The volume is organized in three parts according to the last research works in assembly line subject.

The first part of the book is devoted to the assembly line balancing problem. Assembly line balancing problem (ALBP) consists of a finite set of tasks, where each of them has a duration time and precedence relations, which specify the acceptable ordering of the tasks. One of the problems inherent in organizing the mass production is how to group the tasks to be performed on a workstation so as to achieve the desired level of efficiency. Line balancing is an attempt to locate tasks to each workstation on the assembly line. The basis ALB problem is to assign a set of tasks to an ordered set of workstations, so that the precedence relationships were satisfied, and performance factors were optimized.

The first part includes chapters dealing with different problems of ALBP. We can read about identification of the plant needs and design steps of the line. A knowledge about balancing and sequencing is given. A novel generalized assembly line balancing problem, entitled ASALBP: the Alternative Subgraphs Assembly Line Balancing Problem, is also presented in this part. In this problem, alternative variants for different parts of an assembly or manufacturing process are considered. Each variant is represented by a subgraph that determines the tasks required to process a particular product and the task precedence relations. One of the chapter concludes balancing of assembly line model in garment production by suggesting possible scenarios that eliminate the bottlenecks in the line by various what-if analyses using simulation technique. Different problems of final results estimation are mentioned.

In the second part of the book some optimization problems in assembly line structure are considered. In many situations there are several contradictory goals that have to be satisfied simultaneously. In fact, real-world optimization problems rarely have a single goal. This is the case for the Industrial Car Sequencing Problem (ICSP) on an automobile assembly line. The ICSP consists of determining the order in which automobiles should be produced, taking into account the various model options, assembly line constraints and production environment goals. Also, a multiple objects optimization in mixed model assembly line is shown. Assembly line structure is known since the beginning the 20th century and the mathematical description was first given more than 50 years ago. There are a lot of discussions how to convert this structure to another. One of the idea (conversion from assembly line into production cell) is also discussed in the second section of the book. Optimization, not only of models but also of real equipment (feeders), is underlined as well.

The third part of the book deals with testing problems in assembly line. This section gives an overview on new trends, techniques and methodologies for testing the quality of a product at the end of the assembling line. Collecting meaningful data, exchanging information more smoothly and improving the communication between suppliers, manufacturers, developers and customers are the most important objectives.

The contents of the whole book present us with valuable overview of theory and practice in assembly line production structure and assembly line balancing problem.

The editor would like to express his gratitude to the authors for their excellent work and interesting contributions. We hope that assembly lines structure and assembly line balancing problem become more clear and the book brings closer all readers to the detailed knowledge of described problems.

Waldemar Grzechca
The Silesian University of Technology
Poland

Part 1

Assembly Line Balancing Problem

1

Final Results of Assembly Line Balancing Problem

Waldemar Grzechca
The Silesian University of Technology
Poland

1. Introduction

The manufacturing assembly line was first introduced by Henry Ford in the early 1900's. It was designed to be an efficient, highly productive way of manufacturing a particular product. The basic assembly line consists of a set of workstations arranged in a linear fashion, with each station connected by a material handling device. The basic movement of material through an assembly line begins with a part being fed into the first station at a predetermined feed rate. A station is considered any point on the assembly line in which a task is performed on the part. These tasks can be performed by machinery, robots, and/or human operators. Once the part enters a station, a task is then performed on the part, and the part is fed to the next operation. The time it takes to complete a task at each operation is known as the process time (Sury, 1971). The cycle time of an assembly line is predetermined by a desired production rate. This production rate is set so that the desired amount of end product is produced within a certain time period (Baybars, 1986). In order for the assembly line to maintain a certain production rate, the sum of the processing times at each station must not exceed the stations' cycle time (Fonseca et al, 2005). If the sum of the processing times within a station is less than the cycle time, idle time is said to be present at that station (Erel et al,1998). One of the main issues concerning the development of an assembly line is how to arrange the tasks to be performed. This arrangement may be somewhat subjective, but has to be dictated by implied rules set forth by the production sequence (Kao, 1976). For the manufacturing of any item, there are some sequences of tasks that must be followed. The assembly line balancing problem (ALBP) originated with the invention of the assembly line. Helgeson et al (Helgeson et al, 1961) were the first to propose the ALBP, and Salveson (Salveson, 1955) was the first to publish the problem in its mathematical form. However, during the first forty years of the assembly line's existence, only trial-and-error methods were used to balance the lines (Erel et al,, 1998). Since then, there have been numerous methods developed to solve the different forms of the ALBP. Salveson (Salveson, 1955) provided the first mathematical attempt by solving the problem as a linear program. Gutjahr and Nemhauser (Gutjahr & Nemhauser, 1964) showed that the ALBP problem falls into the class of NP-hard combinatorial optimization problems. This means that an optimal solution is not guaranteed for problems of significant size. Therefore, heuristic methods have become the most popular techniques for solving the problem.

2. Heuristic methods in assembly line balancing problem

The heuristic approach bases on logic and common sense rather than on mathematical proof. Heuristics do not guarantee an optimal solution, but results in good feasible solutions which approach the true optimum. Most of the described heuristic solutions in literature are the ones designed for solving Single Assembly Line Balancing Problem. Moreover, most of them are based on simple priority rules (constructive methods) and generate one or a few feasible solutions. Task-oriented procedures choose the highest priority task from the list of available tasks and assign it to the earliest station which is assignable. Among task-oriented procedures we can distinguish immediate-update-first-fit (IUFF) and general-first-fit methods depending on whether the set of available tasks is updated immediately after assigning a task or after the assigning of all currently available tasks. Due to its greater flexibility immediate-update-first-fit method is used more frequently. The main idea behind this heuristic is assigning tasks to stations basing on the numerical score. There are several ways to determine (calculate) the score for each tasks. One could easily create his own way of determining the score, but it is not obvious if it yields good result. In the following section five different methods found in the literature are presented along with the solution they give for our simple example. The methods are implemented in the Line Balancing program as well. From the moment the appropriate score for each task is determined there is no difference in execution of methods and the required steps to obtain the solution are as follows:

Step 1. Assign a numerical score $n(x)$ to each task x.

Step 2. Update the set of available tasks (those whose immediate predecessors have been already assigned).

Step 3. Among the available tasks, assign the task with the highest numerical score to the first station in which the capacity and precedence constraints will not be violated. Go to STEP 2.

The most popular heuristics which belongs to IUFF group are:
IUFF-RPW Immediate Update First Fit – Ranked Positional Weight,
IUFF-NOF Immediate Update First Fit – Number of Followers,
IUFF-NOIF Immediate Update First Fit – Number of Immediate Followers,
IUFF-NOP Immediate Update First Fit – Number of Predecessors,
IUFF-WET Immediate Update First Fit – Work Element Time.
In the literature we can often find the implementation of Kilbridge & Wester or Moodie & Young methods, too. Both of them base on precedence graph or precedence matrix of produced items.

3. Measures of final results of assembly line balancing problem

Some measures of solution quality have appeared in line balancing problem. Below are presented three of them (Scholl, 1999).
Line efficiency (LE) shows the percentage utilization of the line. It is expressed as ratio of total station time to the cycle time multiplied by the number of workstations:

$$LE = \frac{\sum_{i=1}^{K} ST_i}{c \cdot K} \cdot 100\% \tag{1}$$

where: K - total number of workstations,
c - cycle time.

Smoothness index (SI) describes relative smoothness for a given assembly line balance. Perfect balance is indicated by smoothness index 0. This index is calculated in the following manner:

$$SI = \sqrt{\sum_{i=1}^{K} (ST_{max} - ST_i)^2} \tag{2}$$

where:
ST_{max} - maximum station time (in most cases cycle time),
ST_i - station time of station i.
Time of the line (LT) describes the period of time which is need for the product to be completed on an assembly line:

$$LT = c \cdot (K - 1) + T_K \tag{3}$$

where:
c - cycle time,
K -total number of workstations,
T_K – processing time of last station.
Two-sided assembly lines (Fig. 1.) are typically found in producing large-sized products, such as trucks and buses. Assembling these products is in some respects different from assembling small products. Some assembly operations prefer to be performed at one of the two sides (Bartholdi, 1993).

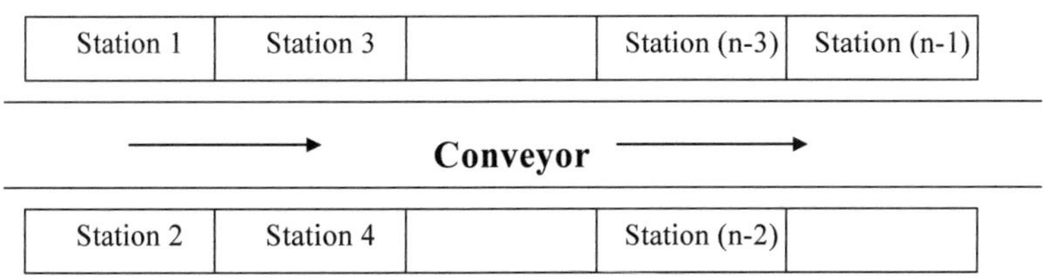

Fig. 1. Two-sided assembly line structure

The final result estimation of two-sided assembly line balance needs some modification of existing measures (Grzechca, 2008).
Time of line for TALBP

$$LT = c \cdot (Km - 1) + Max\{t(S_K), t(S_{K-1})\} \tag{4}$$

where:
Km – number of mated-stations
K – number of assigned single stations
$t(S_K)$ – processing time of the last single station
As far as smoothness index and line efficiency are concerned, its estimation, on contrary to LT, is performed without any change to original version. These criterions simply refer to each individual station, despite of parallel character of the method.

But for more detailed information about the balance of right or left side of the assembly line additional measures will be proposed:
Smoothness index of the left side

$$SI_L = \sqrt{\sum_{i=1}^{K} (ST_{maxL} - ST_{iL})^2} \quad (5)$$

where:
SI_L- smoothness index of the left side of two-sided line
ST_{maxL}- maximum of duration time of left allocated stations
ST_{iL}- duration time of i-th left allocated station
Smoothness index of the right side

$$SI_R = \sqrt{\sum_{i=1}^{K} (ST_{maxR} - ST_{iR})^2} \quad (6)$$

where:
SI_R- smoothness index of the right side of two-sided line,
ST_{maxR}- maximum of duration time of right allocated stations,
ST_{iR}- duration time of i-th right allocated station.

4. Line and station efficiency

Efficiency line was introduced to the assembly line balancing problem by Salveson. It was the optimization goal of ALBP and the best solution was when it achieved 100%. Unfortunately this measure is only useful when number of stations or cycle time are changed. If for many final results we obtain the same number of stations and cycle time the line efficiency does not deliver us the detailed knowledge about quality of the line balance. Example of 12 tasks will be discussed. In Table 1 processing task times are presented. Figure 2 shows the relations between tasks (technology of assembly).

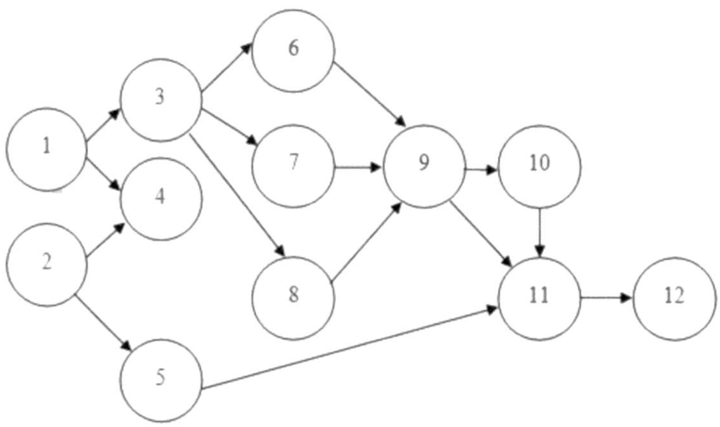

Fig. 2. Precedence graph of 12 tasks numerical example

task	1	2	3	4	5	6	7	8	9	10	11	12
time	20	40	70	10	30	11	32	60	27	38	50	12

Table 1. Operation times for numerical example

Ranked Positional Weight (Halgeson et al, 1961) and Immediate Update First Fit - Working Element Time heuristics for obtaining the balance of assembly line were chosen. As we can see (Fig. 3 ÷ 5) final results are different. For time-oriented balance we got 5 stations and LE for different heuristics is the same. Author proposes an additional station efficiency measurement which allows to find "bottleneck" in the production system and helps to estimate a good feasible line balance. The new measure describes more detailed the efficiency of each workstation and helps to find the worst point in whole assembly structure. Station efficiency (LE_{STi}) shows the percentage utilization of each workstation. It is expressed as ratio of station time to the cycle time:

$$LE_{STi} = \frac{ST_i}{c} \cdot 100\% \tag{7}$$

Measures	RPW heuristic	IUFF-WET heuristic	Measures	RPW heuristic	IUFF-WET heuristic
SI	40,42	59,08	LE_{ST1}	90 %	81 %
LT	462	462	LE_{ST1}	92 %	98 %
LE	80%	80%	LE_{ST1}	65 %	**59 %**
LE_{ST1}	90 %	**100 %**	LE_{ST1}	62 %	62 %

Table 2. Detailed measures of final results of RPW and IUFF-WET heuristics

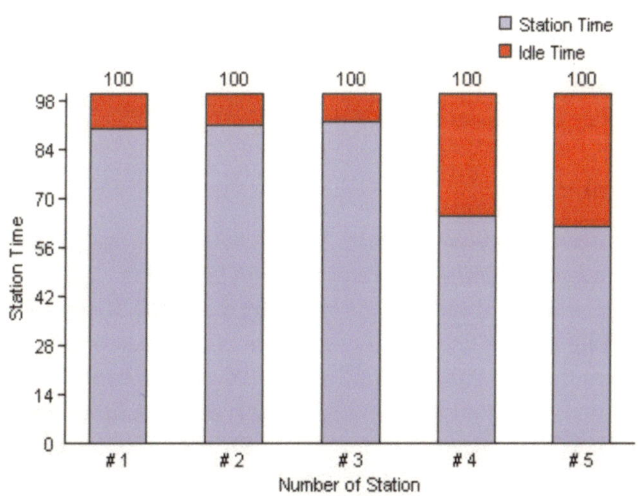

Fig. 3. Final balance of 12 tasks example using RPW heuristic

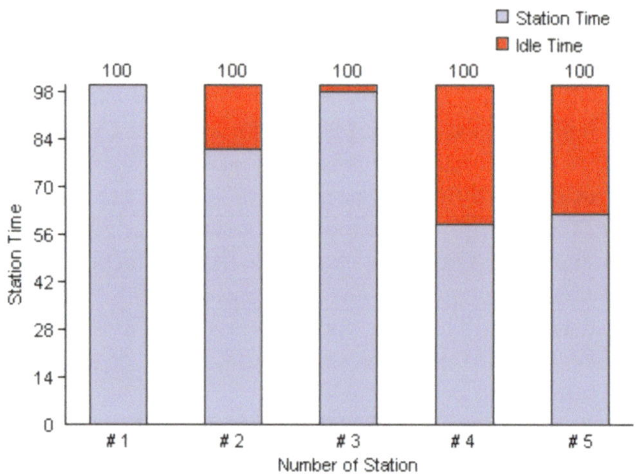

Fig. 4. Final balance of 12 tasks example using IUFF-WET heuristic

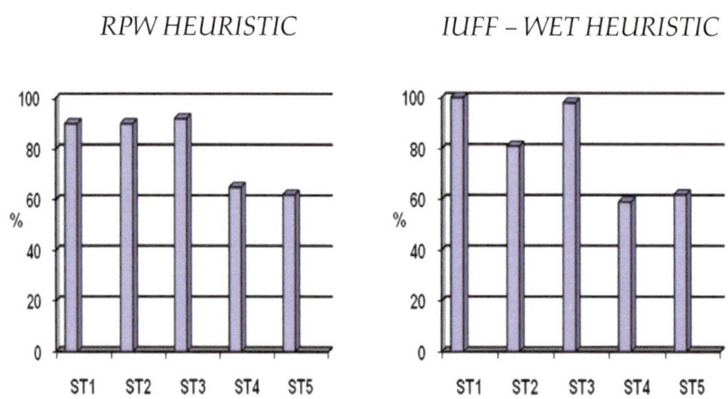

Fig. 5. Station efficiency of 12 tasks example using RPW and IUFF – WET heuristics

5. Last station problem

Below are presented two heuristic solution. We consider the case of single assembly line (Fig. 6.) where all tasks can be assigned to any position (E). The cycle time is 16.
Fig. 7. presents solution where three stations have efficiency 100 % or very close to the value. Last station has an idle time which is a consequence of completed number of tasks. Fig. 8. shows solution where the last station is utilized in 100 %. The contribution of tasks causes idle times of second and third station. As we can observe both solutions are feasible but with different assignment of tasks and different station efficiency. RPW solution is near optimal but the last station idle time is the biggest of all stations. WET solution is feasible and not so close to optimal but smoothness index of this final result is smaller (balancing or equalizing problem).

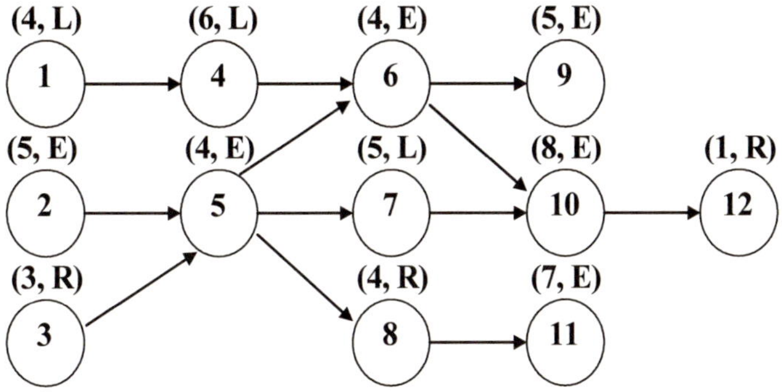

Fig. 6. Numerical example of 12 tasks - time duration and positional constrains are given in brakes (L - left, R - right, E - any position)

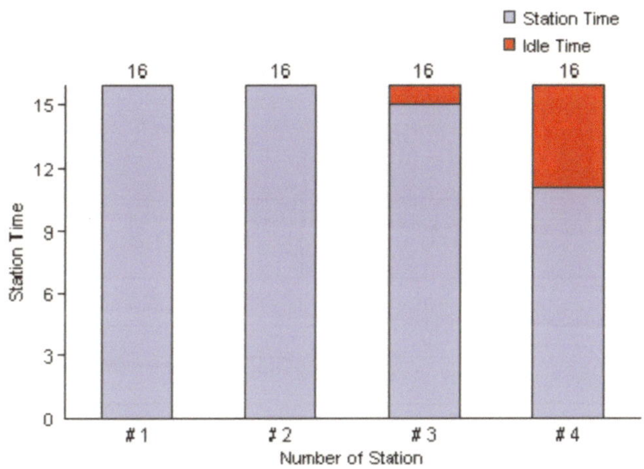

Fig. 7. Final solution of assembly line balancing problem - RPW heuristic

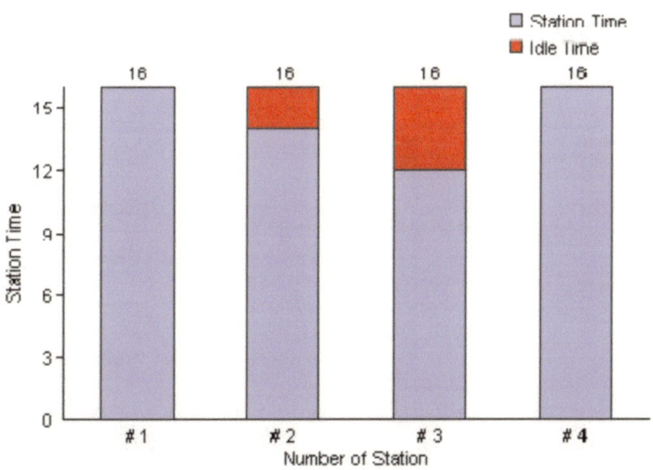

Fig. 8. Final solution of assembly line balancing problem - IUFF - WET heuristic

6. Assembly line structure problem

We consider numerical example from Fig. 8. The positional constrains are respected. As we can observe not only assigning of task becomes a problem (Fig. 9. ÷ Fig. 14.). The structure of assembly line can changed for different cycle times (Grzechca, 2010).

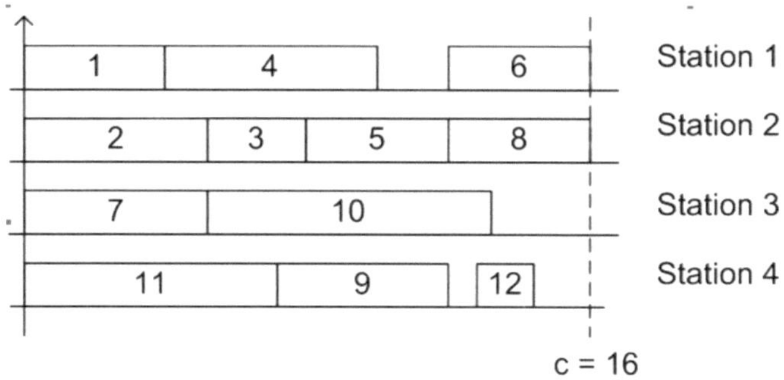

Fig. 9. Assembly line balance (c=16)

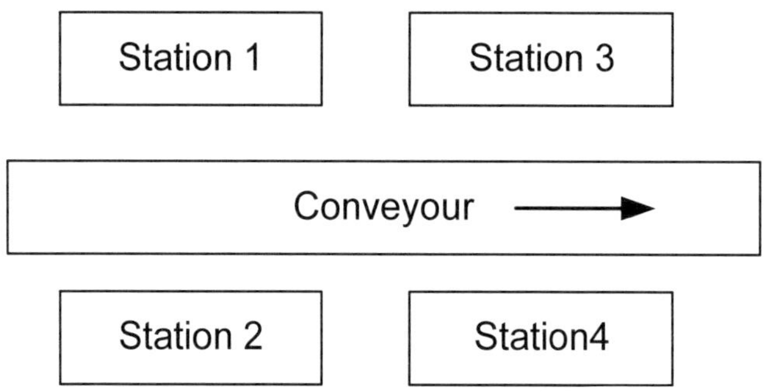

Fig. 10. Assembly line structure (c=16)

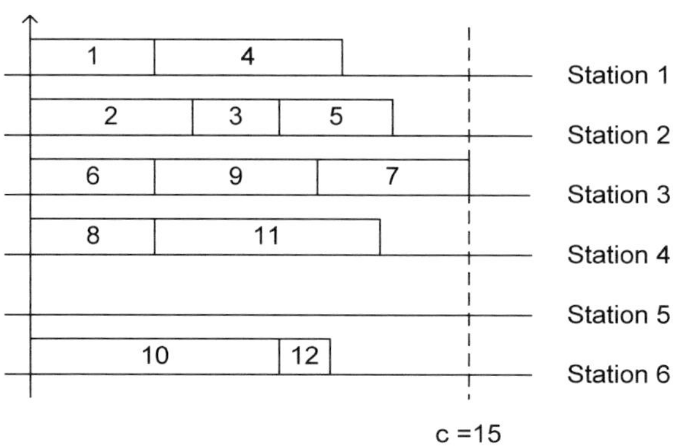

Fig. 11. Assembly line balance (c=15)

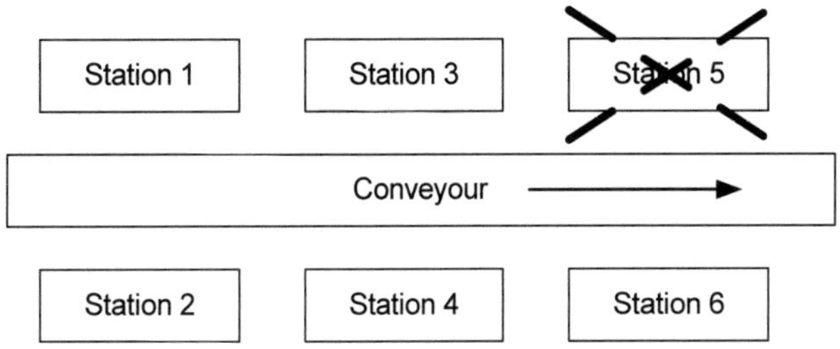

Fig. 12. Assembly line structure (c=15)

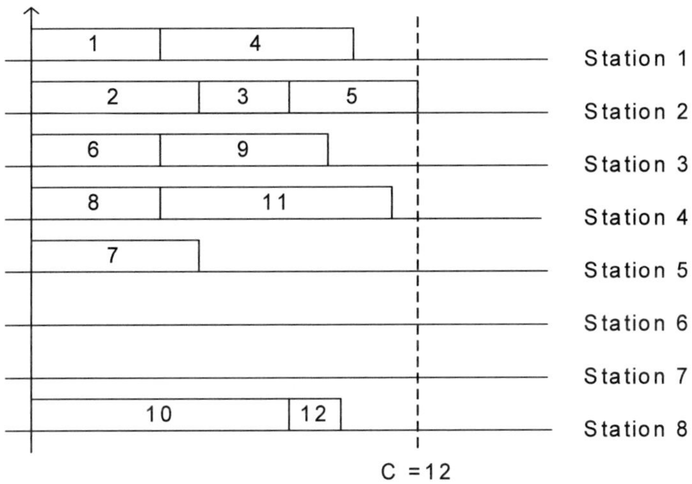

Fig. 13. Assembly line balance (c=12)

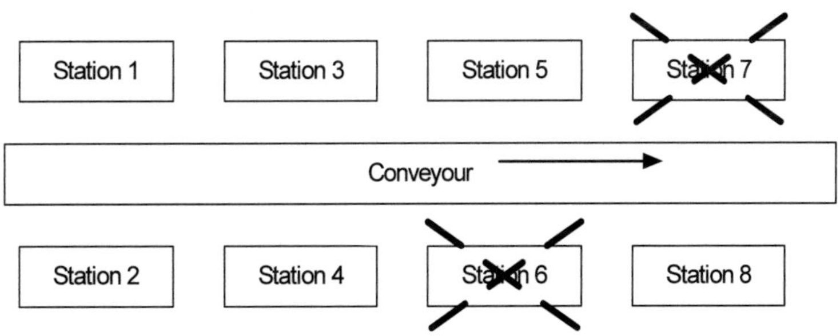

Fig. 14. Assembly line structure (c=12)

In two-sided assembly line balancing problem it is very difficult to obtain a complete station structure. This type of line very hard depends on precedence and position constraints.

7. Conclusions

Assembly lines are a popular manufacturing structure. Assembly line balancing problem is known more than 50 years. There are hundreds exact and heuristic methods. It is very important to obtain the feasible and acceptable results. It is very important to analyze and estimate the final results and to implement the best one. Author of the chapter hopes that the presented knowledge helps to understand the problem.

8. References

Bartholdi, J.J. (1993). Balancing two-sided assembly lines: A case study, *International Journal of Production Research,* Vol. 31, No,10, pp. 2447-2461

Baybars, I. (1986). A survey of exact algorithms for simple assembly line balancing problem, *Management Science,* Vol. 32, No. 8, pp. 909-932

Erel, E., Sarin S.C. (1998). A survey of the assembly line balancing procedures, *Production Planning and Control,* Vol. 9, No. 5, pp. 414-434

Fonseca D.J., Guest C.L., Elam M., Karr C.L. (2005). A fuzzy logic approach to assembly line balancing, *Mathware & Soft Computing,* Vol. 12, pp. 57-74

Grzechca W. (2008) Two-sided assembly line. Estimation of final results. *Proceedings of the Fifth International Conference on Informatics in Control, Automation and Robotics ICINCO 2008, Final book of Abstracts and Proceedings,* Funchal, 11-15 May 2008, pp. 87-88, CD Version ISBN: 978-989-8111-35-7

Grzechca W. (2010) Structure's Uncertainty of Two-sided Assembly Line Balancing Problem. URPDM 2010 Coimbra, CD version

Gutjahr, A.L., Neumhauser G.L. (1964). An algorithm for the balancing problem, *Management Science,* Vol. 11,No. 2, pp. 308-315

Helgeson W. B., Birnie D. P. (1961). Assembly line balancing using the ranked positional weighting technique, *Journal of Industrial Engineering,* Vol. 12, pp. 394-398

Kao, E.P.C. (1976). A preference order dynamic program for stochastic assembly line balancing, *Management Science,* Vol. 22, No. 10, pp. 1097-1104

Lee, T.O., Kim Y., Kim Y.K. (2001). Two-sided assembly line balancing to maximize work relatedness and slackness, *Computers & Industrial Engineering,* Vol. 40, No. 3, pp. 273-292

Salveson, M.E. (1955). The assembly line balancing problem, *Journal of Industrial Engineering,* Vol.6, No. 3. pp. 18-25

Scholl, A. (1999). *Balancing and sequencing of assembly line,* Physica- Verlag, ISBN 9783790811803, Heidelberg New-York

Sury, R.J. (1971). Aspects of assembly line balancing, *International Journal of Production Research,* Vol. 9, pp. 8-14

2

Assembly Line Balancing and Sequencing

Mohammad Kamal Uddin and Jose Luis Martinez Lastra
Tampere University of Technology
Finland

1. Introduction

Assembly line balancing (ALB) and sequencing is an active area of optimization research in operations management. The concept of an assembly line (AL) came to the fact when the finished product is inclined to the perception of product modularity. Usually interchangeable parts of the final product are assembled in sequence using best possibly designed logistics in an AL. The initial stage of configuring and designing an AL was focused on cost efficient mass production of standardized products. This resulted in high specialization of labour and the corresponding learning effects. However, the recent trend gained the insight of the manufacturers of shifting the AL configuration to low volume assembly of customized products, mass customization. The strategic shift took effect due to the diversified customer needs along with the individualization of products. This eventually triggered the research on AL balancing and sequencing for customized products on the same line in an intermix scenario, which is characterized as mixed-model assembly line balancing (MMALB) and sequencing. The configuration planning of such ALs has acquired an important concern as high initial investment is allied with designing, installing and re-designing an AL.

The research carried out in this manuscript aims to contribute to the problem domain of MMALB and sequencing. Balancing refers to objective depended workload balance of the assembly jobs to different workstations. Sequencing refers to find an optimal routing/job dispatching queue considering the demand scenario, available time slots and resources. Primary factors associated to this problem domain includes different assembly plans (e.g. mixed/batch/single model production), variations in processing workstations (e.g. manual/robotic/hybrid stations), physical line layouts (e.g. straight/parallel/U-shaped lines) and varied work transporting methods (e.g. conveyor/pallet-based). These factors are mostly plant specific and must be considered as the design pre-requisites for line balancing and sequencing.

The contribution of this work is twofold. Firstly, a brief review of the problem domain of ALB and sequencing is presented. This includes systematic design approach of an AL and different performance and workstation related indexes which helps the line designer to identify plant specific design factors for line balancing, re-balancing and sequencing. Different heuristics and meta-heuristics based ALB solution strategies, classification of ALB problems, MMALB and sequencing are also addressed (section 2).

Secondly, a logic and mathematical formulation based methodology for solving ALB problem is proposed (section 3), addressed to low volume product customization in shop floor (MMALB). The presented methodology results in optimizing the shift time for any combination of product customization, assembled in an intermix order. It also defines a repetitive job dispatching queue in accordance to the balancing results. The proposed approach is encoded via MATLAB and validated with reference data to prove the optimal conditions. A small scale practical shop floor problem is also analysed with the presented methodology (section 4) to show the optimality conditions. The conclusions are drawn in section 5.

2. Assembly line design

Systematic design of ALs is not an independent and easy task for the manufacturers. Designers need to deal with current physical factory layout in the initial phase. Cost and reliability of the system, complexity of the tasks, equipment selection, ALs operating criteria, different constraints, scheduling, station allocation, inventory control, buffer allocation are the most important area of concern. The development of an approach to design of ALs consisting of seven phases depicted in figure 1.

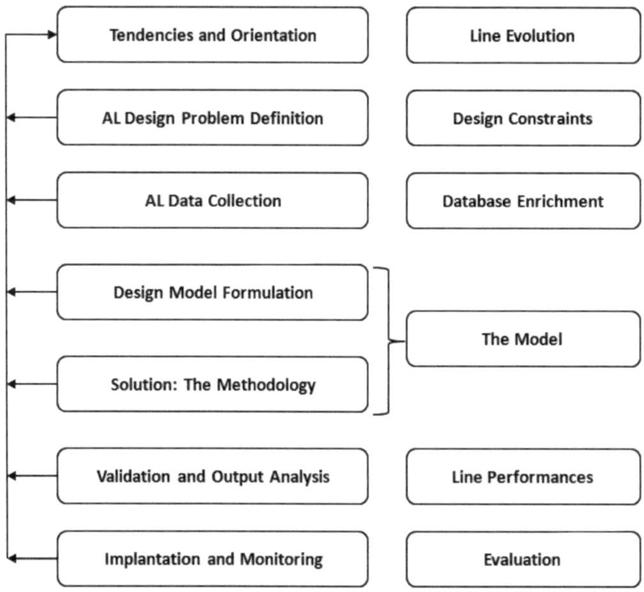

Fig. 1. Development of an approach to AL design (Rekiek & Delchambre, 2006)

Tendencies and orientation of ALs are linked to line evolution. Designers need to collect information in this step about the tendencies of the line to be implemented. Balancing and sequencing problem varies with the types of ALs. For instance, single model line produces a single product over the line. Facility layout, tool changes, workstation indexes remains fairly constant. Batch model lines produce small lots of different products on the line in batches. In mixed-model case, several variations of a generic product are produced at the same time in an intermixed scenario. Consideration of work transport system is also a concern. Apart from manual work transport on the line, three types of mechanized work transport systems are identified as continuous transfer, synchronous transfer (intermittent transfer) and asynchronous transfer (Papadopoulos et al., 1993). Different line orientations need to be identified by the designer as it varies widely according to the production floor layout. Straight, Parallel, U-Shaped lines (Becker & Scholl, 2006) are generally applied.

Various design factors are important to study and integrate with the AL design and balancing. The decisive solution variations depend on the production approaches, objective functions and constraints. Some of the design constraints related to ALB are precedence constraints, zoning constraints and capacity constraints (Vilarinho & Simaria, 2006). Efficient description of line design problem is associated with database enrichment. To collect AL data, knowledge about several performance indexes and workstation indexes are important for a line designer (Table 1).

Performance Indexes	Workstation Indexes
1. Variance of time among product versions	1. Operator skill, motivation
2. Cycle time	2. Tools required
3. No of stations	3. Tools change necessary
4. Traffic problems	4. Setup time
5. Station space	5. Buffer allocation
6. Transportation networks	6. Average station time
7. Communication among the groups	7. Variance of time among product versions (diff. models)
8. Task complexity	8. Ergonomic values (required grip strength)
9. Reliability	9. Need of storage
	10. Working place
	11. Worker absenteeism during operation

Table 1. Performance and workstation indexes for ALB and sequencing

AL design model and solution methodology combine the model stage. Design tools are modelled and formulated after collection and verification of the input data. Design tools modelling include the output data, interaction between different modules and methods to be developed. Wide range of heuristic as Branch and Bound search, Positional weight method, Kilbridge and Wester Heuristic, Moodie-Young Method, Immediate Update First-Fit (IUFF), Hoffman Precedence Matrix (Ponnambalam et al., 1999) and meta-heuristic based solution strategies as Genetic Algorithm GA (Sabuncuoglu et al., 1998), Tabu Search TS (Chiang, 1998) , Ant Colony Optimization ACO (Vilarinho & Simaria, 2006), Simulated Annealing SA (Suresh & Sahu, 1994) for ALB problems are adopted in industrial and research level (figure 2). Validation of the models is a result of performance towards the objectives of that particular line.

Line performances of AL design is a measure of multi-objective characteristics. Varied objective functions are considered for ALB (Tasan & Tunali, 2006). Designer's goal is to design a line considering higher efficiency, less balance delay, smooth production, optimized processing time, cost effectiveness, overall labour efficiency and just in time production (JIT). The aim is to propose a line by exploiting the best of the design methods which will deal in actual fact with user preferences.

Fig. 2. Different solution procedures for ALB

Design evaluation refers to a user friendly developed interface where all necessary AL data is accessible extracted from different database. Validation of different algorithms and methods is integrated with different design packages (Rekiek & Delchambre, 2006).

2.1 Classification of ALB problems

Classification of ALB problem is primarily based on objective functions and problem structure. Different versions of ALB problems are introduced due to the variation of objectives (figure 3).

Objective Function Dependent Problems:

- Type F: Objective dependent problem, it is associated with the feasibility of line balance for a given combination of number of stations and cycle time (time elapsed between two consecutive products at the end of the AL).
- Type 1: This type of problem deals with minimizing number of stations, where cycle time is known.
- Type 2: Reverse problem of type 1.
- Type E: This type of problem is considered as the most general version of ALBP. It is associated with maximizing line efficiency by minimizing both cycle time and number of stations.
- Type 3, 4 and 5: These corresponds to maximization of workload smoothness, maximization of work relatedness and multiple objectives with type 3 and 4 respectively.

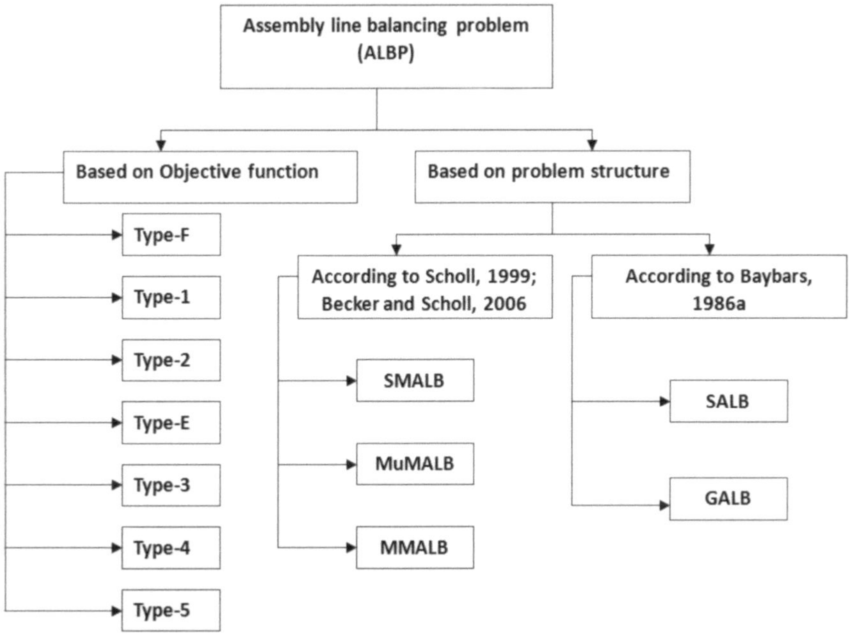

Fig. 3. Classification of ALBP (Scholl & Boyesen, 2006)

Problem Structure Dependent Problems:

- SMALB: This refers to single model ALB problems, where only one product is produced.
- MuMALBP: Multi model ALB problems, where more than one product is produced on the line in batches.
- MMALBP: Mixed model ALB problems, various models of a generic product are produced on the line in an intermixed situation.
- SALBP: Simple ALB balancing problems, simplest version of balancing problems, where the objective is to minimize the cycle time for a fixed number of workstation and vice versa.
- GALBP: A general ALB problem includes those problems which are not included in SALBP. Those are for instance, mixed model line balancing, parallel stations, U-shaped and two sided lines with stochastic task times.

2.2 MMALB and sequencing

Production system planning usually starts with the product design initially. The reason behind this, a great deal of future costs is determined in this phase. Initial configuration and installation of productive units triggers the actual cost of the production planning of assembly system. Resources required by the production process also determines by the configuration planning. Different methodologies are utilized as depicted in figure 2, to support the configuration planning which are included under the term ALB.

In the case of mixed model lines, different models often utilize available capacity in very different intensities. Therefore modification of balancing or line re-balance might be necessary. A family of products is a set of distinguished products (variants), whose main functions are preferably similar, usually produced by mixed-model lines. Mixed model lines are generally employed in the cases (Rekiek & Delchambre, 2006), where

- The cycle time is usually greater than a minute.
- The line price cannot be amortized by a single product model.
- The product must not be delivered in a short time.
- Each product is quite similar to others.
- The same resources are required to assemble the products.
- The set up time of the line needs to be short.

MMALBP occurs when designing or redesigning a mixed-model assembly line. This is subjected to find a feasible assignment of tasks to workstations in such a manner that production demand of different product variants are met within the defined shift times. Minimization of assembly costs, satisfying the constraints are also a concern. Mixed model lines are classified into two different types, which are referred as dual problems.

- MMALBP-1: minimizes the number of workstation for a given cycle time.
- MMALBP-2: Minimizes the cycle time for a given number of workstation.

In type 1 problem cycle time or, the production rate, is pre-specified. That is why; it is more frequently used to design a new AL where demands are forecasted beforehand. Type 2 problem deals with maximization of production rate of an existing AL. This is applied for example when changes in assembly process or in product range require the line to be redesigned.

Mixed model sequencing aims to minimize sequence dependent work overload. Sequencing is based on a detailed model scheduling depending on the operation times, worker movements, necessary tool changes required, station borders and other characteristics of the line. Different models are composed of different product options and thus require different materials and parts, so that the model sequence influences progression of material demand over time. As ALs are commonly coupled with preceding production levels by means of a just in time (JIT) supply of required materials, the model sequence need to facilitate this. An important prerequisite for JIT-supply is the steady demand rate of materials over time, as otherwise advantages of JIT are sapped by enlarged safety stocks that become necessary to avoid stock outs during the peak demand. Accordingly, JIT centric sequencing approaches aim at distributing the material requirements over the planning horizon (Boyesen et al., 2007).

3. Methodology for solving MMALB and sequencing

A logic and mathematical formulation based methodology is proposed for solving MMALB and sequencing. During the development of this approach, a constant speed AL is considered where task transportation, machine setup and tool changing times are taken within the task times. Task time of each model, precedence relations of tasks are known whereas work in progress buffers, station parallelization, assignment restrictions, zoning constraints are not allowed. MMALB problem type 2 is considered. The balancing is achieved in two consecutive stages which are named as first stage and second stage.

3.1 First stage: balancing from equivalent single model

Balancing in this stage finds locally optimized solution in the first stage iteration. Objective of this stage is to find solution(s) with specified number of stations with a minimum cycle time. Solutions are considered as locally optimized as the principle objective is to find a solution which will define a smooth production by minimizing objective function of second stage. The concept of ALBP-1, where the aim is to optimize the number of workstations with a predefined fixed cycle time is utilized in first stage of this proposed approach. The fixed cycle time is considered as the solution lower bound, CT_{min} for finding desired station numbers, CT_{min} is increased by 1 sec per iteration. Solution lower bound is determined with minimum cycle time (Gu et al., 2007) as:

$$CT_{min} = max\left[\frac{1}{S}\sum_{i=1}^{n} t_i \, , max\, t_i\right] \tag{1}$$

Where, t_i is the i_{th} task time and S is the desired number of stations. The flow diagram of first stage is illustrated in figure 4.

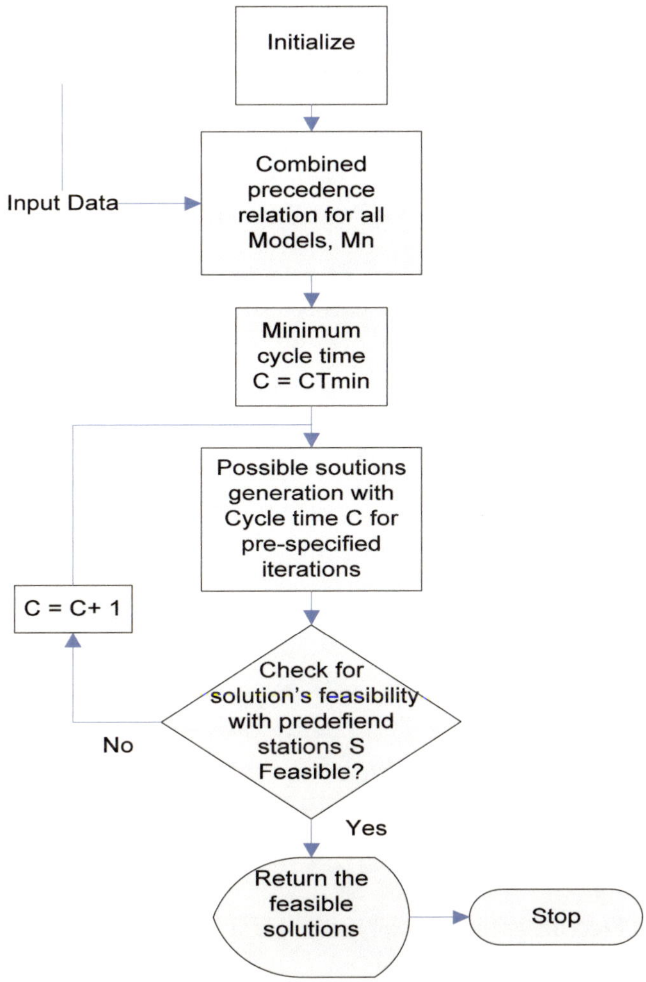

Fig. 4. Flow diagram of first stage iteration

Tasks of different models are first considered as an equivalent single model. Combined precedence diagram alter different models into one equivalent single model. A simple combined precedence relation example is given in figure 5, with 12 tasks, where node containing the task number and the values indicate tasks time.

The following algorithm defined as step by step procedure, generates a number of feasible solutions for equivalent single model. Optimized feasible solutions are stored as the input solutions of second stage.

1. Open a new station S_1 with a cycle time $C = CTmin$.
2. Determine the set of tasks without predecessor, $s = \{i, j \ldots\ldots n\}$
3. Assign randomly one task from s in station S_1.
4. Remove the assigned task from the precedence graph, update station time as the cycle time $C = CTmin - t_i$
5. Update set of tasks without predecessor as $s = \{j, k \ldots\ldots n\}$
6. Assign tasks randomly from s to S_1 until C is positive and update C and s each time.
7. When C is negative or zero for randomly assigned any task from s, check for the other tasks in s to be fitted in S_1.
8. When C is negative or zero for all the tasks in existing s, open a new station S_2 and $C = CTmin$ is restored for S_2.
9. Repeat steps 1 to 8 until the assignment of all tasks.
10. Generate all feasible solutions.
11. Check the solutions with predefined station numbers. If the solutions are not feasible, repeat the above steps with $C = CTmin + 1$ and so on until the desired number of stations are met.
12. When a number of feasible solutions are achieved, store finally updated C as the cycle time. Store and return the workstation based solutions with the station assignment information for next stage.

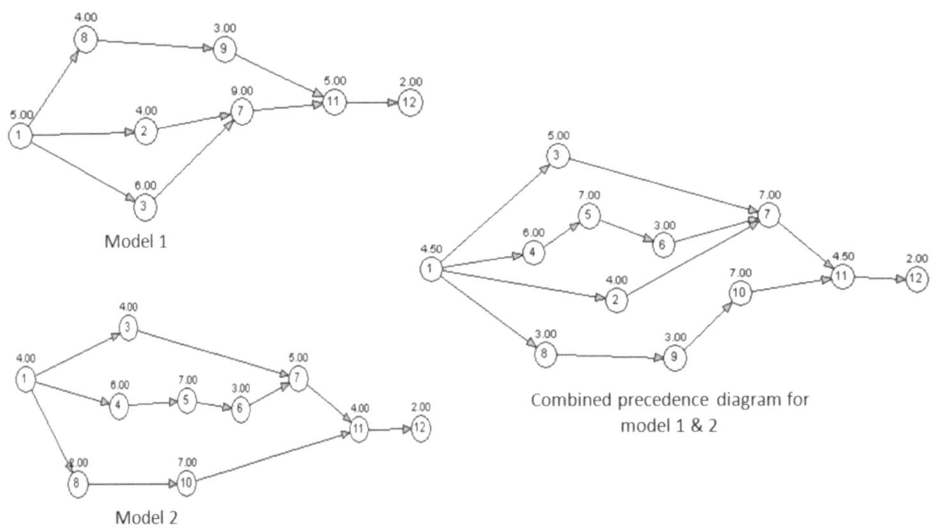

Fig. 5. Combined Precedence diagram for model 1 and 2

3.1.1 First stage experimentation

Benchmarked 'Buxey' data sets of 29 tasks for SMALBP-2 (Scholl, 1993) are tested with first stage balancing approach. Precedence matrix (table 2) defines the precedence constraints associated to the tasks. Precedence task set 1, 2 refers task 2 precedes task 1 in a $\{0, 1\}$ task matrix where column precedes the row. A 1 is placed where there is a precedence relation, otherwise 0. Solution flexibility can be determined from precedence matrix by measuring $F - ratio$ (flexibility ratio). Higher $F - ratio$ indicates less precedence constraints and greater flexibility in generating multiple feasible solutions (Rubinovitz et al., 1995).

$$F - ratio = \frac{2 \times Z}{K(K-1)} \tag{2}$$

Where, Z is the number of zeros above the diagonal and K is the number of task elements. $F - Ratio$ for the combined precedence diagram of figure 5 is 0.78.

Tasks	1	2	3	4	5	6	7	8	9	10	11	12
1		1	1	1	0	0	0	1	0	0	0	0
2	0		0	0	0	0	1	0	0	0	0	0
3	0	0	D	0	0	0	1	0	0	0	0	0
4	0	0	0	I	1	0	0	0	0	0	0	0
5	0	0	0	0	A	1	0	0	0	0	0	0
6	0	0	0	0	0	G	1	0	0	0	0	0
7	0	0	0	0	0	0	O	0	0	0	1	0
8	0	0	0	0	0	0	0	N	1	0	0	0
9	0	0	0	0	0	0	0	0	A	1	0	0
10	0	0	0	0	0	0	0	0	0	L	1	0
11	0	0	0	0	0	0	0	0	0	0		1
12	0	0	0	0	0	0	0	0	0	0	0	

Table 2. Precedence matrix for combined precedence diagram for figure 5

First stage MATLAB program compiled for 'Buxey 29 tasks Problem' (Scholl, 1993) and the task times are shown in table 3.

Task No.	Time, Sec	Task No.	Time, Sec	Task No.	Time, Sec
1	7	11	21	21	1
2	19	12	10	22	9
3	15	13	9	23	25
4	5	14	4	24	14
5	12	15	14	25	14
6	10	16	7	26	2
7	8	17	14	27	10
8	16	18	17	28	7
9	2	19	10	29	20
10	6	20	16	-	-

Table 3. Task times of 'Buxey' benchmarked problem

3.1.2 Experiment results

First stage generates multiple feasible solutions for different number of stations. Tasks assignment is shown below, where S1 to S9 represents predefined 9 stations with assigned tasks. Minimum cycle time achieved 37 seconds which fulfil the benchmarked solution result. Station assignments of the tasks are: S1 {2, 7, 9, 10, 26}, S2 {1, 6, 12, 27}, S3 {3, 4, 5, 14}, S4 {8, 11}, S5 {13, 17, 25}, S6 {15, 16, 20}, S7 {18, 19, 21, 22}, S8 {23, 28}, S9 {24, 29}.

'Buxey' 29 tasks problem			
Benchmarked Results		Stage1 procedure	
Predefined stations, m	*Minimal cycle time*	*Minimal cycle time C*	*CPU run time, sec*
7	47	48	193.83
8	41	41	136.04
9	37	37	105.45
10	34	34	85.45
11	32	32	73.46
12	28	30	50.82
13	27	27	24.42
14	25	25	8.91

Table 4. Comparison between benchmark results and stage1 procedure

Benchmark results and the results obtained by first stage balancing are depicted in table 4. Figure 6 shows line balancing solution for 'Buxey' 9 station problem obtained by first stage balancing procedure.

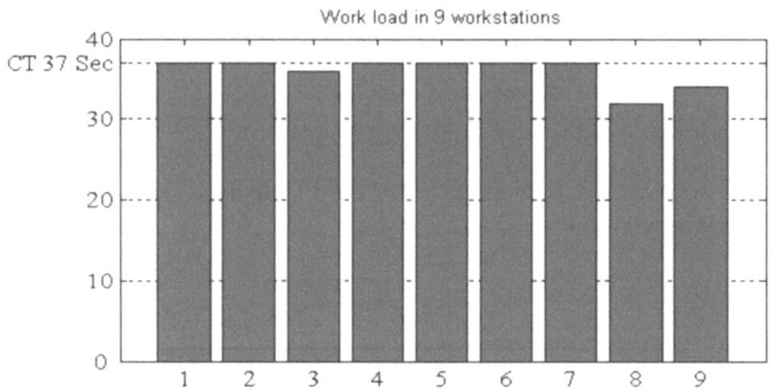

Fig. 6. Workstations Vs Workload (37 sec cycle time)

3.2 Second stage: balancing for mixed-models

This stage finds optimal solutions for mixed-models with the results achieved from first stage. Feasible solutions generated from the first stage are decoded and scaled with second stage objective function. The aim is to obtain the best solutions from first stage in terms of second stage objective which ensures a minimal balance delay. The feasible solutions of first stage are coded as the workstation based solutions. Workstation based solution representation scheme is shown in figure 7.

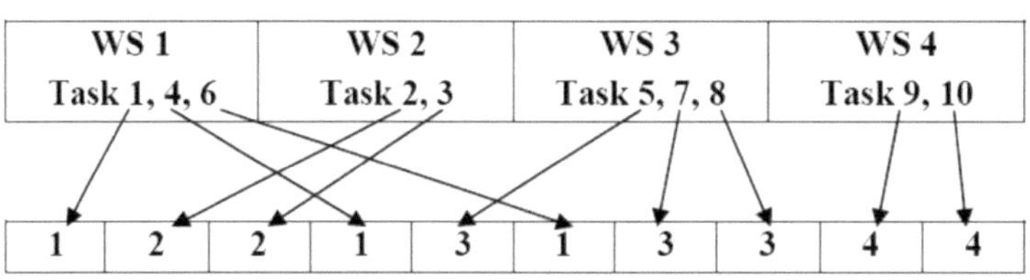

Fig. 7. Workstation based solution representation

Inputs for second stage objective function from the generated first stage solutions are as follows:

1. Number of workstations S, represented by the solution which is the highest numerical number of the solution.
2. No of tasks K in precedence graph as the length of the solution.
3. Tasks assignment in workstations according to the solution representation scheme.
4. The initial problem definition of MMALB-2 describes the inputs to the objective function are number of models to be produced M, production demand for each model N_m, where $m = 1\ to\ M$ and task times for each model t_{km}.

3.2.1 Objective function formulation

Objective function considered for MMALBP-2 to facilitate a smooth workload balance among the stations, while taking smoothed station assignment load into consideration. It also optimizes shift time as cycle time of single model case is replaced by shift time in mixed-model balancing.

Notations:

M	Number of models to be produced.
N_m	Scheduled quantity to be produced for each model where $m = 1$ to M.
T	Shift time period for the scheduled quantity to be produced.
K	Number of total tasks.
CT_{min}	Minimum cycle time.
t_{km}	Task times where $k = 1$ to K and $m = 1$ to M. t_{km} represents the work time of task k on model m.
E_k	Total time required to complete $\sum_{m=1}^{M} N_m$ units in the scheduled period for task k
S	Number of stations.
Q_{sm}	Amount of time that the operator in station s is assigned on each unit of model m
T_s	Station time where $s = 1$ to S.
P_{sm}	Total time assigned to station s on model m.
P_m	Average amount of total work content for all units of model m assigned to each station.

All models of production demand can be summarized as the total products to be produced, where;

$$\text{Total products to be produced} = \sum_{m=1}^{M} N_m \tag{3}$$

The total task times required to complete the production of all models are:

$$E_k = \sum_{m=1}^{M} N_m \times t_{km} \tag{4}$$

In MMAL, operation time is denoted as Q_{sm}; where $s = 1\ to\ S$ and $m = 1\ to\ M$; which refers the amount of time required in station s for each unit of model m. Mixed-model line balancing solutions are obtained here from the single model balancing algorithm of first stage by replacing cycle time C to shift time period T. Total time assigned to station s in period T can be defined as

$$T_s = \sum_{m=1}^{M} N_m \times Q_{sm} \tag{5}$$

Total time assigned to station s on model m in period T is

$$P_{sm} = N_m \times Q_{sm} \tag{6}$$

Now, P_m represents average or desired amount from the total work content for all units of model m assigned to each station and P_m can be presented as

$$P_m = N_m \times \frac{\sum_{k=1}^{n} t_{km}}{S} \tag{7}$$

Hence, minimizing the value of $(P_m - P_{sm})$ points to smooth out or equalize total work load for each model over all work stations. Therefore the objective function Y $(SSAL, Smoothed\ Station\ Assignment\ Load)$, can be abridged as to minimize the following function

$$Y = \min \sum_{s=1}^{S} \sum_{m=1}^{M} [P_m - P_{sm}] \tag{8}$$

3.2.2 Mixed-model line sequencing

Tasks associated to ALs are mostly dealing with the repetitive periodic tasks occurring at a regular interval. A static AL's task sequencing heuristic (Askin & Standridge, 1993) is integrated to the results of MMALB-2 obtained from second stage. The objective of sequencing is to generate a dispatch system which controls the order of entry of the products to the first station.

Let, q_m is the proportion of product type m to be assembled in the line where $m = 1\ to\ M$. The initial step is to develop an AL balance for the weighted average product. Let t_{km} is the task time for k of model m and S_s is the set of tasks assigned to station s where $s = 1\ to\ S$. In that case if the cycle time is CT, the average feasibility condition can be stated as:

$$\sum_{k \in S_s} \sum_{m=1}^{M} q_m t_{km} \leq CT \tag{9}$$

This condition indicates the averaged across all items produced in the long term, no workstation is overloaded. According to the feasibility condition, one single product ALB problem needs to be solved. Due to this, task time of task k can be summarized as:

$$t_k = \sum_{m=1}^{M} q_m t_{km} \tag{10}$$

For each model m, Q_m amount need to be produced. If r be the greatest common denominator of all Q_m a repeating cycle comprised of $N_m = Q_m/r$ units should suffice where the models are from $m = 1\ to\ M$. The cycle would be repeated r times to satisfy the period demand. In that case, $N = \sum_{m=1}^{M} N_m$ items are produced in each cycle.

The objective of designing such cycle is to define a smooth production rate of each model type. This will also prevent the excessive idle time at the workstation due to the mix-induced starving of workstations. A workstation is starved if on completion of all the defined tasks, there are no tasks available for it to work on because the next task has not been completed in the prior station. This is even more crucial in the bottleneck stations. That is why, the maintaining of a smooth flow of the parts to those stations is necessarily important. The bottleneck stations are the stations with maximal total work or equivalently average work load per cycle. If a partial sequence overloads this workstation with respect to average cycle time CT, the subsequent stations are starved. If a partial sequence under loads the bottleneck station, the initial output rate from the line will be too high which will result in accumulating the inventory. In case of the relative workload of station s is C_s, it workload can be defined as:

$$C_s = \sum_{k \in S_s} t_k \quad (11)$$

The bottleneck station S_b is the station with maximum workload or equivalently or average workload per cycle. Hence,

$$S^b = \text{argmax}_s \, C_s \quad (12)$$

Let, X_{mn} is the value equals to 1 if model m is placed in the n_{th} position and 0 otherwise. In that case, $m(n)$ will indicate the type of model placed in n_{th} position in the assembly sequence. Now, the approach is to select the n_{th} model to be started to insert in the line to optimize as following:

$$\text{minimize maximum}_{1 \le n \le N} \left[\sum_{m=1}^{M} \sum_{k \in S_{s^b}} t_{km(n)} - nC_{s^b} \right] \quad (13)$$

Sequencing is done in two consecutive steps:
Step 1: Initialization, create a list of all products to be assigned during the cycle and named as list A.
Step 2: Assign a product. For $n = 1 \; to \; N$ from list A, create a list B of all product types that could be assigned without violating any constraints. From list B select the product type m' that minimizes the objective function of equation 13. Add model type m' to the n_{th} position. Remove a product type m' from list A and if $n < N$, go to step 2. C_{s^b} defines the time accumulated in bottleneck stations.
Aim of this sequencing heuristic is to create a list of unassigned products first, which is then reduced first to a list of feasible assignable products and to the single best feasible products. The assumption of this heuristic is that the operator in manual workstations can intermix to a slight degree to keep the line moving even if the station is temporarily overloaded.

4. Case study

A modified practical problem definition of WXYZ Company (Askin and Standridge, 1993) is considered here for the implementation of the addressed integrated approach for MMALB-2 and sequencing. The problem defines assembly of web cameras of four different models. A constant speed, conveyor based, straight AL is considered where tasks contains no zoning constrains, capacity constraints or assignment restrictions. Average demands per shift for four different types of cameras are 20 units of model 1, 30 units of model 2, 40 units of model 3 and 10 units of model 4. Aim is to balance the line for mixed-model assembly system with optimized shift time. Assembly module has four fixed workstations (MMALB-2) where they have decided to place one operator in each station. Each workstation is capable of performing the same set of operations on all four model types. Task times (sec) for each model are shown in table 5.
Now, following the proposed methodology, the aim is to find:

- Optimal cycle time accounting for workstation availability considering combined task relationships for all models (first stage).
- Distributing the tasks of all four models to four different workstations maintaining an overall workload balance, *i.e. SSAL* as the objective of mixed-model balancing considered here and also to find out optimized overall shift timing for assembly of all models according to demand (second stage).
- Finally, constructing a repetitive lot planning through model sequencing (mixed-model sequencing).

Tasks	Model M1	Model M2	Model M3	Model M4	Wt. Avg.	Immediate predecessors
Op 1	14	34	15	10	19	-
Op2	12	15	11	17	14	Op 1
Op 3	39	47	40	51	45	Op2
Op 4	3	4	4	7	5	Op 1
Op 5	11	13	10	9	11	Op 3
Op 6	19	29	21	21	23	Op 4
Op 7	11	14	9	10	11	Op 5
Op 8	21	38	28	32	30	Op 3, Op 6
Op 9	13	19	15	17	16	Op 5, Op 8
Op 10	33	41	42	43	40	Op 7, Op 9
Total	176	254	195	216	234	-

Table 5. Tasks time per model

Ten different tasks or operations are identified for completing the assembly of each model. Task times are different depending on the models. Combined precedence diagram for four models are shown in figure 8.

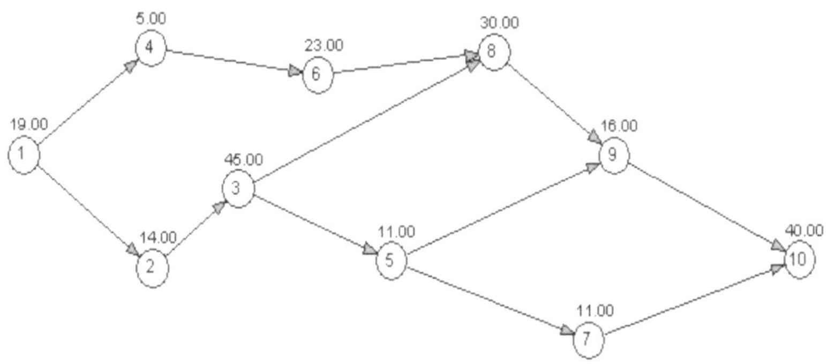

Fig. 8. Precedence diagram of the case problem

Proposed first stage generates two feasible solutions considering minimum cycle time for the case problem. Cycle times of both workstation based solutions are 59 seconds. Next step is to decode and scale the optimized solutions to achieve the best one considering overall SSAL. Feasible solutions represented in figure 9, decoded in table 6, 7.

1	2	2	1	3	1	3	3	4	4

1	2	2	1	3	1	4	3	3	4

Fig. 9. Feasible solutions of the case problem

Work Station	Assigned Tasks	Station Time
1	Op 1, Op 4, Op6	47
2	Op2, Op3	59
3	Op5, Op7, Op8	52
4	Op9, Op10	56

Table 6. Decoded first solution from figure 9

Work Station	Assigned Tasks	Station Time
1	Op 1, Op 4, Op6	47
2	Op2, Op3	59
3	Op5, Op8, Op9	57
4	Op7, Op10	51

Table 7. Decoded second solution from figure 9

Two feasible solutions are explored and scaled with the objective function of second stage. Results obtained are illustrated in table 8. Overall SSAL are 22 and 23.9 for solution 1 and 2. Therefore solution 1 has the better smoothed stations assignment load.

Feasible Solutions	Stations	St. Time(Hr) per shift for mixed-models	SSAL (Y value of the objective function)
Solution 1	1	1.31	7.85
	2	1.56	4.15
	3	1.44	1.65
	4	1.54	5.35
Solution 2	1	1.31	7.85
	2	1.55	4.15
	3	1.58	4.25
	4	1.42	3.55

Table 8. Shift times and SSAL values for generated solutions of figure 9

Production ratio of four models is 2:3:4:1 according to demand. Therefore, a repetitive lot of 10 units need to be considered. As a consequence of demand fluctuation, the ratio may vary but the goal is to find feasibility of a long run path with demand ratio (Askin and Standridge, 1993). The feasible solution of the mixed-model balancing indicates station 2 as bottleneck station as the cycle time of 59 sec is fully consumed. Bottleneck station load per model are 51, 62, 51 and 68 seconds.

According to this sequencing heuristic, initially all models are eligible since the cumulative production level deficit is below one for all models. The sequencing is shown in table 9. M2 is selected to minimize the maximum deviation of actual to desired production for any assignable product. If the presence of bottleneck stations are multiple, larger of the deviation are chosen for constructing the model sequencing. Selection of M2 puts the production schedule [1 – 0.3] or 0.7 ahead of the schedule for M2 and 0.2, 0.4 and 0.1 behind for M1,

M3 and M4. In second step ($n = 2$), selection of M2 is not eligible because its assignment will place M2 $[1 - (-0.4)]$ or 1.4 ahead of the schedule. Following this heuristic, a recurring lot planning of 10 units where 2 units of model 1, 3 units of model 2, 4 units of model 3 and 1 unit of model 4 are achieved for the case problem where the sequence of mixed-models would be M2-M3-M4-M1-M3-M2-M3-M2-M1-M3 with a shift time of 1.56 hours.

Step	**Models If Selected**				**Selected Model**	**WS2 Load (Bottleneck)**
	M1	M2	M3	M4		
1	8,0.2	3, 0.3	8, 0.4	9, 0.1	M2	62(3)
2	5, 0.4	6, -0.4	5, 0.8	12, 0.2	M3	113(-5)
3	13, 0.6	2, -0.1	13, 0.2	4, 0.2	M4	181(4)
4	4, 0.8	7, 0.2	4, 0.6	-	M1	232(-4)
5	8,0	3,0.5	8,1	-	M3	287(-8)
6	16, 0.2	5,0.8	16,0.4	-	M2	359(5)
7	13, 0.4	2,0.1	13, 0.8	-	M3	400(-13)
8	21, 0.6	10,0.4	21,0.2,	-	M2	482(10)
9	2, 0.8	-	2, 0.6	-	M1	529(-2)
10	-	-	0,1	-	M3	590(0)

Table 9. Mixed-model sequencing for the case problem

Most solutions for ALB problems look for one final optimized solution. However, it is fairly important to explore the alternative solutions (Boysen, 2006). This integrated approach facilitates such necessary diversity of the solutions. If 8 station 'Buxey' data sets are focused, three feasible solutions are generated with 41 seconds minimal cycle time. As in the case of mixed-model balancing, the objective function is measured from the solutions obtained by the joint precedence graph, feasible solutions need to satisfy the performance indexes of the line. Such performance indexes are the line efficiency, station smoothness index and the overall balance delay. Generated workstation based solutions are depicted in figure 10.

1	1	1	2	3	3	2	3	2	4	4	2	5	4	5	5	6	6	5	7	5	6	7	8	2	2	4	8	8
1	1	1	2	2	2	2	3	2	3	5	3	3	4	4	4	6	6	5	7	6	6	7	8	4	2	5	8	8
2	1	3	3	3	1	2	4	2	2	5	4	3	2	4	5	6	6	5	7	5	6	7	8	2	1	1	8	8

Fig. 10. Generated 3 feasible Solutions with the first stage approach for 8 Station 'Buxey' problems

As a consequence of the generated balancing solutions, corresponding station load and utilization of the stations for three solutions are depicted in figure 11.

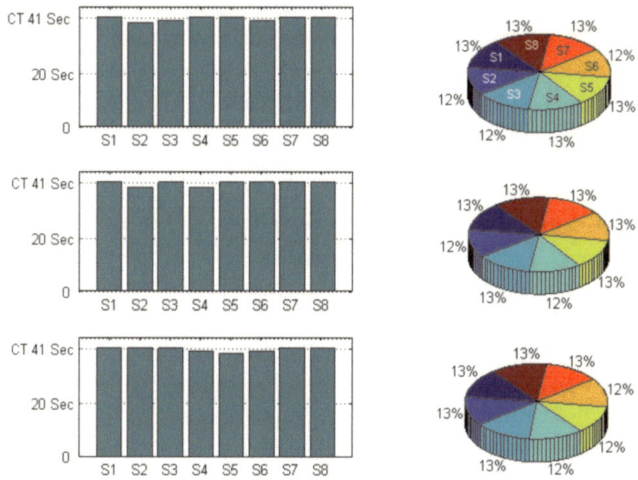

Fig. 11. Station Load over 41 sec Cycle time and consecutive station utilization% for 3 solutions

Line efficiency *LE* gives an impact of percentage of utilization of the line (Boysen et al., 2006). *LE* for generated 3 feasible solutions is 98.2 % as most of the stations are fully utilized with equivalent single model case. Smoothness index (SX) is measured to indicate the relative smoothness of the AL balance (Ponnambalam, et al., 1999). A smoother line results in reducing process inventory as well as smoothed workload balance. SX for all the generated 3 solutions is 2, which indicates the solutions are feasible and having less balance delay.

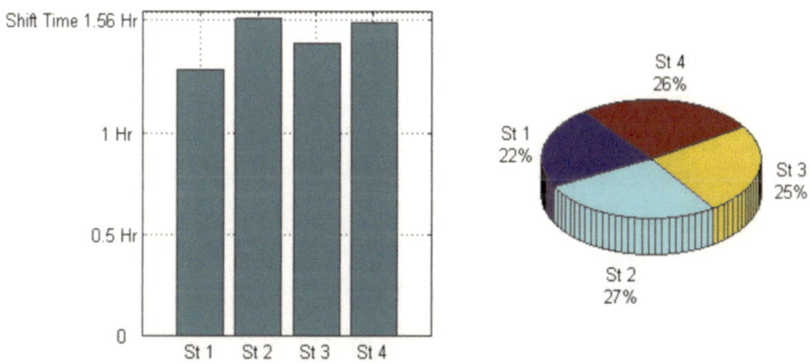

Fig. 12. Shift timing and station utilization of the case problem of MMALB-2

In mixed-model balancing, work elements are assigned to different workstations on a daily basis or an entire shift instead of cycle time basis as is done in single model case. The objective function considered here is to distribute evenly the total entire workloads within the shift time. As in example case problem, selected optimal solution for mixed-model case is solution 1. Station per SSAL values are 7.85, 4.15, 1.65, 5.35 and overall SSAL for the solution is 22. The solution having a smallest value of SSAL indicates optimality of workload balance among the stations. For assembling the entire 100 units of 4 different models, generated optimal solution indicates shift timing of 1.56 hours. Station 2 is fully utilized where as station 1 having the idlest time during an entire shift. Shift timing and station utilization are illustrated in figure 12.

5. Conclusions and future works

Systematic design and balancing of ALs is somewhat complicated, especially for the large scale product customization due to the uneven nature of tasks times of different models. This is the parallel rationale of having a difficulty to a smooth workload balance among workstations. But, in terms of not finding a good balancing structure supported by a proper sequencing of the mixed models, the interim performances of the line will be poor which obstruct the overall mixed-model AL-based production scenario.

The research carried out in this manuscript helps to identify the critical design parameters associated to ALB and sequencing. It also briefly addresses the overall problem domain of ALB and sequencing. The methodology for MMALB and sequencing addressed in this work distributes workloads of mixed-models to predefined workstations considering smoothed station assignment load. This results in optimizing the shift timing of AL for any combination of various models and defines a repetitive production lot planning from model sequencing. The end result can be summarized as maximization of production rate.

It can be concluded from the experimental results that the addressed two stage balancing and sequencing methodology ensures a smooth and optimal production with varied demand for MMALB-2 in ideal conditions. Whereas, the first stage procedure can also be implemented for Single model ALB problems. Proposed approach is shown to perform well as the optimized solution generation scheme is converged from the different feasible solutions. Integrated sequencing approach of this work also imparts an understanding of a smooth production of the mixed-models by defining a repetitive production schedule. Overall, the results of this work are important when designing and balancing an AL layout from the scratch or redesigning for product customization.

In a more complex assembly environment, there might be several constrains like equipment restrictions, facility layout restrictions, buffer allocation and stations length which essentially differ from plant to plant. For an overall understating of extensive performances of MMALB and sequencing, all those plant and line oriented constraints should be taken into account within the balancing methodology and this is considered to be the future extension of this work.

6. Appendix: MATLAB codes for the case study

Weighted task times representation: Cost Function

```
function c= cost()
c = [19 14 45 5 11 23 11 30 16 40];
```

Precedence matrix Representation: Graph Function

```
function g= graph()
g = [0 1 0 1 0 0 0 0 0 0
    0 0 1 0 0 0 0 0 0 0
    0 0 0 0 1 0 0 1 0 0
    0 0 0 0 0 1 0 0 0 0
    0 0 0 0 0 0 1 0 1 0
```

```
0 0 0 0 0 0 0 1 0 0
0 0 0 0 0 0 0 0 0 1
0 0 0 0 0 0 0 0 1 0
0 0 0 0 0 0 0 0 0 1
0 0 0 0 0 0 0 0 0 0];
```

The proposed approach for assigning the tasks:

```
function [stInfo loadInfo stationCount ] = assign(tTime, nOfTask, Ctmin, tMatrix);
  global readyTaskList;
  global readyTaskCount;
  global taskCount;
  global parentCount;
  global taskMatrix;
  global taskTime;
  global cTime;
  global stCount;
  global stID;
  global loadID;
  global minStationID;
    stCount = 0;
  cTime = Ctmin;
  taskTime = tTime;
  taskCount = nOfTask;
  taskMatrix = tMatrix;
  parentCount = buildParentCount();
  readyTaskCount = 0;
  buildReadyTaskList();
  minStationID = ones( 1, taskCount);
    while readyTaskCount > 0.5
    t = getReadyTask();
    loadTask(t);
    updateReadyTaskList(t);
  end
    stInfo = stID;
  loadInfo = loadID;
  stationCount = stCount;
  function loadTask(t);
  global taskTime;
  global stTimeLoad;
  global stTaskLoad;
  global stID;
  global loadID;
  global minStationID;
    reqTime = taskTime(t);
  minSID = minStationID(t);
  st = getStation( reqTime, minSID);
  stTaskLoad( st) = stTaskLoad( st) + 1;
  stTimeLoad(st) = stTimeLoad(st) + reqTime ;
  stID(t) = st;
```

```
  loadID(t) = stTaskLoad( st) ;
function st = getStation( reqTime, mnstID);
  global stCount;
  global stTimeLoad;
  global stTaskLoad;
  global cTime;
  st = -1;
  for i= 1:stCount
    if stTimeLoad(i) +  reqTime < cTime + 0.5 && mnstID < i + 0.5
      if st > 0
        if stTimeLoad(i) > stTimeLoad(st)
          st = i;
        end
      else
         st = i;
      end
    end
  end
  if st < 0
     stCount = stCount +1;
     stTimeLoad(stCount) = 0;
     stTaskLoad(stCount) = 0 ;
     st = stCount;
  end
function  pCount = buildParentCount();
  global taskMatrix
  pCount = sum(taskMatrix, 1);
function buildReadyTaskList();
  global parentCount;
  global taskCount;
  for t = 1:taskCount;
    if parentCount(t) == 0
      addReadyTask(t);
    end
  end
function  updateReadyTaskList(t);
   global parentCount;
  global readyTaskList;
  global readyTaskCount;
  global taskCount;
  global taskMatrix;
  global minStationID;
  global stID;
  for i = 1:taskCount;
    if taskMatrix(t,i) > 0.5 % if dependency exist
      parentCount(i) = parentCount(i)-1;
       if minStationID(i) < stID(t)
         minStationID(i) = stID(t);
      end
```

```
        if parentCount(i) < 0.5 % if no parent exist
          addReadyTask(i);
        end
      end
    end
function rTask = getReadyTask();
    global readyTaskList;
    global readyTaskCount;
    ind = getRandIndex(readyTaskCount);
    rTask = readyTaskList( ind );
    readyTaskList( ind ) = readyTaskList( readyTaskCount );
    readyTaskCount = readyTaskCount -1;
function addReadyTask(t);
    global readyTaskList;
    global readyTaskCount;
    readyTaskCount = readyTaskCount + 1 ;
    readyTaskList( readyTaskCount ) = t;
function ind = getRandIndex(readyTaskCount);
    ind = rand;
    ind = ind * readyTaskCount;
    ind = round(ind);
    if ind < readyTaskCount;
      ind = ind +1;
    end
```

Solution Generation: Schedule Generator

```
close all;
clear all;
stCount = inf;
maxStation = 4
G_RAPH =graph();
C_OST = cost();
[temp, taskCount] = size(C_OST);
%ctMin = max(C_OST);
TTM = [14 12 39 3 11 19 11 21 13 33 % task times t1:t10 for model 1
    34 15 47 4 13 29 14 38 19 41 % task times t1:t10 for model 2
    15 11 40 4 10 21 9 28 15 42 % task times t1:t10 for model 3
    10 17 51 6 9 21 10 32 17 43]; % task times t1:t10 for model 4
[nom,not] = size (TTM) % no of models, no of tasks
%NOS = 4 % predefined number of stations
% weighted average
tw = ceil(sum(TTM(1:nom,1:not))/4)
max_tw = max(max(tw)) % maximum task time
cmin = floor(sum (tw)/maxStation)
ctMin = max(cmin, max_tw) % minimum cycle time, lower bound
%C = CTmin
solutionCount = 0;
while(stCount > maxStation)
    for i=1:300 % no of iteration
```

```
      [stInfo ldInfo stationCount ] = assign(C_OST, taskCount, ctMin, G_RAPH);
      if stCount > stationCount
         stCount = stationCount;
      end
      if stationCount <= maxStation
         flag = 0;
         j =1;
         while j<=solutionCount && flag<0.5
            %v1 =  sum(abs(loadInfo(j,1:taskCount) - ldInfo(1:taskCount) ));
            v2 =  sum(abs(sationInfo(j,1:taskCount) - stInfo(1:taskCount) ));
            if v2 < 0.5 % v1 <0.5 &&
               flag = 1;
            end
            j = j+1;
         end
         if flag < 0.5
            solutionCount = solutionCount + 1;
            sationInfo(solutionCount,:) = stInfo(1:taskCount);
            loadInfo(solutionCount,:) = ldInfo(1:taskCount);
            NumberOfStation(solutionCount) = stationCount;
         end
      end
   end
  if stCount > maxStation
     ctMin = ctMin +1;
  end
end
sationInfo
loadInfo
Updated_Final_CYCLE_TIME = ctMin
```

Mixed Model Scaling for optimized SSAL value:

```
s1 = [1 4 6] % Task Assigned to Station 1
s2 = [2 3]
s3 = [5 7 8]
s4 = [9 10]
time_s1 = sum(tw(s1))
time_s2 = sum(tw(s2))
time_s3 = sum(tw(s3))
time_s4 = sum(tw(s4))
st = [s1 s2 s3 s4];
Nm= [20 30 40 10] % demand for each model
E = (Nm*TTM); % total time required to complete all Nm (100) units
%T = sum(E(st(1:end)))
T1 = sum(E(s1(1:end))) %Station 1 work load for all models
T2 = sum(E(s2(1:end))) %Station 2 work load for all models
T3 = sum(E(s3(1:end))) %Station 3 work load for all models
T4 = sum(E(s4(1:end))) %Station 4 work load for all models
ST_TIME = [T1 T2 T3 T4]/3600 % In hour
 Station_time= [T1 T2 T3 T4]
```

```
L1 = TTM(1,:);
L2 = TTM (2,:);
L3 = TTM(3,:);
L4 = TTM(4,:)
W = [sum(L1) sum(L2) sum(L3) sum(L4)];
Qsm1 = [(sum(L1(s1(1:end)))) (sum(L2(s1(1:end)))) (sum(L3(s1(1:end)))) (sum(L4(s1(1:end))))];
Qsm2 = [(sum(L1(s2(1:end)))) (sum(L2(s2(1:end)))) (sum(L3(s2(1:end)))) (sum(L4(s2(1:end))))];
Qsm3 = [(sum(L1(s3(1:end)))) (sum(L2(s3(1:end)))) (sum(L3(s3(1:end)))) (sum(L4(s3(1:end))))];
Qsm4 = [(sum(L1(s4(1:end)))) (sum(L2(s4(1:end)))) (sum(L3(s4(1:end)))) (sum(L4(s4(1:end))))];
Psm1 = Nm.*Qsm1;
Psm2 = Nm.*Qsm2;
Psm3 = Nm.*Qsm3;
Psm4 = Nm.*Qsm4;
Psm = [Psm1 Psm2 Psm3 Psm4];
Pm1 = (Nm.*W)/maxStation;
SSAL1 = sum(abs(Pm1 - Psm1)); % SSAL in Station 1
SSAL2 = sum(abs(Pm1 - Psm2));
SSAL3 = sum(abs(Pm1 - Psm3));
SSAL4 = sum(abs(Pm1 - Psm4));
% Overall SSAL
SSAL = [SSAL1 SSAL2 SSAL3 SSAL4]/100
```

Solutions obtained for first stage balancing

```
maxStation = 4
nom = 4
not = 10
tw =   19   14   45   5   11   23   11   30   16   40
max_tw = 45
cmin = 53
ctMin = 53
sationInfo =
   1  2  2  1  3  1  3  3  4  4 and  1  2  2  1  3  1  4  3  3  4
Updated_Final_CYCLE_TIME = 59
```

MMALBP second stage balancing

```
s1 =   1   4   6
s2 =   2   3
s3 =   5   7   8
s4 =   9   10
time_s1 = 47
time_s2 = 59
time_s3 = 52
time_s4 = 56
Nm = 20   30   40   10
T1 = 4700
T2 = 5600
T3 = 5200
T4 = 5600
ST_TIME = 1.3056   1.5586   1.4444   1.5386
SSAL = 7.8500   4.1500   1.6500   5.3500
```

7. References

Askin, R.G. & Standridge, C.R. (1993). Modelling and analysis of manufacturing systems; John Wiley and Sons Inc, ISBN 0-471-51418-7

Baybars, L. (1986). A survey of exact algorithms for the simple assembly line balancing problems. *Management science,* Vol. 32, No. 8, (August, 1986), pp. (909-932)

Becker, C. & Scholl, A. (2006). A survey on problems and methods in generalized assemblyline balancing, *European journal of operational research,* Vol. 168, Issue. 3 (February, 2006), pp. (694–715), ISSN 0377-2217

Boysen, N., Fliedner, M. & Scholl, A. (2006). A classification of assembly line balancing problems. *European journal of operational research,* Elsevier, Vol 183, No. 2 (December, 2007), pp. (674–693)

Gu, L., Hennequin, S., Sava, A., & Xie, X. (2007). Assembly line balancing problem solved by estimation of distribution, Proceedings of the 3rd Annual IEEE conference on automation science and engineering scottsdale, AZ, USA

Papadopoulos, H.T; Heavey, C. & Browne, J. (1993). Queuing Theory in Manufacturing Systems Analysis and Design; Chapman & Hall, ISBN 0412387204, London, UK

Ponnambalam, S.G., Aravindan, P. & Naidu, G.M. (1999). A comparative evaluation of assembly line balancing heuristics. *International journal of advanced manufacturing technology,* Vol. 15, No. 8 (July 1999), pp. (577-586), ISSN: 0268-3768

Rekiek, B. & Delchambre, A. (2006). Assembly line design, the balancing of mixed-model hybrid assembly lines using genetic algorithm; Springer series in advance manufacturing, ISBN-10: 1846281121, Cardiff, UK

Rubinovitz, J. & Levitin, G. (1995). Genetic algorithm for assembly line balancing, *International Journal of Production Economics, Elsevier,* Vol. 41, No. 1-3 (October, 1995), pp (343-354), ISSN 0925-5273

Sabuncuoglu, I., Erel, E. & Tanyer, M. (1998). Assembly line balancing using genetic algorithms. *Journal of intelligent manufacturing,* Vol. 11, No. 3 (June, 2000), pp. (295-310), ISSN: 0956-5515

Scholl, A. (1993). Data of assembly line balancing problems. Retrieved from http://www.wiwi.uni-jena.de/Entscheidung/alb/, last accessed: 07 February 2008

Scholl, A. (1999). Balancing and sequencing of assembly lines, Second edition, Heidelberg: Physica, 318S, DM98,00, ISBN: 3790811807

Suresh, G. & Sahu, S. (1994). Stochastic assembly line balancing using simulated annealing, *International journal of production research,* Vol. 32, No. 8, pp. (1801-1810), ISSN: 1366-588X (electronic) 0020-7543 (paper)

Tasan, S.O. & Tunali, S. (2006). A review of current application of genetic algorithms in assembly line balancing, *Journal of intelligent manufacturing,* Vol. 19, No. 1 (February, 2008), pp. (49-69), ISSN: 0956-5515

Vilarinho, P.M. & Simaria, A.S. (2006). ANTBAL: An ant colony optimization algorithm for balancing mixed-model assembly lines with parallel workstations, *International journal of production research,* Vol 44, Issue 2, pp. 291–303, ISSN ISSN: 1366-588 0020-7543

3

A Metaheuristic Approach to Solve the Alternative Subgraphs Assembly Line Balancing Problem

Liliana Capacho[1] and Rafael Pastor[2]
[1]*University of Los Andes*
[2]*Technical University of Catalonia*
[1]*Venezuela*
[2]*Spain*

1. Introduction

Nowadays, assembly line balancing problems are commonly found in most manufacturing and production systems. In its basic form, an assembly line balancing problem consists of finding an assignment of tasks to a group of workstations in such a way that the precedence constraints among the tasks are maintained and the sum of the times of the task assigned to each workstation does not exceed the maximum workstation time (i.e. the cycle time). According to the objective considered, two variants of the problem are distinguished: (1) the problem aims at minimizing the number of workstations for a given cycle time and (2), given the number of workstations, the problem seeks to minimize the cycle time. Over the last years a significant amount of research work has been done towards solving assembly line balancing problems efficiently. Finding the best solution is a crucial task for maintaining the competitive advantage of industries and, in some cases, for their survival. Falkenauer (2005), pg. 360, argues that the efficiency difference between an optimal and a sub-optimal assignment can yield economies reaching millions of dollars per year. However, solving real life problems is a very difficult task for decision makers and practitioners since even the simple case is NP-hard by nature. For this reason, most assembly line balancing problems involve only a few aspects of the real systems (see, for example, (Becker & Scholl, 2006)).

In order to deal with the complexity of industrial problems, a great variety of problem definitions (i.e. generalized assembly line balancing problems) have arisen, which consider other restrictions apart from the technological ones. Most common, these include mixed models, multiple products, different line layouts, parallel workstations and multiple objectives. However, real problems require tackling many of those generalizations simultaneously (Falkenauer, 2005). Such a consideration must also be taken into account when alternatives processes are involved. Alternatives may appear when, for example, new technologies are taking place in a production system, in which different procedures are available to complete a production unit, or when the processing order affects the processing times of certain tasks; i.e., the realization of one task facilitates, or makes more difficult, the completion of other tasks (see, for example, (Scholl *et al.*, 2008) and (Das & Nagendra, 1997)).

A novel generalized assembly line balancing problem, entitled ASALBP: the Alternative Subgraphs Assembly Line Balancing Problem, is addressed here. In this problem alternative variants for different parts of an assembly or manufacturing process are considered. Each variant is represented by a subgraph that determines the tasks required to process a particular product and the task precedence relations. Thus, alternative assembly sub-processes for a sub-assembly may involve completely different set of tasks. Consequently, in addition to cycle time or workstations requirements, subgraph constraints must also be taken into account to ensure that tasks belonging to a particular subassembly are processed considering its corresponding assembly subgraph. Furthermore, it is also considered that task processing times are not fixed, but instead are dependent on the assembly subgraph. Therefore, total processing time may vary from one processing alternative to another (even when the alternatives involve the same set of tasks). Similarly to the simple case, the ASALB problem aims at minimizing the number of required workstations for a given bound on the cycle time (i.e., ASALBP-1), or minimizing the cycle time for a given number of workstations (i.e., ASALBP-2). This work focuses on ASALBP-1. As previously discussed, to solve the ASALBP efficiently, two problems must be solved simultaneously: the decision problem, which involves selecting a single assembly subgraph for each subassembly that allows alternatives, and the balancing problem to assign the tasks to the workstations.

In practice, due to the complexity of assembly problems, a two-stage approach is usually used to solve assembly balancing problems that involve alternatives. In a first stage, an assembly variant is selected considering a given criterion, such as, for example, shortest total processing time. When the production alternative has been selected (i.e., the complete assembly process has been defined) and a single precedence graph is available, the problem is then balanced in a second stage. (Capacho & Pastor, 2008) illustrated, by means of numerical examples, how by selecting a priory an alternative, it cannot be guaranteed that an optimal solution of the global problem will be obtained, because the best solution of a problem can be discarded if it does not meet the selection criteria. For instance, it was shown that the alternative assembly variant with the longest processing time required the smallest number of workstations for a given cycle time.

The Alternative Subgraphs Assembly Line Balancing Problem was introduced by (Capacho & Pastor, 2006) and was optimally solved by means of two mathematical programming models. The computational experiment carried out with the models showed that only small- and medium-scale problems could be solved optimally in significantly small computing times. Other attempts have also been done to solve the ASALBP optimally, see for example (Scholl *et al.*, 2006). In order to solve ASALBP for practical sizes, (Capacho *et al.*, 2006, 2009) proposed a group of heuristic methods based on adapting well-known priority rules (e.g. (Talbot *et al.*, 1986), most of which are based on the precedence relations and the cycle time. Solutions of good quality were obtained for problems involving up to 305 tasks and 11 assembly subgraphs. In this work the use of weighted, rather than nominal, values for the priority rules is explored. In particular, it is considered an adaptation of a class of metaheuristic methods, namely GRASP (Greedy Randomized Adaptive Search Procedure), which has been successfully applied to hard combinatorial problems (Delorme *et al.*, 2004).

A GRASP overview and literature reviews can be found in (Festa & Resende, 2008a, 2008b). The former research work discusses the algorithmic aspects of this type of procedures; the second one presents GRASP applications covering a wide range of optimization problems, including scheduling, routing, graph theory, partitioning, location, assignment, and manufacturing. Other examples of GRASP can be found in (Dolgui *et al.*, 2010), which propose an algorithm for balancing transfer lines with multi-spindle machines; (Andrés *et al.*, 2008) and (Martino & Pastor, 2007), both of which tackle problems involving setup times. Basically, a GRASP is an iterative process in which each iteration consists of two phases: the construction phase, which generates a feasible solution, and the search phase that uses a local optimization procedure to find a local optimum in the neighbourhood of the constructed solution (Resende & Ribeiro, 2003). Feasible solutions are generated by iteratively selecting the next element to be incorporated to a partial solution. The best element is selected by ordering all candidate elements according to a greedy function. In this work, the adaptation to the ASALBP of thirteen well-known priority rules are used for selecting tasks and three priority rules are use for selecting the subgraphs (see (Capacho *et al.*, 2009)). This resulted in a total of 39 construction methods, which are used to generate the initial feasible solutions. Furthermore, the application of two neighbourhood search strategies produces a total of 78 GRASP algorithms that are proposed, implemented and tested here. The performance of such algorithms is evaluated by considering a set of 150 medium- and large-scale problem instances; and compared with the results obtained in (Capacho *et al.*, 2009) and with known optimal solutions (see (Capacho & Pastor, 2006, 2008)).

The remainder of this chapter is organized as follows: Section 2 describes the metaheuristic procedures (i.e., GRASP) that are designed and implemented here; Section 3 presents the computational experiment carried out to evaluate and compare the proposed algorithms; Section 4 presents the conclusions and proposes future research work; and, finally, Section 5 lists the References.

2. Grasp procedures for solving the ASALBP

As mentioned above, a GRASP involves a construction phase and a local search phase. In the proposed procedures (see (Capacho, 2007)), the construction phase generates a feasible solution by applying a construction method in which both the subgraphs and the assembly tasks are randomly selected following probability distributions based on weighted priority rule values. Weighted values are proportional or inversely proportional (when using a maximizing or minimizing criterion, respectively) to the values obtained considering a given priority rule. The local search phase generates and evaluates all neighbours of the current solution and maintains the best neighbour solution (i.e., the one that requires the fewest number of workstations). This process is repeated for a given length of time. The best overall solution generated is the final solution.

A solution for the Alternative Subgraphs Assembly Line Balancing Problem consists of a task sequence, a number of required workstations and a set of assembly subgraphs (i.e., one subgraph for each subassembly that allows alternatives).

2.1 The construction phase

To construct an initial solution, one subgraph is randomly selected for each available subassembly following a distribution (also referred to as an assembly route), following a probability function based on a priority rule for subgraphs. Once the subgraphs have been chosen, the set of available assembly tasks (AVT) is defined. The set of available tasks is formed with the tasks that belong to the selected subgraphs and those tasks that do not allow assembly variants. The assignable tasks are determined to form the list of candidate tasks (LCT). A task is assignable if all of its predecessors have already been assigned and the sum of its time and the time of the tasks assigned to the current workstation does not exceed the cycle time. For each task in the candidate list LCT, the priority rule value is computed to construct a probability distribution, which is then used to randomly select the next task to be assigned to the current workstation. Once a task has been assigned it is removed from AVT. New lists LCT are generated and tasks are systematically assigned until all assembly tasks have been assigned (i.e., the sets AVT and LCT are empty) and the solution has been completed. The probability distributions used for selecting subgraphs and tasks are built using weighted values of the following priority rules.

2.1.1 Priority rules for subgraphs

The priority rules used for selecting subgraphs are the following:

a. **Minimum NP**: this rule ranks the subgraphs of the same subassembly according to ascending number of precedence relations involved in each subgraph, which is the total number of arcs entering into and within the subgraph.
b. **Minimum TT**: subgraphs are ranked according to ascending total processing time (i.e., the sum of the times of all tasks belonging to the subgraph).
c. **Minimum NT:** subgraphs are ranked according to ascending number of tasks.

Let consider, for example, a subassembly of a given manufacturing process that allows three alternative subgraphs, S1, S2 and S3, with total processing time of 30, 35 and 35 time units, respectively. Considering the priority rule TT, the cumulative probability distribution for selecting a subgraph s is as follows ($r \in [0, 1)$ is a random value):

$$s = \begin{cases} S1 \text{ if } 0 \le r < 0.30 \\ S2 \text{ if } 0.30 \le r < 0.65 \\ S3 \text{ if } 0.65 \le r < 1 \end{cases} \qquad (1)$$

2.1.2 Priority rules for tasks

Table 1 lists the priority rules considered to build the probability distribution to select the next tasks to be assigned. It is valid to mention that priority rules 3, 4, 5, 6 and 13 of Table 1 are minimization rules while all others are maximization ones. These rules are thirteen well-known priority rules that have been adapted to the ASALBP (see (Capacho *et al.*, 2009)) and random choice assignment. Basically, these rules are determined by considering the cycle time and task precedence relations and by measuring tasks processing times.

The following notation is considered:

n	Number of tasks
ct	Cycle time
m_{max}	Upper bound on the number of workstations
t_{ir}	Duration of task i when processed through route r ($i = 1,\ldots,n$; $r \in R_i$)
P_{ir}	Set of immediate predecessors of task i, if processed through route r ($i=1,\ldots,n$; $r \in R_i$)
S_{ir}	Set of all successors of task i, if it is processed through route r ($i=1,\ldots,n$; $r \in R_i$)

Once the set of selected routes SR is known, the following values can be defined:

$sub(i)$	Subgraph chosen for task i ($\forall i \in AVT$); in this way it is possible to know the duration of task i ($t_{i,sub(i)}$).
Ei,Li	Earliest and latest workstation to which task i can be assigned, respectively ($\forall i \in AVT$).
SIi, Si	Set of immediate and total successors of task i, respectively ($\forall i \in AVT$).

No.	Label	Priority rule	Computation procedure
1	RPW	Rank Positional Weight	$RPW_i = t_{i,sub(i)} + \sum_{j \in S_{i,sub(i)}} t_{j,sub(j)}$ (2)
2	T	Task Time	$t_{i,sub(i)}$
3	EW	Earliest Workstation	$EW_i = \left\lceil (t_{i,sub(i)} + \sum_{j \in P_{i,sub(i)}} t_{j,sub(j)}) / ct \right\rceil$ (3)
4	LW	Latest Workstation	$LW_i = m_{max} + 1 - \left\lceil (t_{i,sub(i)} + \sum_{j \in S_{i,sub(i)}} t_{j,sub(j)}) / ct \right\rceil$ (4)
5	N	Task Number	i
6	Sk	Slack	$Sk_i = LW_i - EW_i$ (5)
7	TLW	Task Time divided by Latest Workstation	$TL_i = t_{i,sub(i)} / LW_i$ (6)
8	IS	Number of Immediate Successors	$IS_i = \lvert SI_i \rvert$ (7)
9	TS	Number of Total Successors	$TS_i = \lvert S_i \rvert$ (8)
10	TTS	Task Time plus Total Number of Successors	$TTS_i = t_{i,sub(i)} + TS_i$ (9)
11	STS	Average Time of Successors	$STS_i = (\sum_{j \in S_{i,sub(i)}} t_{j,sub(j)}) / TS_i$ (10)
12	TSSk	Number of Total Successors divided by the Slack	$TSSk_i = TS_i / \left(Sk_i + 1\right)$ (11)
13	LWTS	Average Latest Workstation	$LWTS_i = LW_i / \left(TS_i + 1\right)$ (12)

Table 1. Priority rules used to form the probability distributions for tasks

The combination of the resulting probability distributions, based on the various priority rules used for tasks and subgraphs, produced 39 constructive methods, which are presented in Table 2.

2.2 The local search phase

Two different neighbourhood search strategies based on task exchange movements are considered. At this point, it is valid to recall that a solution to the problem is represented by a sequence of tasks.

The following notation is used to describe such strategies:

m_k Number of workstations required for a given sequence (solution) k

IS Initial tasks sequence generated in the construction phase

WS Working sequence (the first WS is IS)

SS Stored sequence (the first SS is IS)

NS Neighbour sequence

Slk_j Slack of workstation j (i.e., cycle time minus workstation time)

α Weight parameter

Weighted rules for tasks	Weighted rules for subgraphs					
	NP		*TT*		*NT*	
	No.	Heuristic Name	No.	Heuristic Name	No.	Heuristic Name
RPW	1	W-NP_RPW	14	W-TT_RPW	27	W-NT_RPW
T	2	W-NP_T	15	W-TT_T	28	W-NT_T
EW	3	W-NP_EW	16	W-TT_EW	29	W-NT_EW
LW	4	W-NP_LW	17	W-TT_LW	30	W-NT_LW
N	5	W-NP_N	18	W-TT_N	31	W-NT_N
Sk	6	W-NP_Sk	19	W-TT_Sk	32	W-NT_Sk
TLW	7	W-NP_TLW	20	W-TT_TLW	33	W-NT_TLW
IS	8	W-NP_IS	21	W-TT_IS	34	W-NT_IS
TS	9	W-NP_TS	22	W-TT_TS	35	W-NT_TS
TTS	10	W-NP_TTS	23	W-TT_TTS	36	W-NT_TTS
STS	11	W-NP_STS	24	W-TT_STS	37	W-NT_STS
TSSk	12	W-NP_TSSk	25	W-TT_TSSk	38	W-NT_TSSk
LWTS	13	W-NP_LWTS	26	W-TT_LWTS	39	W-NT_LWTS

Table 2. Heuristic methods used in the construction phase

The proposed local optimization procedures generate the neighbourhood of the working sequence *WS* by using a transformation or exchange movement (which are described in what follows). Each exchange generates a neighbour sequence *NS,* which is evaluated and compared with the stored sequence *SS.* If *NS* improves the stored sequence *SS* (i.e., it requires fewer workstations) the neighbour sequence becomes the stored sequence *SS*.

When a neighbour sequence requires the same number of workstations as the store sequence (situation that may frequently occur in line balancing problems), a secondary objective function (13) is used as tie-breaker. This function creates solutions by attempting to load the first workstations at maximum capacity and the last ones at minimum capacity. To achieve this objective, the weight parameter α of f has been set to 10. It was verified that equivalent results can be obtained by using $\alpha=10^e$, being e an integer greater than 1.

$$min\ f = \sum_{j=1}^{m_k} \alpha^j \cdot Slk_j \qquad (13)$$

The local search ends when, for each task in *WS*, all feasible exchanges have been made; i.e., all neighbours have been generated and evaluated. For the next iteration, the stored sequence *SS* is assigned to the working sequence *WS*. The entire procedure is repeated until a predetermined computing time has been completed. The final solution is the best of all solutions that have been generated.

Exchange Movements

An adaptation of two classical transformations (see (Armentano & Basi, 2006)) has been considered to generate the neighbourhood of a given solution, as follows.

a. The exchange of the positions in WS of a pair of tasks

In this case, the exchange movement tries to exchange the position, in the working sequence *WS,* of any two tasks i and k, provided it is feasible; i.e., the precedence relations amongst the tasks are maintained. Furthermore, both task i and task k should have been assigned to different workstations. When task i and task k belong to the same subgraph s, new neighbour sequences are searched by interchanging s with each one of the remaining subgraphs available for such tasks.

b. The movement of task *i* to another position of the working sequence WS

In this type of movement a task is yielded to a different workstation. A task i can be moved to the position of task k when the tasks precedence relations are maintained and when task k and task i have been assigned to different workstations. In this case, all tasks between the positions of task i and k, including task k itself, are moved in the sequence one position backwards. Furthermore, for each movement, neighbour sequences are generated by interchanging the alternative subgraphs available for the moved task.

When a movement exchange type a is used in the local search phase, the local optimization procedure is regarded as LOP-1; otherwise, it is regarded as LOP-2.

Examples of exchange movements

Let consider an example (see (Capacho *et al.*, 2009)) of an ASALB problem involving 11 tasks and 7 alternative subgraphs, which represent the assembly variants that are allowed for three parts of a manufacturing system: alternative S1 and S2, for subassembly one; S3 and S4, for the second part; and S5, S6 and S7, for the third subassembly. Then, if the *NR* (i.e., minimum number of tasks) is the rule applied, subgraphs S1, S3 and S5 are selected for subassembly 1, 2 and 3, respectively. By choosing such subgraphs, the precedence relations presented in Table 3 are determined.

task	1	2	3	4	5	6	7	8	9	10	11
Immediate predecessors	-	-	-	1, 2	3	2, 5	4	6	7	8	9

Table 3. Resulting precedence relations by considering rule *NT* for selecting subgraphs.

Then, by applying rule *RPW* for selecting the next task to be assigned, the solution presented in Table 4 is obtained. Table 4 includes the number of required workstations m, the tasks assigned to each workstation, and the corresponding workstation time wt.

m=6	workstation					
	I	II	III	IV	V	VI
Tasks	2, 1	3	4, 5	6, 7	8, 10	9, 11
wt	11	17	16	16	15	15

Table 4. Task assignment to the workstations by applying rule *RPW*.

As can be observed in Table 4, tasks are assigned to the workstations in a particular order, which defines the tasks sequence that is used as the initial sequence *ISq*, as follows:

ISq = 2, 1, 3, 4, 5, 6, 7, 8, 10, 9, 11.

If a transformation type a is applied, the neighbour solution of Figure 1 is obtained by interchanging the positions of task 2 and task 3. This movement is possible since, as can be seen in Table 3, neither task 2 nor task 1 are predecessors of task 3, and neither task 1 nor task 3 are successors of task 2 (i.e., the precedence constraints are maintained). Moreover, as shown in Table 4, the tasks are assigned to different workstations: task 2 is assigned to workstation I and task 3 is assigned to workstation II.

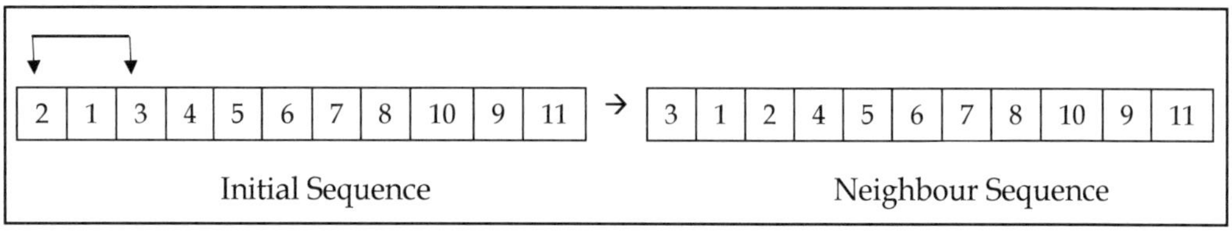

Fig. 1. Generation of a neighbour sequence applying transformation *a*.

If transformation *b* is considered, the neighbour sequence is generated by moving task 2 to the position of task 3; and by moving task 1 and task 3 one position backwards in the sequence. The resulting neighbour sequence is as follows.

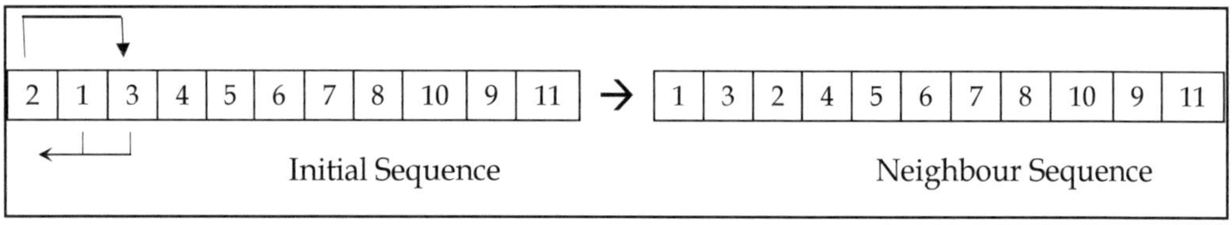

Fig. 2. Generation of a neighbour sequence applying transformation *b*.

3. Computational experiment

To evaluate and compare the performance of the proposed GRASP procedures described in the previous section, a computational experiment was carried out considering medium- and large-scale ASALBP. The data sets (see Table 5) used in the computational experiment are base on the following 10 benchmark problems: Gunther, Kilbrid, Hann, Warnecke, Tongue, Wee-Mag, Lutz3, Arcus2, Bartholdi and Scholl; with 35, 45, 53, 58, 70, 75, 89, 111, 149 and 297 tasks, respectively. Each benchmark problem is subdivided into two, three and four subassemblies; involving five, eight and eleven subgraphs, respectively. For each problem instance five cycle time values were considered. Benchmarks are available online at the web page for assembly line balancing research: www.assembly-line-balancing.de.
A total of 150 problem instances, involving from 37 to 305 tasks, were solved considering a computation time of 0.1 second on a Pentium IV, 3GHZ CPU with 512 Mb of RAM with each of the 78 proposed algorithms. All heuristic methods were implemented using C++.

4. Analysis of the results

To present the results obtained in the computational experiment, the following notation is used: *NI*, number of the tested instances; *CT*, computation time; *NBS*, number of best solutions obtained; *PBS*, percentage of best solutions; Δ_{max} , Δ_{av} , Δ_{min}, maximal, average and minimal deviation from the best solution, respectively. For each problem instance, the relative deviation from the best solution *BS*, of each heuristic solution *HS*, is computed as follows:

$$\Delta = \frac{HS - BS}{BS} \times 100 \qquad (14)$$

Problem	Cycle time values					Number of subgraphs		
						5	8	11
	ct_1	ct_2	ct_3	ct_4	ct_5	Number of tasks		
Bowman	20	-	-	-	-	10	-	-
Mansor	48	62	94	-	-	11	-	-
Mitchell	14	21	35	-	-	21	21	-
Buxey	30	36	54	-	-	29	29	-
Gunther	41	44	49	61	81	37	37	37
Kilbrid	57	79	92	138	184	45	46	48
Hahn	2004	2338	2806	3507	4676	56	56	63
Warnecke	54	62	74	92	111	63	63	67
Tonge	160	176	207	251	320	73	75	75
Wee-Mag	28	33	39	46	56	77	81	83
Lutz3	75	83	97	118	150	93	98	101
Arcus2	5785	6540	7916	9400	11570	115	121	125
Bartholdi	403	470	564	705	805	151	157	160
Scholl	75	83	97	118	150	299	302	305

Table 5. Data Sets

The overall performance of all methods is evaluated by considering the number of best solutions provided by the methods. The best solution for a problem instance, the basis for the comparative analysis, is the best of all solutions found by the proposed heuristic methods. The results are also compared with previous results obtained by methods using nominal, rather than weighted, values of the priority rules (e.g., Capacho *et al.*, 2009).

Table 6 shows the results obtained by applying all proposed procedures to solve all data sets. As observed in Table 6, better results were obtained by methods using local procedure LOP-2, which in most cases outperformed methods applying LOP-1; i.e., 24 methods performed better, in two cases both performed the same, and for 3 procedures it behave worse that LOP-1. On the other hand, all methods using LOP-2 provided best solutions in more than 54% of the cases.

The best performance was recorded for the method that employed W-NT_TSSk in the construction phase and applied LOP-2; which generated best solutions in 85,3% of the cases (this represents a 3.3% of improvement comparing with the same method when applying LOP-1). Furthermore, method W-NT_TSSk yielded a comparatively small value of Δ_{max} (16.7%), and the smallest value of Δ_{av} (1%).

Regarding LOP-1, the method that performed the best was W-NT_TSSk, which generated best solutions in 82,7% of the cases. Method W-NT_TSSk performed the same, generating best solutions in 82.7% of the cases regardless of the local optimization procedure applied.

Other methods that also performed well are those that employed W-TT_LWTS, W-TT_LWTS, W-NT_TS, W-TT_TS and applied LOP-2, all of which generated the best solutions in 78.7% of cases. Table 6 also shows that for most methods Δ_{max} is significantly high.

As it can be observed in Table 6, priority rule *EW* has a very poor performance, regardless of the rule used for subgraphs and the local optimization procedure applied; generating best solutions, at best, in 56 % of the cases.

	Method	LOP-1				LOP-2			
		NBS	*PBS*	Δ_{max}	Δ_{av}	*NBS*	*PBS*	Δ_{max}	Δ_{av}
1	W-NP_RPW	112	74.7	16.7	1.9	114	76.0	16.7	1.8
2	W-NP_T	92	61.3	33.3	3.1	94	62.7	33.3	2.9
3	W-NP_EW	75	50.0	33.3	4.0	85	56.7	33.3	3.7
4	W-NP_LW	74	49.3	33.3	3.8	85	56.7	33.3	3.6
5	W-NP_N	85	56.7	25.0	3.2	90	60.0	25.0	2.9
6	W-NP_Sk	74	49.3	33.3	3.7	83	55.3	33.3	3.4
7	W-NP_TLW	96	64.0	33.3	3.0	98	65.3	33.3	2.8
8	W-NP_IS	87	58.0	33.3	3.3	90	60.0	33.3	3.1
9	W-NP_TS	111	74.0	16.7	1.7	113	75.3	16.7	1.6
10	W-NP_TTS	93	62.0	25.0	2.7	94	62.7	25.0	2.6
11	W-NP_STS	76	50.7	33.3	3.8	81	54.0	33.3	3.5
12	W-NP_TSSk	109	72.7	16.7	1.8	109	72.7	16.7	1.7
13	W-NP_LWTS	111	74.0	16.7	1.9	112	74.7	16.7	1.7
14	W-TT_RPW	113	75.3	16.7	1.4	116	77.3	16.7	1.4
15	W-TT_T	99	66.0	33.3	2.5	100	66.7	33.3	2.3
16	W-TT_EW	80	53.3	33.3	3.3	84	56.0	33.3	3.0
17	W-TT_LW	85	56.7	33.3	3.2	88	58.7	33.3	3.0
18	W-TT_N	94	62,7	25.0	2.6	96	64.0	25.0	2.3
19	W-TT_Sk	83	55.3	33.3	3.3	89	59.3	33.3	3.0
20	W-TT_TLW	102	68.0	33.3	2.3	104	69.3	33.3	2.1
21	W-TT_IS	97	64.7	25.0	2.3	100	66.7	25.0	2.0
22	W-TT_TS	116	77.3	16.7	1.3	118	78.7	16.7	1.1
23	W-TT_TTS	105	70.0	25.0	2.0	106	70.7	25.0	1.8
24	W-TT_STS	88	58.7	33.3	2.8	90	60.0	33.3	2.6
25	W-TT_TSSk	124	82.7	16.7	1.0	124	82.7	16.7	1.0
26	W-TT_LWTS	119	79.3	16.7	1.3	118	78.7	16.7	1.3
27	W-NT_RPW	116	77.3	16.7	1.4	113	75.3	16.7	1.4
28	W-NT_T	99	66.0	33.3	2.5	100	66.7	33.3	2.3
29	W-NT_EW	79	52.7	33.3	3.4	83	55.3	33.3	3.2
30	W-NT_LW	85	56.7	33.3	3.2	88	58.7	33.3	3.0
31	W-NT_N	93	62.0	25.0	2.6	95	63.3	25.0	2.4
32	W-NT_Sk	83	55.3	33.3	3.3	89	59.3	33.3	3.0
33	W-NT_TLW	102	68.0	33.3	2.3	104	69.3	33.3	2.1
34	W-NT_IS	96	64.0	33.3	2.5	99	66.0	33.3	2.2
35	W-NT_TS	116	77.3	16.7	1.3	118	78.7	16.7	1.1
36	W-NT_TTS	105	70.0	25.0	2.0	106	70.7	25.0	1.8
37	W-NT_STS	88	58.7	33.3	2.8	90	60.0	33.3	2.6
38	W-NT_TSSk	123	82.0	16.7	1.0	128	85.3	16.7	1.0
39	W-NT_LWTS	119	79.3	16.7	1.3	118	78.7	16.7	1.3
Δ_{min}= 0 in all cases									

Table 6. Results of applying the GRASP procedures (*NI* = 150)

On the other hand, Table 6 also shows that methods employing different local search procedures behave very similarly when the same heuristic method is used to build the initial solution. This means that when a constructive method performs well when applying LOP-1, it also performs well (or better) when applying LOP-2. This behaviour can also be observed in Figure 3, which shows the percentage of improvement (in one workstation) of the proposed procedures comparing with other multi-pass methods.

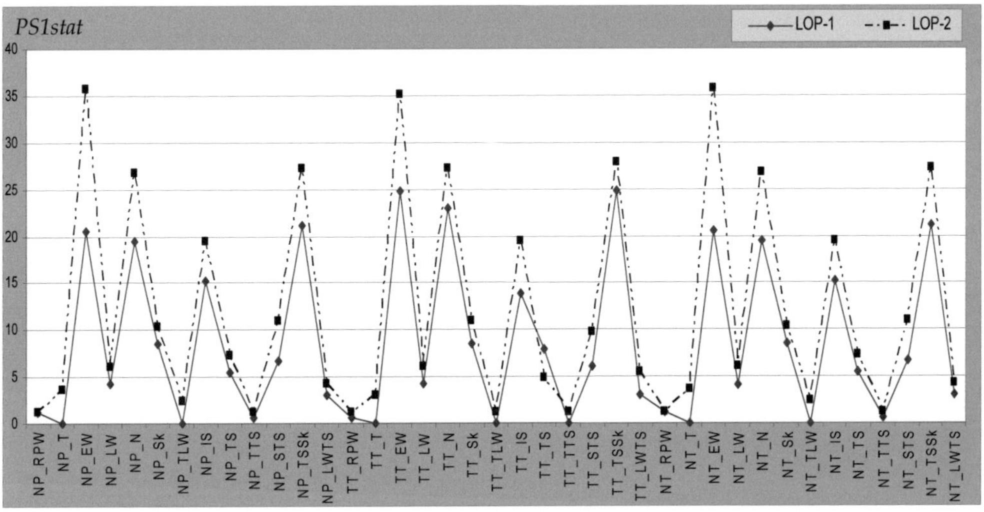

Fig. 3. Performance of GRASP procedures

The results obtained with the proposed GRASP methods were also compared with the results obtained by multi-pass methods, in which the assembly subgraphs are randomly selected and tasks are assigned according to nominal values of the priority rules of Table 1. Table 7 shows the percentages of improvement in the solution provided by the proposed GRASP methods comparing with multi-pass ones (e.g., Capacho *et al.*, 2009), considering a CT=0.1 seconds. Data in Table 7 includes Imp_{max}, Imp_{av}, Imp_{min}, maximal, average, and minimum percentage of improvement, respectively. It also shows the percentage of cases in which the best solution found requires 1, 2 or 3 workstations less (%1*ws*, %2*ws*, %3*ws*), respectively, than the best result provided by the corresponding multi-pass methods. As can be seen in Table 7, the best results were obtained when LOP-2 was applied. This strategy provided an additional 12.9% of best solutions, getting in some cases up to 35.6%. Table 7 also reveals that both types of procedures were able to generate improved solutions in which up to three fewer workstations were required; i.e., in 0.1% and 0.3% of the cases with LOP-1 and LOP-2, respectively.

5. Conclusions and further research work

In this chapter, the metaheuristic approach GRASP was used to solve the Alternative Subgraphs Assembly Line Balancing Problem (ASALBP). Thirty-nine construction methods, based on weighted priority rule values, and two local search strategies (LOP-1 and LOP-2) were proposed, implemented and evaluated.

Local method	Imp_{max}	Imp_{av}	Imp_{min}	% 1 *ws*	% 2 *ws*	% 3 *ws*
LOP-1	24.9	6.8	0.2	2.2	0.3	0.1
LOP-2	35.6	12.9	2.7	3.6	0.6	0.3

Table 7. Comparison of nominal- and weighted- rule based methods

The results obtained showed that methods that used LOP-2 performed better than those that used LOP-1, achieving best solutions in up to 85.3% of all cases, considering all the proposed construction methods. Nevertheless, some methods applying LOP-1 (i.e., W-NT_TSSk and W-TT_TSSk) also performed well, providing best solutions in up to 82.7%.

The results also showed that a significant improvement can be obtained in comparison to previous results obtained using multi-pass methods based on single priority rule values and using random choice for subgraphs. Improved solutions were obtained in which the number of workstations required was reduced by one, two or even three, which represent the best results obtained with any method developed up to now to solve the ASALBP. Thus, all proposed methods that used LOP-2 could be applied to solve an Alternative Subgraphs Assembly Line Balancing Problem to select the best overall solution.

Further research involves exploring other methods to solve the ASALBP. The growing interest on using Evolutionary Algorithms to solve optimization problems in industry makes the use of such procedures an attractive approach, which, in addition, has been successfully applied to complex assembly line balancing problems. On the other hand, in order to increase the practicality of the problem, its definition can be extended by including new features such as, for example, stochastic processing times, setups, and different line layouts.

6. Acknowledgment

Supported by the University of Los Andes, Mérida - Venezuela.

7. References

Andrés, C., Miralles, C. & Pastor, R. (2008). Balancing and scheduling tasks in assembly lines with sequence-dependent setup times. *European Journal of Operational Research*, Vol. 187, No. 3, pp. 1212-1223.

Armentano, A. & Bassi, O. (2006). Graph with memory-based mechanisms for minimizing total tardiness in single machine scheduling with setup Times. *Journal of Heuristics*, Vol. 12, pp. 427-446.

Becker, C. & Scholl, A. (2006). A survey on problems and methods in generalized assembly line balancing. *European Journal of Operational Research,* Vol. 168, pp. 694-715.

Capacho, L. (2007). *ASALBP: the Alternative Subgraphs Assembly Line Balancing Problem. Formalization and resolution procedures.* PHD thesis. Tecnical Universtity of Catalunya. Barcelona, Spain.

Capacho L. & Pastor, R. (2006). The ASALB problem with processing alternatives involving different tasks: definition, formalization and resolution. *Lecture Notes in Computer Science,* Vol. 3982, pp. 554-563.

Capacho, L. & Pastor, R. (2008). ASALBP: The alternative subgraphs assembly line balancing problem. *International Journal of Production Research,* Vol. 46, pp. 3503-3516.

Capacho, L., Pastor, R., Dolgui, A. & Guschinskaya, O. (2009). An evaluation of constructive heuristic methods to solve the alternative subgraphs assembly line balancing problem. *Journal of Heuristics,* Vol. 15, pp. 109-132.

Capacho, L., Pastor, R., Guschinskaya, O. and Dolgui, A. (2006). Heuristic Methods to Solve the Alternative Subgraphs Assembly Line Balancing Problem. *Proceedings of the IEEE Conference on Automation Science and Engineering - CASE 2006,* pp. 501-506, ISBN: 1-4244-0310-38-11, Shanghai-China. October 7-10, 2006.

Das, S. & Nagendra, P. (1997). Selection of routes in a flexible manufacturing facility. *International Journal of Production Economics,* Vol. 48, pp. 237–247.

Delorme, X., Gandibleux, X. & Rodriguez, J. (2004). GRASP for set packing problems, *European Journal of Operational,* Vol. 153, pp. 564-580.

Dolgui, A., Eremeev, O. & Guschinskaya, O. (2010). MIP-based GRASP and genetic algorithm for balancing transfer lines. *Matheuristics: Hybridizing Metaheuristics and Mathematical Programming,* Maniezzo, V., Stutzle, T., Voss, S. (eds.), Annals of Information Systems, Springer, 10, 189-208.

Falkenauer, E. (2005). Line Balancing in the Real World. *Proceedings of the International Conference on Product Lifecycle Management* PLM′05, 360 - 370.

Festa, P. & Resende, M. (2008a). An annotated bibliografy of GRASP. Part I: Algoritms. Technical Report. AT&T Labs Research. Available online at www.optimization-online.org/DB_FILE/2008/06/2011.pdf. Visited: 01.03.11.

Festa, P. & Resende, M. (2008b). A*n annotated bibliografy of GRASP. Part II: Aplications. Technical Report.* AT&T Labs Research. 01. 03.11. Available online at: www2.research.att.com/~mgcr/doc/gabib-appl.pdf. Visited: 01.03.11.

Martino, L. & Pastor, R. (2007). *Heuristic procedures for solving the General Assembly Balancing Problem with Setups (GALBPS).* Tecnical Report, IOC-DT-P-2007-15. Tecnical Universtity of Catalunya.

Resende, M. & Ribeiro, C. (2003). Greedy Randomized Adaptive Search Procedures. Handbook of Metaheuristics. *International Series in Operations Research & Management Science,* Vol. 57, pp. 219-249, DOI: 10.1007/0-306-48056-5_8

Scholl, A., Boysen, N. & Fliedner, M. (2008). The sequence-dependent assembly line balancing problem. *Operations Research Spectrum,* Vol. 30, pp. 579-609.

Scholl, A., Boysen, N. & Fliedner, M. (2009). Optimally solving the alternative subgraphs assembly line balancing problem. *Annals of Operations Research,* vol. 172, No.1, pp. 243-258.

Talbot, F.B, Patterson, J.H. & Gehrlein, W.V. (1986). A comparative evaluation of heuristic line balancing techniques. *Management Science,* Vol. 32, pp. 431-453.

4

Model Sequencing and Worker Transfer System for Mixed Model Team Oriented Assembly Lines

Emre Cevikcan and M. Bulent Durmusoglu
Istanbul Technical University
Turkey

1. Introduction

Assembly lines are the most commonly used method for a mass production environment. Their main purpose is to increase efficiency by maximizing the ratio between throughput and required costs. However, in the last few decades, market demands have changed enormously. The emphasis is now on shorter lead times, larger product variety, higher quality and more customized options. At the same time, socioeconomic conditions have improved and workers have a greater interest in work satisfaction. Increasing the importance of quality and flexibility of the assembly system, while providing a satisfying work environment, these changes lead to the utilization of teamwork for assembly line design. Unlike conventional assembly lines, team-oriented assembly lines consist of multimanned workstations, where workers' groups simultaneously perform different assembly works on the same product and workstation. In addition, it is also different from installing parallel (multiple) stations, where individual products are distributed among several workers who perform the same tasks but on different products (Dimitriadis, 2006). Team-oriented assembly systems in which workers are organized in groups outperform traditional assembly systems in terms of cost, lead time, flexibility, quality, and worker satisfaction. However, in such lines, performance is affected in a negative way on the condition that high level of variability among station times result in utility and idle times.

Some research in the literature states the convenience of team oriented assembly systems for current market needs when compared to conventional assembly lines the stations of which consists of one worker. For instance, Wild, R. (1975) examined the types, basic design considerations and benefits of teamwork. The author stated that implementing teamwork yields good results in quality development and workers may benefit via increased confidence through the development of social skills in a teamwork environment. Groover, M. P. (2001) maintained that compared with workers on a conventional line, the members of an assembly team achieved a greater level of personal satisfaction at having accomplished a major portion of product. Bukchin, J. et al. (1997) provided a straightforward study applying teamwork approach to assembly line design. According to the authors, team-oriented assembly system approach should be used to overcome the disadvantages of classical assembly design: low quality, poor working environment, long flow time and high costs of material handling. In spite of the fact that teamwork approach is applied in assembly lines frequently, very few reported studies have utilized. Johnson, R. V. (1991) discussed the fact that labor cost increases with both the number of the teams and the percentage of tasks that can be allocated to one team only when using disconnected teams for balancing. Bukchin J. and Masin M. (2004) presented a multi objective design of team oriented assembly system requiring bill of materials of the product. They proposed both the optimal solution procedure based on a backtracking branch and bound algorithm and a heuristic algorithm based on the optimal algorithm for large-scale problems. More recently, in Dimitriadis (2006), a two-level heuristic for single model team-oriented ALBP has been proposed. However, this heuristic attempts to solve balancing problem for only single model assembly line.

What is more, the capability of synchronous assembly of products an important matter when considering current market needs. However, in mixed model assembly lines (MMAL), complication exists as result of congestion and starvation caused by the arrival of different models to the line, having different assembly time requirements at each station. In this context, an effective model sequencing for mixed model assembly line (i.e. determining the order in which products have to be introduced into the assembly line) increases the performance of such a line.

Sequencing is usually carried out with the two primary goals such as levelling the workload (total operation time) at each workstation on the assembly line to reduce line inefficiencies such as idleness, utility work, work deficiency and work congestion (Thomopoulos, 1967; Macaskill, 1973, Bolat and Yano, 1992; Xiaobo and Ohno, 1997; Hyun et al., 1998; Sarker and Pan, 2001; Erel et. al., 2007), keeping a uniform parts usage (Xiaobo, et al., 1999; Tamura, et al., 1999; Jin and Wu, 2002; Kurashige, et al., 2002; Ventura and Radhakrishnan, 2002). What is more, some attempts have been made to solve multi-objective MMAL sequencing problem via heuristics (Kotani et al., 2004; Aigbedo and Monden, 1997; Korkmazel and Meral, 2001), metaheristics (Tavakkoli-Moghaddam and Rahimi-Vahed, 2006; Akgunduz and Tunali, 2010) and mixed integer linear programming models (Giard and Jeunet, 2010).

When considering the large body of literature, it is revealed that there has not been any published study addressing both worker transfer and sequencing in MMALs. That being the case, in this chapter, a model sequencing and worker transfer systematic for team oriented assembly lines is developed. Utility time is considered in the determination of bottleneck station(s) via simulation. Then, model sequencing with four method alternatives and worker transfers between sequential stations are utilized as the two techniques to reduce utility time. Last but not least, the proposed systematic has the potential of being applied to real-life industrial sized MMALs. In this context, a real life mixed model tractor assembly system is presented to demonstrate the application of the proposed system.

The remaining sections of the chapter are organized as follows. In Section 2, the description of the problem is presented. In Section 3, proposed model sequencing and worker transfer systematic given.. Section 4 contains industrial application of the proposed methods to a real life assembly system and finally in Section 5, conclusions are provided.

2. Description of the problem

The following are the characteristics of the mixed model assembly line in which the problem arises and problem assumptions:

- The assembly of the products is performed when they are moving on the conveyor system with a constant speed through the assembly line which consists of K work stations.
- An assembly team work in each station. The members of a team simultaneously perform different assembly works on the same product and workstation.

- Consecutive products are launched on the line from the first station at a constant time interval.
- The products enter each work station in the same sequence.
- There is no buffer between work stations.
- Each product model may have different assembly times which are assumed to be deterministic.
- The product moves on a conveyor which has a constant speed.
- Each work station has upstream and downstream boundaries.
- The operator returns to the upstream boundary of the station or to the next product, whatever is reached first, in zero time after finishing the work load on the current product due to the fact that the speed of conveyor is much slower than the walking speed of the workers.
- When utility work occurs (i.e. the operator can not complete the assembly tasks of a product within his/her allowable work zone), utility workers are utilized. Whenever the operator finds that he/she might fail to complete the operation within his/her work zone, he/she calls the utility worker who additionally assists such that the work can be completed on time.
- Each utility work can be performed on time (i.e. a constraint in terms of utility worker does not exist.)

An example of operators' movements in their work stations is shown in Figure 1. The lines with arrowheads represent assembly times and dotted lines represent the movement of the operators. The conveyor moves from the left to the right. Physical length of workstation "*k*" is denoted by PL_k. The line consists of four work stations the boundaries of which are represented by vertical lines. Tem sizes of stations are two, one, three and two, respectively. Two utility workers exist to assist operators in the work stations when they fail to complete their work in their work zone. The first utility worker is responsible for performing utility work occurred in work stations 1 and 2 (i.e. Line Segment 1). Similarly, the second utility worker performs utility work occurred in work stations 3 and 4 (i.e. Line Segment 2). To reach the product in position 7 in work station 1, the operator can not go beyond the upstream station boundary and he/she has to wait ot_1 time units. In work station 2, the operator can not complete the jobs on products in the positions of 3 and 7 until downstream boundary. Therefore, a work overload of ut_1 time units are performed on product 3, and ut_4 time units on product 7 is performed by utility worker 1. Similarly, in work station 3, a work overload of ut_2 time units are performed on product 4 is performed by utility worker 2.

3. Proposed methodology

Proposed methodology consists of five steps as shown in Figure 2. Each of these steps will be explained in this section. As mentioned above, the chapter focuses on model sequencing and worker transfer. These two concepts are the last two stages of a five-stage methodology developed by Cevikcan et al. (2009) that can be seen for a detailed discussion of the first three stages. Due to the fact that this chapter focuses on model sequencing and worker transfer system, first three steps of the methodology have not been explained in detail so as not to expand the study in scope.

The following notation is used within this chapter:

UT_t= total utility time for team combination "*t*"

C_m = cycle time for model "*m*"

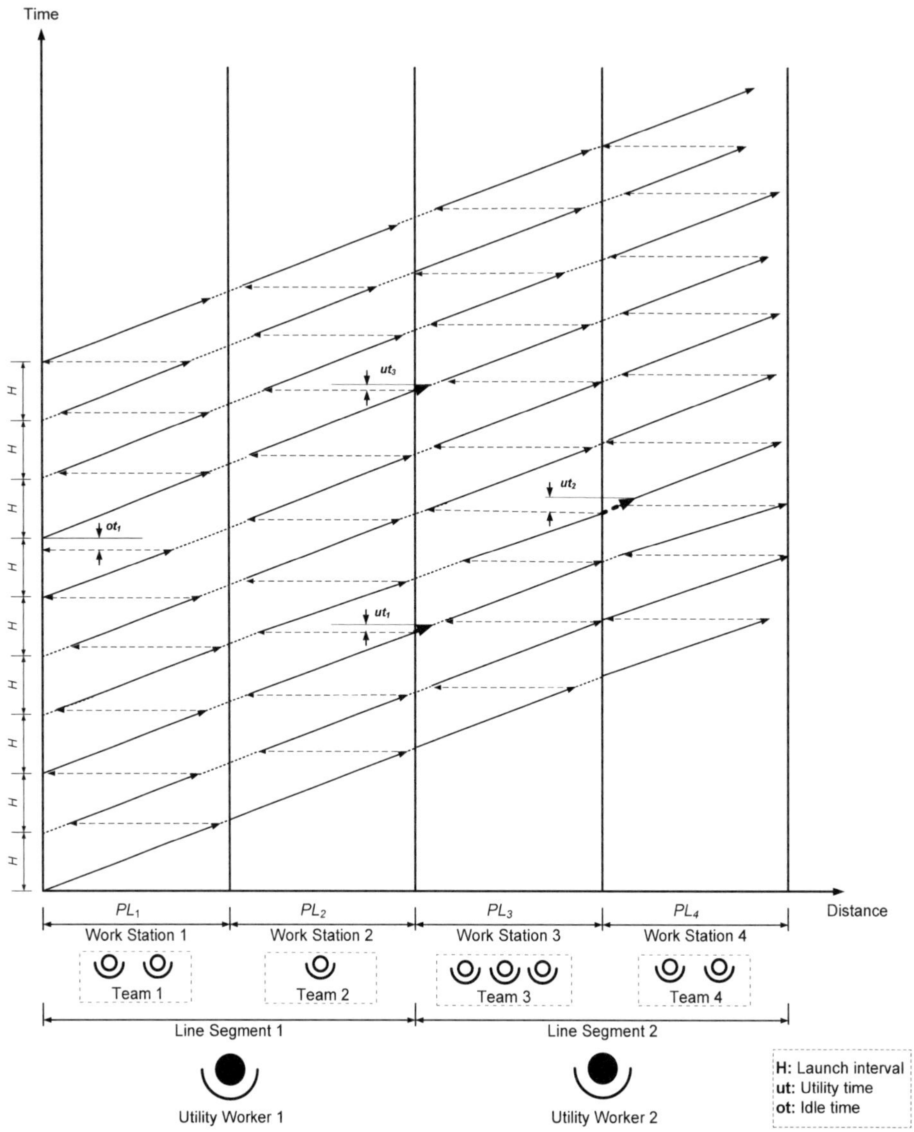

Fig. 1. Operator movement diagram

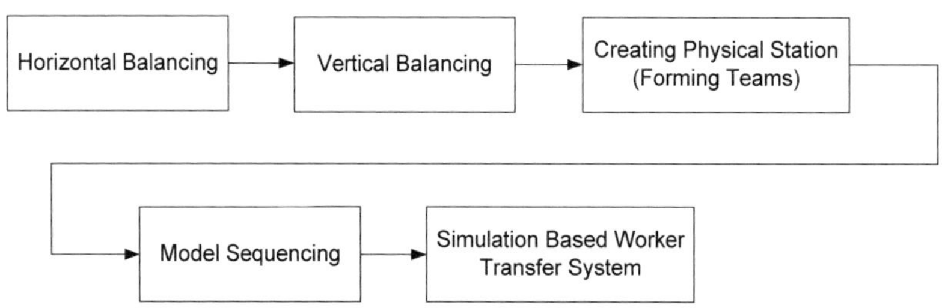

Fig. 2. Proposed methodology

CW = weighted average cycle time
d_m = demand for model "m"
D= total demand
$DD(m,p)$ =Priority value of p_{th} product of model "m"
M = the number of models
$m(j)$ = the model in the j_{th} sequence
m = index for model
m_k = the size of k_{th} team before worker transfer
m'_k =the size of k_{th} team after worker transfer
p_m = p_{th} product of model "m"
sr = parameter value of search range
t = index for team combination

3.1 Horizontal balancing

Horizontal balancing is an efficiency improvement process that groups the tasks into families such that of each task family takes the time as close as the required cycle time (or takt time) of each model or model family without exceeding this cycle time and violating the precedence relation between tasks. When the five steps of the methodology are considered, the horizontal balancing will have a marginal effect on the whole system performance. Therefore, the algorithms which seek the optimal solutions, were not comprehensively analyzed in the literature, a practical and fast algorithm was decided to be chosen for industrial cases. The chosen algorithm is "regional approach" heuristic which is suggested by Bedworth and Bailey (1987).

3.2 Vertical balancing

Although the result of the horizontal balancing satisfies the cycle time constraints, it could be unfair task assignments between workers. Therefore a vertical balancing procedure is performed (Merengo et al., 1999). This procedure tries to reassign some of the tasks of the workers who have more workload than the average, to the workers who have less, under the constraints of precedence relations and cycle time.

3.3 Creating physical station

In traditional assembly lines, only one worker per station is assumed, however this is not the actual situation on the shop floor in today's assembly systems. Different number of workers works together as a team in one place/station. Figure 3 shows the algorithm developed for forming teams.Team formation approach in this chapter, is based on the heuristic algorithm which is developed for project scheduling under limited resources (Elsayed and Boucher, 1985). Here, a modified version of this algorithm is used for a different field. Re-allocating tasks to workers (resources) under assembly time (ASTIM) criterion are performed using this algorithm.

The maximum allowable number of workers for a physical station, i.e. the maximum size for teams is an important external constraint for the methodology. The team has to be large enough in order to enable team dynamics and allow a variety of ideas and skills, yet small enough to enable cohesiveness of the team members.

Human interactions are added to the algorithm, considering the different design parameters of a real life assembly system. The user compares physical station assembly time with model cycle times for each team size and eliminates each the team size alternative even if its physical station assembly time exceeds the pre-determined tolerance of cycle time for a single model. The different design parameters of a real life assembly system is considered during this stage. The determination of the team size in the considered station could depend on the location of the station and the location depends on the task sets being assigned to the teams until the considered station. In a team oriented assembly line, some work elements assigned to a worker of a specific workstation can be delayed by the work elements assigned to some other worker of the same workstation, hence physical station assembly times are allowed to exceed model cycle times within a tolerance (0-20%).

The step of creating the physical station begins with the first worker, and proceeds consecutively with respect to the sequence of workers. For instance, if he/she chooses three as the size of the first team, the algorithm starts to schedule team size alternatives such that the first worker of the second team is worker 4. The algorithm stops creating the physical station process when the last worker is assigned to the last team. The following section describes the scheduling of assembly tasks for a team of workers.

3.4 Model sequencing

At this step, the sequence with which the models of the product will enter the assembly line will be determined. Figure 4 shows the model sequencing flow diagram of the methodology. While evaluating each sequencing algorithm, total utility time is regarded as the performance measure. The sequencing algorithm with the lowest utility time is preferred among alternative methods for the current scheduling period.

According to Figure 4, on the condition that a mixed model sequence is developed instead of batch sequence, the utilization of cycle time as a parameter is concerned. When cycle time of each model is used for model sequencing, cycle time based sequencing algorithm is applied. If demand values of models are considered as a sequencing parameter, demand value based sequencing algorithm is focused. What is more, an algorithm is included to the systematic so as to improve the performance of demand value based sequencing algorithm in terms of utility time.

3.4.1 Cycle time based sequencing algorithm (CTBSA)

That models with longer station times should not enter the line consequently is considered in model sequencing because of the fact that in the sequence, consecutive positions of models with long assembly times trigger utility time. On the other hand, the assembly times are proportional to cycle time due to cycle time constraint in assembly line balancing problem. Therefore, regarding cycle time as an indicator of a work content for a product model, CTBSA utilizes the cycle time with the aim of workload levelling among stations. CTBSA attempts to position models with longer cycle time and models with shorter cycle time consecutively with the aim of decreasing utility time. The steps of the cycle time based sequencing algorithm are as follows:

1. Calculate the weighted average cycle time (*CW*).

$$CW = \sum_{m=1}^{M} C_m * d_m \tag{1}$$

2. Assign the value of 1 to *j* (sequence index).

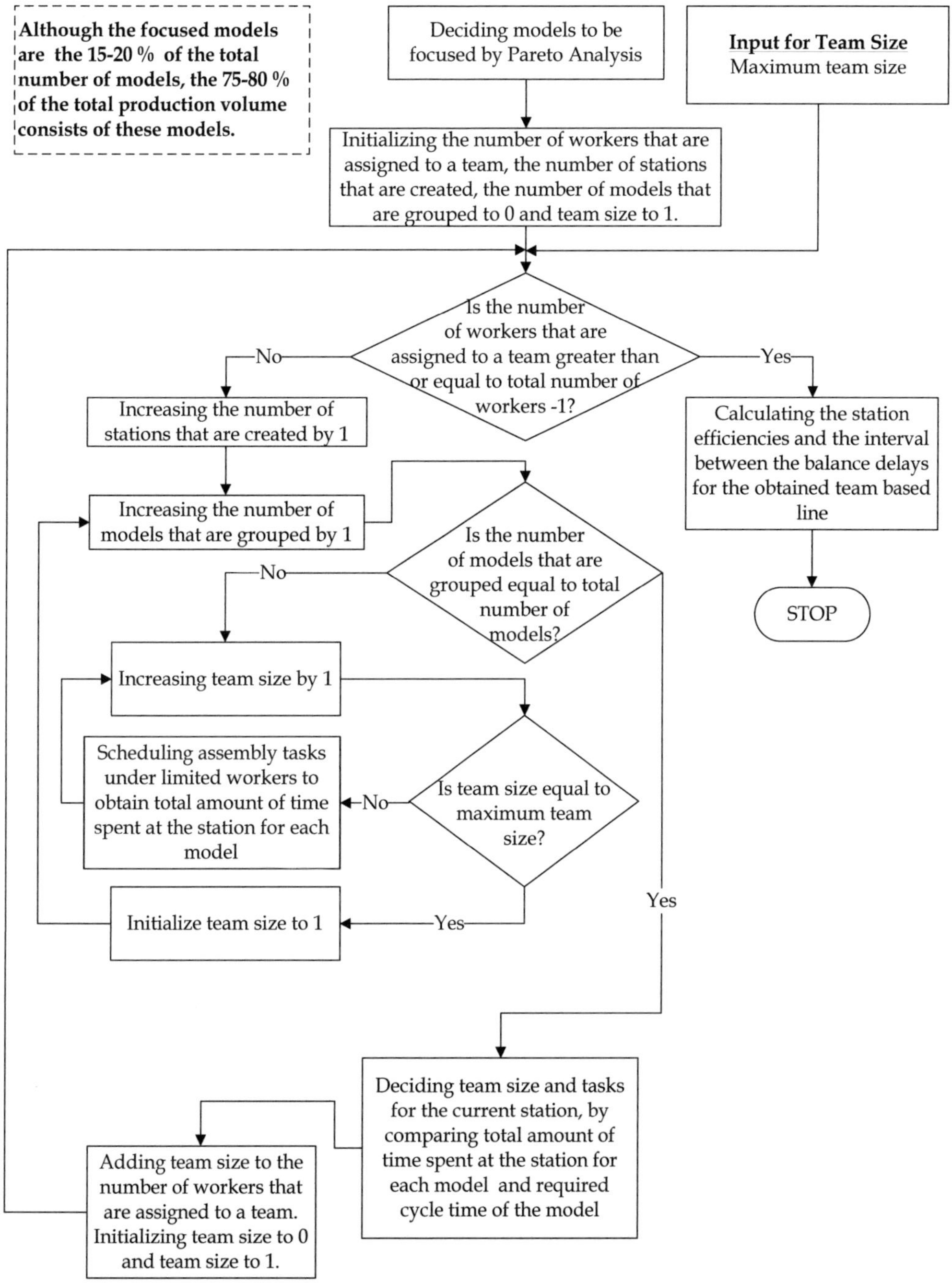

Fig. 3. Algorithm for team forming

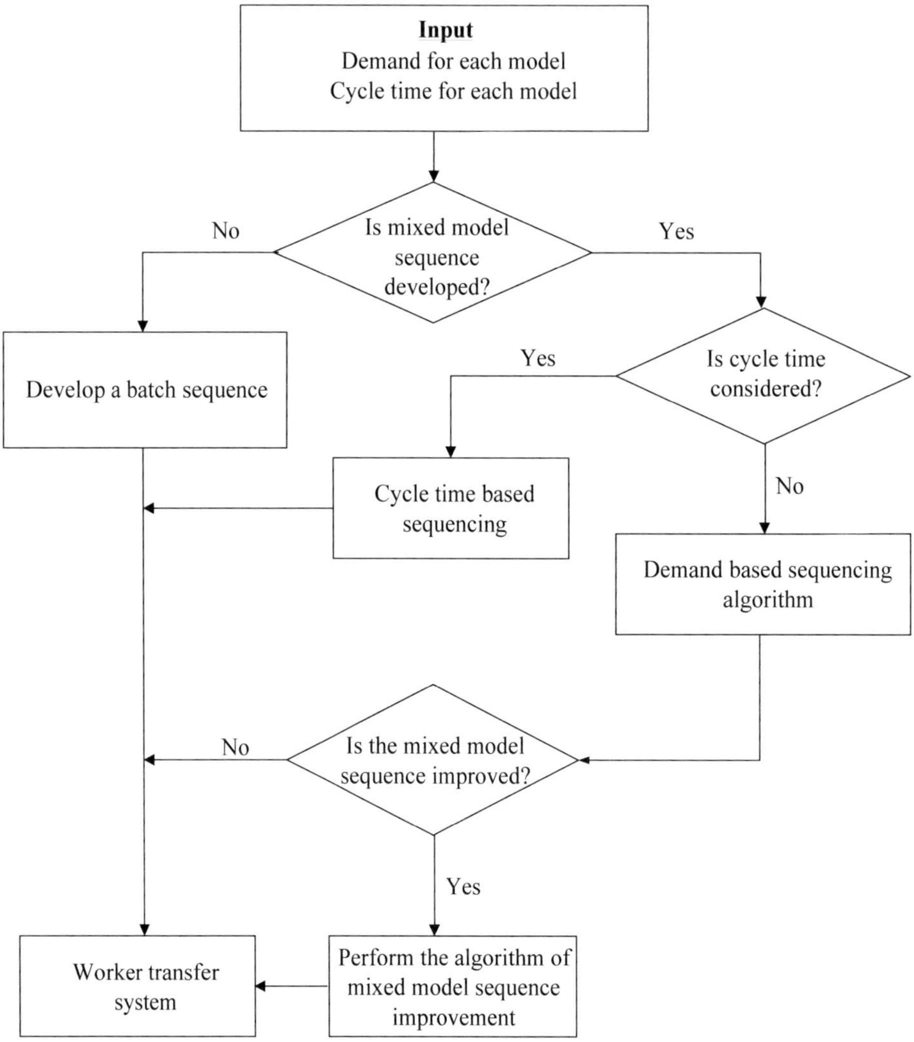

Fig. 4. The model sequencing flow diagram

3. Calculate binary average cycle time of j_{th} model and each of the following models in the sequence.
4. Calculate the absolute value of the difference between the weighted average cycle time and each binary average cycle time.
5. Find the following model with the lowest absolute value other than itself.
6. Assign the value of "j+1" to the sequence number of this model.
7. Increase "j" by one.
8. Go to step 9 if j=D-1, otherwise go to step 3.

$$D = \sum_{m=1}^{M} d_m \tag{2}$$

9. Stop.

3.4.2 Demand based sequencing algorithm (DBSA)

The data requirement is a critical issue for the methods for production systems. Therefore, data requirement affects the adoption of the method in real life production systems. For instance, when BOM (Bill of Materials) data for assembly is required for a MMAL sequencing method, the potential of applying sequencing method reduces because of the fact that it is very difficult to update BOMs in real life assembly systems. That being the case, for real life assembly systems, the sequencing methods with simple data requirement is preferred and developed within this chapter. As an indicator of this issue, demand based sequencing algorithm, the parameter of which is only demand value of each model, is included to the study. The following simple formula is suggested for DBSA (Merengo et al., 1999):

$$DD(m,p) = \left(p - \frac{1}{2}\right)\frac{D}{d_m} \tag{3}$$

This value is calculated for each model type and p, and then the sequence is generated by sorting these values in ascending order. Since DBSA does not include time-based parameters, it can be enhanced in terms of workload leveling via the algorithm the steps of which are given below:

1. Calculate the weighted average cycle time.
2. Assign the value of 1 to j (sequence index).
3. Determine "*pc*" value (the number of pair-wise comparisons).

$$pc = \left(\frac{D}{d_{m(j)}}\right)^{+} + sr \tag{4}$$

4. Calculate binary average cycle time of jth model and each of the following "*pc*" models in the sequence seperately.
5. Calculate the absolute value of the difference between the weighted average cycle time and each binary average cycle time.
6. Find the following model with the lowest absolute value other than itself.
7. Assign the value of "j+1" to the sequence number of this model.
8. Increase "j" by one.
9. Go to step 10 if j=D-1, otherwise go to step 3.
10. Stop.

3.5 Simulation based worker transfer system

In order to eliminate the problems arising from the system unexpected events such as noise factors, the worker transfer (WT) system is built between neighbor stations by the help of computer simulation after model sequencing. The bottleneck and idle stations are determined with respect to the status of the assembly system (i.e. changing the model sequence, the model demands etc.) before the day starts. The worker transfer strategy is established based on the output of the simulation. The proposed methodology is a dynamic design methodology. Horizontal/vertical balancing and creating physical station are performed at the time for only adding or extracting the models. The frequency of model sequencing and worker transfer system is once or more than once every day if necessary. Developed procedure to eliminate bottlenecks is as follows:

1. Determine current team combination as the result of the third step of the methodology.
2. Determine the bottleneck station "R" with the highest total utility time in current team combination via simulation.
3. Generate team combinations by decreasing the size of station "R" by one when increasing the size of an upstream station of station "R" by one.

$$m'_R = m_R - 1 \tag{5a}$$

$$m'_k = m_k + 1 \qquad k=R-1, R-2, \ldots, 1 \tag{5b}$$

4. Generate team combinations by increasing the size of station "R-1" by one when decreasing the size of an upstream station of "R-1" by one.

$$m'_{R-1} = m_{R-1} + 1 \tag{6a}$$

$$m'_k = m_k - 1 \qquad \forall k : m_k \neq 1 \wedge k < R-1 \tag{6b}$$

5. Determine assembly time of stations of each generated team combination.
6. If there exists one or more team combination satisfying cycle time constraint with respect to assembly times for each model, go to step 7; otherwise go to step 11.
7. Calculate total utility time for each feasible (satisfying cycle time constraint) generated team combination.
8. If there exists one or more team combination(s) which has/have lower utility time than current team combination, go to step 9; otherwise go to step 11.
9. Determine the team combination with the best performance by regarding total utility time.
10. Make the team combination which is determined in step 9 be the current team combination and go to step 2.
11. Make the current team combination be the final team combination and stop.

4. Industrial application

The proposed methodology is applied to a chosen segment of a tractor assembly system. The tractor assembly system consists of ten segments and each segment includes many interrelated tasks (Figure 5). Among these segments, the fourth segment of the main line is chosen for the analysis of the proposed methodology. Brake pedal, motor, bumper support assembly is performed in this segment. Table 1 shows the cycle time of each model.

Model	Code	Cycle time (seconds)
T240	1	355
T266	2	576
T3075	3	576
T3085	4	672
T431	5	456
T461	6	486

Table 1. Cycle time of each model

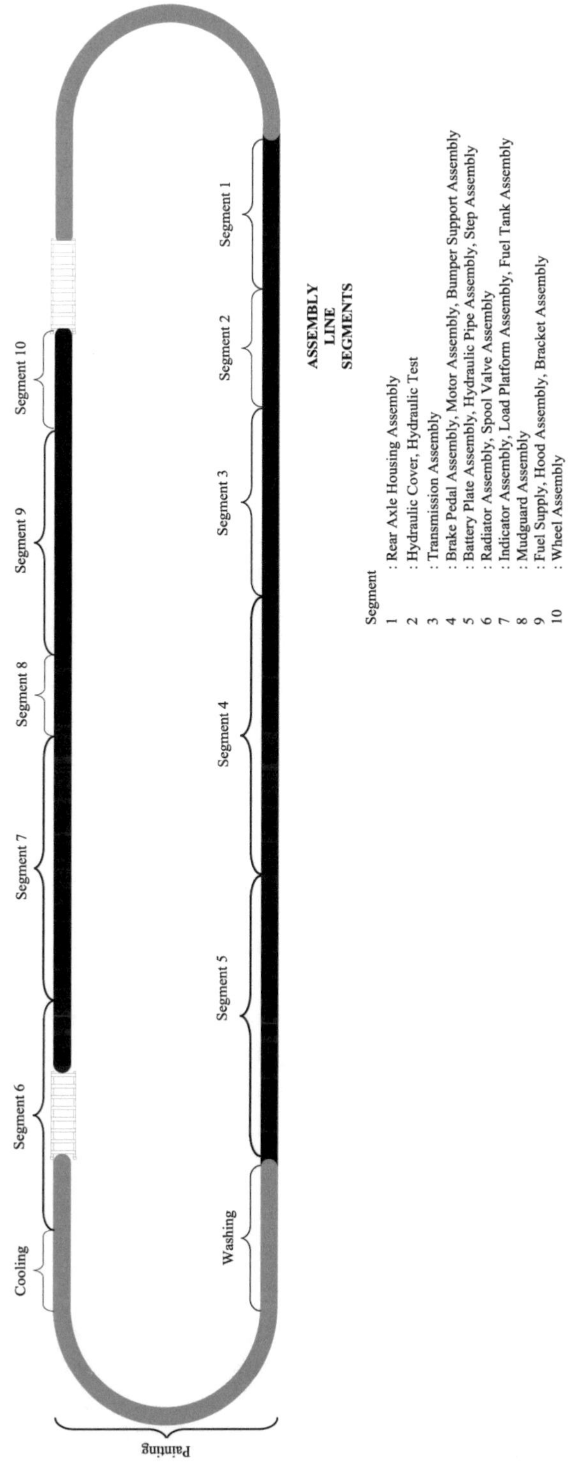

Fig. 5. The Segments of Tractor Assembly Line

In the step of physical station creating, 7 is regarded as the maximum allowable number of workers for a physical station. In this step, the computer program gives user team size alternatives, their assembly time and % deviations of physical station assembly time from cycle time for each model. In this context, the user decides a team size for each physical station. In this application, the user eliminates each the team size alternative when its physical station assembly time exceeds % 15 of cycle time for a model. Fourteen workers are grouped in seven assembly teams as the output of this step. The sizes and assembly times of teams are given in Table 6.

	Station Number						
	1	2	3	4	5	6	7
Team Size	3	1	2	3	2	2	1
Models	Assembly Time (seconds)						
1	355	301	265	286	246	321	285
2	400	398	393	556	483	234	285
3	335	330	393	402	448	341	285
4	335	228	393	414	395	363	285
5	342	264	265	285	293	345	285
6	342	345	265	317	339	345	285

Table 6. The sizes and assembly times of assembly teams

When determining the most appropriate model sequence, four different alternatives are considered which are listed below. Furthermore, the relevant data about sequencing methods are given in Table 7.

Model	Code	Demand Weight	Daily Demand	Number of pairwise comparisons
T240	1	0.492	39	3
T266	2	0.255	21	5
T3075	3	0.042	3	25
T3085	4	0.061	5	17
T431	5	0.114	9	10
T461	6	0.038	3	27
sr=1				
D=80				
CW=456				

Table 7. Relevant data for model sequencing

- Batch sequence (model batch with the lowest average assembly time first)-BS
- Cycle time based model sequence-CTBSA
- Demand based model sequence-DBSA
- Improved demand based model sequence-IDBSA

Assembly line is simulated under each model sequence via MS Excel. Total utility time, is considered with the aim of evaluating the performance of model sequences. Table 8 shows the performance of model sequences. As can be seen in Table 8, improved demand based model sequence yields best results in terms of total utility time.

Sequence	Total Utility Time (minutes)
BS	322
CTBSA	183
DBSA	215
IDBSA	162

Table 8. Performance of model sequences

Procedure for eliminating the relevant bottleneck is performed as below.
Step 1 of the procedure: Current team combination=3-1-2-3-2-2-1
Step 2 of the procedure: Station 4 is determined as bottleneck station due to the fact that it has the highest total utility time (8.4 minutes) with respect to IDBSA.
Step 3 of the procedure: The following team combinations are generated by decreasing the size of station 4 by one when increasing the size of an upstream station of station 4. Note that team sizes written in bold are changed.
3-1-**3**-**2**-2-2-1
3-**2**-2-**2**-2-2-1
4-1-2-**2**-2-2-1
Step 4 of the procedure: The following team combinations are generated by increasing the size of station 3 by one when decreasing the size of an upstream station of 3 by one.
2-1-**3**-3-2-2-1
Steps 5&6&7 of the procedure: Assembly time of stations of each generated team combination is determined. The performance of feasible team combinations are obtained and listed in Table 9.
Step 8 of the procedure: Team combinations of (3-1-3-2-2-2-1) and (2-1-3-3-2-2-1) have lower utility time than current team combination has (24.6 minutes).
Step 9 of the procedure: Between (3-1-3-2-2-2-1) and (2-1-3-3-2-2-1), (3-1-3-2-2-2-1) has the lowest total utility time (19.4 minutes).
Step 10 of the procedure: (3-1-3-2-2-2-1) is assigned as the current team combination and step 2 is directed.
Step 2 of the procedure: Station 3 is determined as bottleneck station due to the fact that it has the highest total utility time (6.8 minutes) with respect to IDBSA.
Step 3 of the procedure: The following team combinations are generated by decreasing the size of station 3 by one when increasing the size of an upstream station of station 3.
3-**2**-**2**-2-2-2-1
4-1-**2**-2-2-2-1
Step 4 of the procedure: The following team combinations are generated by increasing the size of station 2 by one when decreasing the size of an upstream station of 2 by one.
2-**2**-3-2-2-2-1
Steps 5&6&7 of the procedure: Assembly time of stations of each generated team combination is determined. Table 10 gives the feasibility and performance of feasible team combinations.
Step 8 of the procedure: The only feasible generated team combination (2-2-3-2-2-2-1) have higher total utility time (29.3 minutes) than current team combination has (19.4 minutes).
Step 11 of the procedure: (3-1-3-2-2-2-1) is determined as the final team combination.

	Team Combination	Cycle Time Constraint Feasibility	Total Utility Time (minutes)
	3-1-2-3-2-2-1 (*)		24.6
	3-1-3-2-2-2-1	Feasible	19.4
Step 3	3-2-2-2-2-2-1	Infeasible	-
	4-1-2-2-2-2-1	Infeasible	-
Step 4	2-1-3-3-2-2-1	Feasible	22.7

(*) Current team combination

Table 9. Feasibility and performance of generated team combinations

	Team Combination	Cycle Time Constraint Feasibility	Total Utility Time (minutes)
	3-1-3-2-2-2-1 (*)		19.4
Step 3	3-2-2-2-2-2-1	Infeasible	19.4
	4-1-2-2-2-2-1	Infeasible	-
Step 4	2-2-3-2-2-2-1	Feasible	29.3

(*) Current team combination

Table 10. Feasibility and performance of generated team combinations

5. Conclusions

Teamwork is a flexible, quick response production system consisting of self organised, self motivated, multi-skilled operators who work collectively in stable teams, making joint decisions and sharing responsibility for the team's output in terms of both quality and quantity. Teamwork, however, adds another dimension, moving the emphasis away from just a production method towards a change in the philosophy and culture of a company. In parallel, the emphasis begins to focus on 'reducing non value added activities' ; that is, the cost and time associated with factors that add nothing to the value of the product.

Reflecting the advantages of teamwork to assembly lines, team-oriented assembly systems provides higher level of effectiveness in terms of cost, lead time, flexibility and worker satisfaction when compared to traditional assembly systems. However, in team-oriented assembly systems, output from an assembly line may be severely restricted if high level of variability among station times lead to bottleneck and idle stations. Team-oriented assembly systems are in accordance with the objectives of modern assembly systems such as system flexibility and product quality while creating a more satisfactory working environment. In this context, determining the most effective assembly team combination is a critical success factor of team oriented assembly system design.

Simulation based worker transfer system for team oriented mixed model assembly lines is introduced in this chapter. Meanwhile, the suitability of model sequencing and worker transfer systematic for industrial sized problems is demonstrated by the application in a real life mixed model assembly line. Fourteen operators are grouped into seven assembly teams in order to provide synchronized implementation of parallel assembly tasks for shorter lead times in a chosen segment of the assembly line. Then demand oriented model sequence is decided as the most appropriate model sequence among four model sequences via

simulation. After bottleneck station is determined, worker transfer procedure to eliminate bottleneck is performed. Total utility time is decreased by 21.1% via worker transfer system according to the team combination obtained as the result of the step of team forming for the chosen segment. Thus, this systematic is suggested for decreasing non-value added times in assembly lines.
Future work is focused on worker transfer between different assembly line segments. The proposed systematic can be improved for unsteady demand structure. Furthermore, uniform parts usage may be included as a performance measure for determining the most appropriate model sequence.

6. References

Aigbedo, H. & Monden, Y. (1997). A Parametric procedure for multicriterion sequence scheduling for just- in-time mixed-model assembly lines, International Journal of Production Research, Vol.35, No.9, (December 1996), pp. 2543-2564, ISSN 0020-7543

Akgunduz, O., S. & Tunali, S. (2010). An adaptive genetic algorithm approach for the mixed-model assembly line sequencing problem, *International Journal of Production Research*, Vol.48, No.17, (November 2009), pp. 5157-5179, ISSN 0020-7543

Bedworth, D.D. & Bailey, J. (1987). Integrated Production Control Systems, Analysis, Design, *John Wiley & Sons*, ISBN 0471062235, New York, USA.

Bolat, A. & Yano, C.A. (1992). Scheduling algorithms to minimize utility work at a single station on paced assembly line, *Production Planning and Control*, Vol.3, No.7, (April 1991), pp. 393–405, ISSN 1381-1231.

Bukchin J., Darel, M. & Rubinovitz, J. (1997). Team-oriented assembly system design: a new approach, *International Journal of Production Economics*, Vol.51, No.2, (May 1996), pp. 47-57, ISSN 0360-8352

Bukchin, J. & Masin, M. (2004). Multi-objective design of team-oriented assembly systems, *European Journal of Operational Research*, Vol.156, No.7, (August 2003), pp. 326-352, ISSN 1094-6136.

Cevikcan, E., Durmusoglu, M. B. & Unal, M. E. (2009). A team-oriented design methodology for mixed model assembly systems, *Computers & Industrial Engineering*, Vol.56, No.5, (February 2008), pp. 576-599, ISSN 0360-8352.

Dimitriadis, S. G. (2006). Assembly line balancing and group working: A heuristic procedure for workers' groups operating on the same product and workstation, *Computers & Operations Research*, Vol.33, No.5, (January 2006), pp. 2757–2774, ISSN 0020-7543.

Elsayed, E.A. & Boucher, T.O. (1985). Analysis and Control of Production Systems, *Prentice Hall*, ISBN 0073377856, New Jersey, USA.

Erel., E., Gocgun, Y. & Sabuncuoğlu, İ. (2007). Mixed-model assembly line sequencing using beam search, *International Journal of Production Research*, Vol.45, No.22, (February 2006), pp. 5265-5284, ISSN 0020-7543.

Giard, V. & Jeunet, J. (2010). Optimal sequencing of mixed models with sequence-dependent setups and utility workers on an assembly line, *International Journal of Production Economics*, Vol.123, No.2, (March 2010), pp. 290-300, ISSN 0925-5273.

Groover, M. P. (2001). Automation, Production Systems, and Computer Integrated Manufacturing, 2e, *Prentice Hall*, ISBN 849385962, New Jersey, USA.

Hyun, C. J., Kim, Y. & Kim, Y. K. (1998). A genetic algorithm for multiple objective sequencing problems in mixed model assembly lines, *Computers & Operations Research*, Vol.25, No.7, (May 2007), pp. 675–690, ISSN 0305-0548.

Jin, M. & Wu, S. D. (2002). A new heuristic method for mixed model assembly line balancing problem, *Computers & Industrial Engineering*, Vol.44, No.5, (December 2001), pp. 159-169, ISSN 0360-8352.

Johnson, R. V.(1991). Balancing assembly lines for teams and work groups, *International Journal of Production Research*, Vol.29, No.6, (September 1990), pp. 1205-1214, ISSN 0020-7543.

Korkmazel, T. & Meral, S. (2001). Bicriteria sequencing for the mixed model assembly line in just-in-time production systems, *European Journal of Operational Research*, Vol.131, No.7, (June 2000), pp. 188-207, ISSN 0377-2217.

Kotani, S., Ito, T. & Ohno, K. (2004). Sequencing problem for a mixed-model assembly line in the toyota production system, *International Journal of Production Research*, Vol.42, No. 23, (March 2003), pp. 4955-4974, ISSN 0020-7543.

Kurashige, K., Yanagawa, Y., Miyazaki, S. & Kameyama, Y. (2002). Time-based goal chasing method for mixed-model assembly line problem with multiple work stations, *Production Planning & Control*, Vol.13, No.8, (November 2001), pp. 735-745, ISSN 1381-1231.

Macaskill, J. L. C. (1973) Computer simulation for mixed-model production lines, *Management Science*, Vol.20, No.2, (May 1972), pp. 341-348, ISSN 1526-5501.

Merengo, C., Nava, F. & Pozzetti, A. (1999). Balancing and sequencing manual mixed-model assembly lines, *International Journal of Production Research*, Vol.37, No.12, (September 1998), pp. 2835- 2860, ISSN 0020-7543.

Sarker, B. R. & Pan, H. (2001). Design configuration for a closed-station, mixed-model assembly line: a filling cabinet manufacturing system, *International Journal of Production Research*, Vol.39, No.5, (October 2000), pp. 2251-2270, ISSN 0020-7543..

Tamura, T., Long, H. & Ohno K. (1999). A sequencing problem to level part usage rates and workloads for a mixed-model assembly line with a bypass subline, *International Journal of Production Economics*, Vol 60, No.7, (March 1998), pp. 557-564, ISSN 0925-5273.

Tavakkoli-Moghaddam, R. & Rahimi-Vahed, A. R. (2006). Multi-criteria sequencing problem for a mixed-model assembly line in a JIT production system, *Applied Mathematics and Computation*, Vol.181, No.8, (April 2005), pp. 1471-1481, ISN 0096-3003.

Thomopoulos, N. T. (1967). Line balancing-sequencing for mixed-model assembly, *Management Science*, Vol. 14, No.2, (December 1966), pp. 59-75, ISSN 1526-5501.

Ventura, J. A. & Radhakrishnan, S. (2002) Sequencing mixed model assembly lines for a just-in-time production system, *Production Planning & Control*, Vol.13, No.2, (May 2001), pp. 199-210, ISSN 1381-1231.

Wild, R. (1975). Work Organization: a Study of Manual Work and Mass Production, *John Wiley & Sons*, ISBN 0 471944068, New Jersey, USA.

Xiaobo, Z. & Ohno, K. (1997). Algorithms for sequencing mixed models on an assembly line in a JIT production system, *Computers &Industrial Engineering*, Vol.32, No.1, (November 1996), pp. 47-56, ISSN 0360-8352.

Xiaobo, Z., Zhou, Z. & Asres A. (1999). A note on Toyota's goal of sequencing mixed models on an assembly line, *Computers & Industrial Engineering*, Vol.36, No.10, (June 1998), pp.57-65, ISSN 0360-8352.

5

Assembly Line Balancing in Garment Production by Simulation

Senem Kurşun Bahadır
Istanbul Technical University, Faculty of Textile Technologies and Design, Istanbul Turkey

1. Introduction

Construction of a quality garment requires a great deal of know-how, a lot of coordination and schedule management. Clothing manufacturing consists of a variety of product categories, materials and styling. Dealing with constantly changing styles and consumer demands is so difficult. Furthermore, to adapt automation for the clothing system is also so hard because, beside the complex structure also it is labor intensive. Therefore, garment production needs properly rationalized manufacturing technology, management and planning (Glock et. al, 1995; Caputo, et. al., 2005).

In garment production, until garment components are gathered into a finished garment, they are assembled through a sub-assembly process. The production process includes a set of workstations, at each of which a specific task is carried out in a restricted sequence, with hundreds of employees and thousands of bundles of sub-assemblies producing different styles simultaneously (Chan et al, 1998). The joining together of components, known as the sewing process which is the most labor intensive part of garment manufacturing, makes the structure complex as the some works has a priority before being assembled (Cooklin, 1991). Furthermore, since sewing process is labor intensive; apart from material costs, the cost structure of the sewing process is also important. Therefore, this process is of critical importance and needs to be planned more carefully (Tyler, 1991). As a consequence, good line balancing with small stocks in the sewing line has to be drawn up to increase the efficiency and quality of production (Cooklin, 1991; Tyler, 1991; Chuter, 1988).

An assembly line is defined as a set of distinct tasks which is assigned to a set of workstations linked together by a transport mechanism under detailed assembling sequences specifying how the assembling process flows from one station to another (Tyler, 1991). In assembly line balancing, allocation of jobs to machines is based on the objective of minimizing the workflow among the operators, reducing the throughput time as well as the work in progress and thus increasing the productivity. Sharing a job of work between several people is called division of labor. Division of labor should be balanced equally by ensuring the time spent at each station approximately the same. Each individual step in the assembly of product has to be analyzed carefully, and allocated to stations in a balanced way over the available workstations. Each operator then carries out operations properly and the work flow is synchronized. In a detailed work flow, synchronized line includes short distances between stations, low volume of work in process, precise of planning of production times, and predictable production quantity (Eberle et al, 2004).

Overall, the important criteria in garment production is whether assembly work will be finished on time for delivery, how machines and employees are being utilized, whether any station in the assembly line is lagging behind the schedule and how the assembly line is doing overall (Glock & Kutz, 1995; Hui & Ng, 1999). To achieve this approach, work-time study, assembly line balancing and simulation can be applied to apparel production line to find alternative solutions to increase the efficiency of the sewing line (Kursun & Kalaoglu, 2009).

This chapter deals with assembly line balancing in garment production by simulation. In this chapter, to analyze the structure of garment assembly, a sewing line will be focused on. Firstly, work flow of sewing line and the chronological sequence of assembly operations needed to transform raw materials into finished garment will be described in detail. Then, a detailed work and time study along the sewing line will be summarized considering the precedence constraints. After time study, real-data taken from factory floor will be discussed for distribution fit and goodness of fit. The chapter goes on creation of model of the sewing line by simulation. To set-up the model, all fitted data and allocation of operations to the operators will be transferred to simulation model. Model will be verified by comparing the actual system. Chapter then addresses how simulation model can be used to analyze assembly line's problems such as bottlenecks. Simulation model will be compared with the ones of the actual system according to model statistics; number of current and average content in workstations in the system, cycle time, server utilization percentage, average staying time of jobs, average output, throughput values of workstations.. etc., Hence, this chapter concludes balancing of assembly line model in garment production by suggesting possible scenarios that eliminate the bottlenecks along the line by various what-if analyses using simulation technique. Throughout this chapter how assembly line balancing in garment production can be done by using simulation will be understood.

2. Experimental

In the production of garment, at first garment model is designed. Then, according to model requirements, a sort of fabrics are cut as well as classified due to their sewing sequences. Then, cut fabrics are sewn and assembled in order to form garment. After the sewing and pressing process, garment is controlled for eliminating sewing faults, and finally it is sent to package and expedition.

In this chapter, to analyze the structure of garment assembly processes, a trousers sewing line was considered. The first step performed in this study was to understand trousers sewing processes' components and sewing line problems. The objective was to have a clear idea on how a trousers production-sewing process line flows and then, how the line can be balanced as well as the performance of production line can be increased.

2.1 Sewing line flow

The whole trousers manufacturing cycle includes a sequence of different phases of assembly operations. In Fig. 1, a set of assembly operations to transform raw materials (cut fabrics-accessories) into finished product of trousers is shown.

In the production of trousers, there are mainly four sequence of phases namely (i) pre-preparation of pockets, fly and labels, (ii) production of back of trousers, (iii) production of front of trousers, and (iv) assembling of fabric parts. As seen in Fig.1., at first pockets, fly and labels are prepared in order to be ready for insertion to fabric parts. Then, both back and front pockets are inserted to back and front fabric of the trousers, respectively. Fly is sewn on the front fabric. Front and back fabrics of trousers are prepared individually. Then all fabric parts are assembled in order to form trousers sequentially: Back and front fabrics are assembled. Zipper is attached and, belt and waistband are attached and sewn as well. Finally, hems, pockets, belt loop bartack seams are done and, by this way the sewing process of trousers is finished.

Trousers' sewing line flow

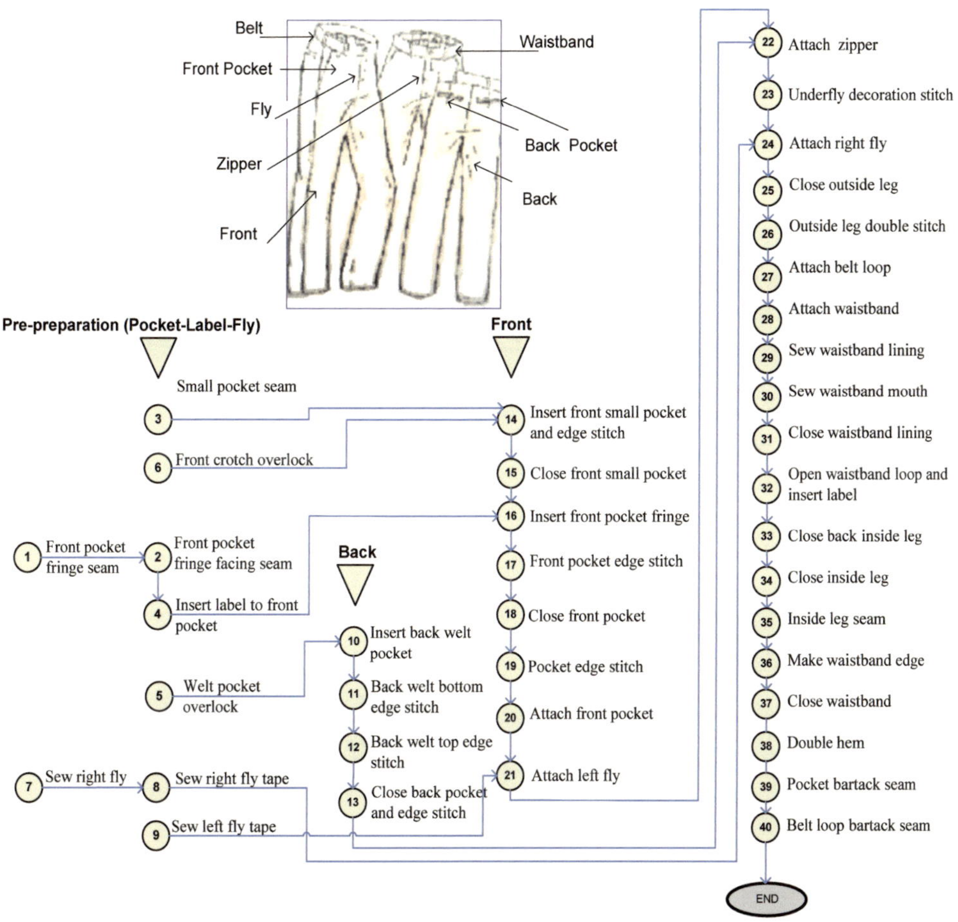

Fig. 1. Trousers' sewing line flow

2.2 Time study

In order to balance the sewing line as well as to increase the efficiency of the line, at first a detailed work and time study was carried out to find the task durations (Niebel, 1976). However, the time required to complete a task depends on a lot of factors such as the task, the operator, the properties of fabric and sub materials, working environment, quality level of the product, the hour of the day, psychology of the operator etc. (Fozzard et al,1996).

Therefore to calculate the approximate real process time of a task, 20 measurements were taken for each task and operator working on the line. Time study was performed along the line by chronometer. Each operation was measured in seconds and recorded. Then the data gathered from job floor was tested for firstly independency. It should be noted here that data taken from job floor should be independent. Then gathered data was tested for distribution fit and goodness of fit.

2.3 Distribution fit and goodness of fit

To estimate the relevant distribution fit of the data gathered, histogram of each process was obtained firstly. For instance; histograms of process 1-front pocket fringe seam, process 3-small pocket seam, process 18-close front pocket, process 19-pocket edge stitch are shown in Fig. 2. The estimated distributions for the processes mentioned above were obtained as *Logistic (42.20,3.47), Weibull (23.95,7.10), Uniform (43.76,60.24), Lognormal (36,1.17,0.972)*, respectively. The red lines in the figures show the estimated distributions. Similarly, the distributions estimated for all tasks were calculated.

After the estimation of the fit distribution, to validate the goodness of fit Chi Squared test, Kolmogorov Smirnov test and Anderson Darling test can be applied. While the Chi Squared test is asymptotic, which is valid only as the number of data points gets larger, it might not be appropriate for this study as 20 measurements were taken for each operator. Since the Kolmogorov Smirnov test is not a limited distribution, being appropriate for any sample size, it was chosen to test the goodness of fit. In order to do the tests, an SPSS program was used. The level of significance was set at 0.05 (95% confidence interval) for the Kolmogorov Smirnov test (Law & Kelton, 2000; Brunk, 1960) and, consequently all the goodness of fit distributions estimated were validated. Table 1 summarizes the estimated fit distributions for all processes.

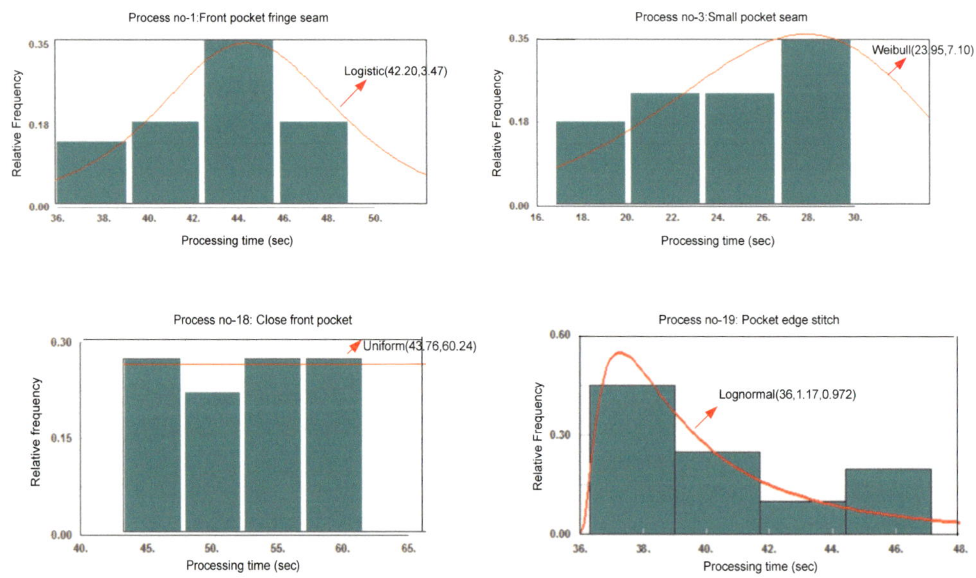

Fig. 2. Examples of estimated distributions for some processes

No:	Process Name	Fit Distribution (sec)	No:	Process Name	Fit Distribution (sec)
1	Front pocket fringe seam	Logistic (42.20,3.47)	21	Attach left fly	Uniform (31.34,46.86)
2	Front pocket fringe facing seam	Uniform (34.15,49.15)	22	Attach zipper	Uniform (48.88,73.62)
3	Small pocket seam	Weibull (23.95,7.10)	23	Underfly decoration stitch	Uniform (45.28,63.22)
4	Insert label to front pocket	Uniform (23.74,36.86)	24	Attach right fly	Normal (11.68,2.02)
5	Welt pocket overlock	Uniform (2.68,6.12)	25	Close outside leg	Normal (77.23,4.37)
6	Front crotch overlock	Uniform (17.92,28.28)	26	Outside leg double stitch	Normal (53.28,4.95)
7	Sew right fly	Lognormal (6,1.34,0.775)	27	Attach belt loop	Uniform (47.78,61.72)
8	Sew right fly tape	Uniform (2.56,4.64)	28	Attach waistband	Lognormal (93,2.07,0.972)
9	Sew left fly tape	Lognormal (2,0.034,0.81)	29	Sew waistband lining	Normal (44.88,3.96)
10	Insert back welt pocket	Normal (57.29,5.66)	30	Sew waistband mouth	Uniform (71.06,89.54)
11	Back welt bottom edge stitch	Normal (31.50,2.48)	31	Close waistband lining	Normal (64.96,3.43)
12	Back welt top edge stitch	Uniform (13.43,25.67)	32	Open waistband loop and insert label	Logistic (28.10,4.23)
13	Close back pocket and edge stitch	Uniform (51.78,70.22)	33	Close back inside leg	Uniform (34.36,51.24)
14	Insert front small pocket and edge stitch	Lognormal (37,1.22,0.98)	34	Close inside leg	Normal (45.80,4.79)
15	Close front small pocket	Uniform (30.53,48.27)	35	Inside leg seam	Uniform (13.00,22.90)
16	Insert front pocket fringe	Normal (52.42,4.14)	36	Make waistband edge	Logistic (61.80,3.79)
17	Front pocket edge stitch	Logistic (42.10,3.54)	37	Close waistband	Uniform (79.27,95.53)
18	Close front pocket	Uniform (43.76,60.24)	38	Double hem	Uniform (34.11,47.49)
19	Pocket edge stitch	Lognormal (36,1.17,0.972)	39	Pocket bartack seam	Uniform (54.39,72.31)
20	Attach front pocket	Uniform (65.26,80.54)	40	Belt loop bartack seam	Normal (64.46,4.53)

Table 1. Estimated distributions for processes

For instance; as seen in Table 1, the estimated distribution for the processes 10, 11, 16, 24, 25, 26, 29, 31, 34, 40 were found as *normal* distribution. In order to test if *normal* distribution is appropriate for the input data or not, Kolmogorov Smirnov test results were evaluated. With reference to Table 2 results, it was confirmed that the estimated *normal* distributions for these processes are appropriate for the input data. As seen in Table 2, the asymptote significant (2-tailed) values of the mentioned processes were found to be greater than the level of significance (0.05) for the Kolmogrov Smirnov test. Thus, these results can permit us to state that *normal* distributions for the processes mentioned above are appropriate and herewith, the distribution fit was validated. Similarly, each estimated distribution for each process was validated for goodness of fit by Kolmogorov-Smirnov test and it was found that all estimated distributions are appropriate for the input data so that they are ready to be transformed in simulation model of sewing line.

One-Sample Kolmogorov-Smirnov Test

		Processes (No)									
		10	11	16	24	25	26	29	31	34	40
N		20	20	20	20	20	20	20	20	20	20
Normal Parameters[a]	Mean	57.29	31.50	52.42	11.68	77.23	53.28	44.88	64.96	45.80	65.46
	Std. Deviation	5.66	2.48	4.14	2.020	4.371	4.95	3.96	3.43	4.79	4.53
Most Extreme Differences	Absolute	0.182	0.112	0.125	0.189	0.097	0.150	0.159	0.134	0.126	0.089
	Positive	0.163	0.093	0.125	0.108	0.084	0.150	0.159	0.122	0.075	0.071
	Negative	-0.182	-0.112	-0.079	-0.189	-0.097	-0.116	-0.112	-0.134	-0.126	-0.089
Kolmogorov-Smirnov Z		0.814	0.500	0.559	0.846	0.434	0.673	0.709	0.599	0.563	0.398
Asymp. Sig. (2-tailed)		0.522	0.964	0.913	0.471	0.992	0.756	0.696	0.865	0.909	0.997

a. Test distribution is Normal.

Table 2. Kolmogorov-Simirnov test results

2.4 Setting up the simulation model

Simulation is a technique to model a real-life or hypothetical situation on a computer so that it can be used for analyzing the behavior of system. By changing variables predictions can be made on system behavior. It provides predictions on the performance of an existing system. Moreover, by suggesting possible scenarios on system alternative solutions can be compared. Therefore, it is a very useful engineering technique to suggest investment strategies to companies for a particular design problems.

If the operations in the system are based on chronological sequence of events, then it is called as discrete-event simulation. Since our sewing line consists of a sequence of different phases of assembly operations, it is an example of discrete-event simulation. Therefore, to set up the model, ENTERPRISE DYNAMICS ® simulation program (Student Version) from Incontrol Simulations Solutions, which is a software program for discrete event simulation, was used (Incontrol Simulation Solutions, 2003).

Before setting up model of sewing line, it is necessary to identify the components of model. With reference to sewing line flow seen in Fig. 1, 40 processes were considered to be assigned to 40 operators, and these operators with their machines including queues, materials, assembly operations with precedence constraints were determined as components of model.

2.4.1 Model building basics

In the creation of a model, the decision of right atom at a right place is critical issue. An atom can be a machine, a counter, a queue or a product etc. To create our sewing line model, mainly six different types of atoms were used as shown in Fig. 3. The explanations of atoms that were used in our model are as follows:

- Product: This atom represents the products/customers/raw materials that comes into an atom through an input channel and leave the atom through an output channel
- Source: The function of this atom is to produce products into the model
- Queue: This atom represents the waiting area for customers or products.
- Server: This atom corresponds to a machine or a counter. Atoms coming to a server undergo a process and stay in this atom for a certain time (the process time).
- Assembler: The atom which is used for assembly operation
- Sink: The products or customers leave the model through this atom and finishes the schedule (Incontrol Simulation Solutions, 2003).

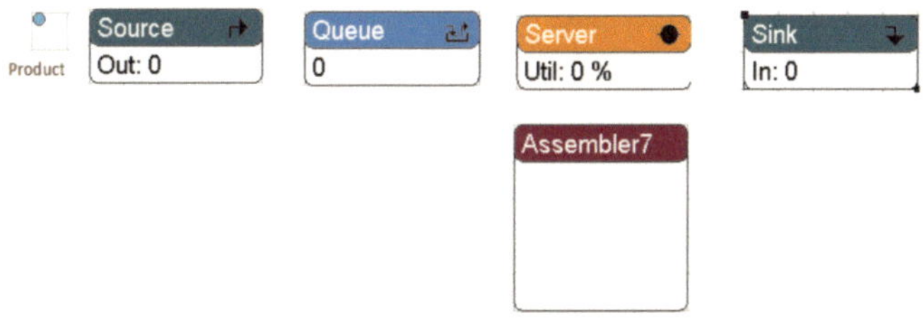

Fig. 3. Type of atoms used in our simulation model

By using these atoms, the model of sewing line was formed as shown in Fig. 4.

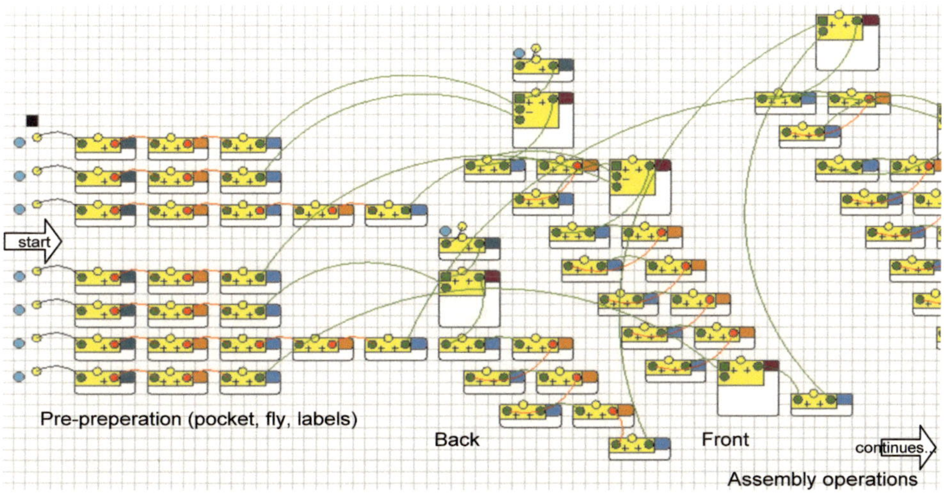

Fig. 4. Simulation model preview of sewing line

Arrows in the figure show the connections as well as the relations between atoms. Before and after server and assembly atoms or in other words before and after sewing processes, products always wait in queues for being processed like real system. Here, the data for model were entered by considering precedence constraints. Data in Table 1, as explained in the proceeding section, was transformed into simulation model for each operation individually. Also, the interval arrival time of raw material feeding in the system was obtained as an exponential distribution and directly transformed into the model as well.

Unfortunately, to analyze the real system and create the model, some conjectures were considered:

- The 8 hour working day of the system.
- Only one worker is at each machine.
- Allowances are not taken into consideration.
- Delay times (machine breakdowns, changing apparatus) are not taken into consideration.
- There is no energy problem in the system.
- Fabric loss is not taken into consideration.
- Raw material is unlimited.
- The supervisor's job on the line was ignored.

2.5 Model verification and validation

By considering the conjectures, simulation model was run. Verification of model was done step by step comparing with actual system. The model statistics; number of current and average content in atoms in the system, cycle time, server utilization percentage, waiting time of jobs, average output, throughput values of atoms.. etc., were compared with those of the actual system, and in all cases there were no significant differences between the model and the actual system.

3. Results

To analyze the results of the system, three performance measures were considered:

- average staying times of jobs in queues,
- average content of jobs in machines,
- quantity of the average daily output

Since our system is an example of a nonterminating simulation, it was evaluated in two stages to consider the effect of the warm-up period . Firstly, to find the warm-up period, the simulation model was run for 800 hours (5 months as a working day) at a 95% confidence level. Nevertheless, with these results, the average output quantity of the system for a day cannot be evaluated. To find the quantity of the average daily output, the system was run 100 times, each run consisting of 8 hours of simulated time, taking into account the warm-up period (Law & Kelton, 2000).

3.1 Results based on reference layout model

Results of the reference layout model are summarized in Table 3 according to the performance measures. As seen in Table 3, it can be observed that the average number of finished trousers in a day is 295, the average content of jobs in machines is 28 and the average staying time of jobs in queues is 260 sec. The state diagrams of the performance measures for 100 observations (5 months) are shown in Fig. 5-7. When these results were compared with those of the actual system, it was also found that the actual system and the reference model results were alike.

To increase the efficiency of the line, firstly bottlenecks were determined, and then possible scenarios were tried by what-if analysis. As a result four scenarios were developed for the production of trousers.

Performance measures	Average	St. Deviation	L-bound (95%)	U-bound (95%)	Min.	Max.
Average content of jobs in machines	28.34	0.04	28.33	28.35	28.26	28.42
Average staying of jobs in queues	260.66	1.07	260.45	260.87	258.10	264.03
Average daily output	295.12	0.48	295.03	295.21	294.00	296.00

Table 3. Results based on reference layout model

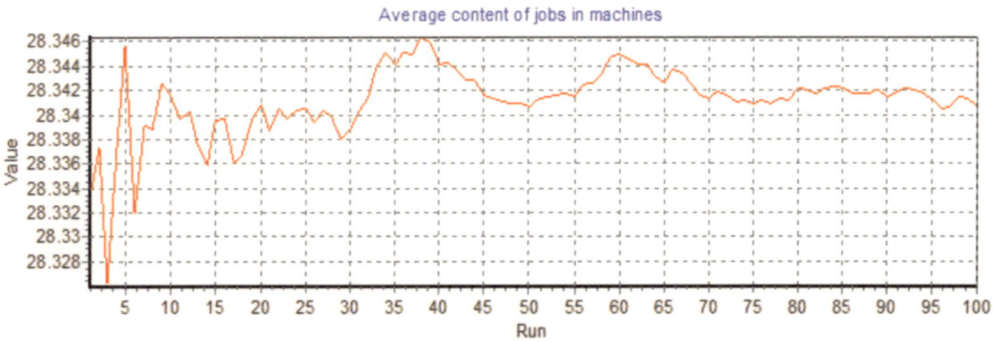

Fig. 5. Reference layout model: Average content of jobs in machines for 100 observations (5 months)

Fig. 6. Reference layout model: Average staying of jobs in queues for 100 observations (5 months)

In order to determine bottlenecks in the reference layout model; number of current and average content in atoms in the system, cycle time, server utilization percentage, waiting time of jobs, average output, throughput values of atoms.. etc. were taken into account. It was observed that process 28: Attach waistband with a higher processing time blocks the system. The server utilization status of process 28 is shown in fig. 8. As seen in the figure, machine is busy with 97.7 percentage of total time. Therefore, in reference layout model, process 28 was identified as bottleneck. By this way, the first scenario was developed by adding one extra machine to the system in order to overcome process 28's bottleneck problem.

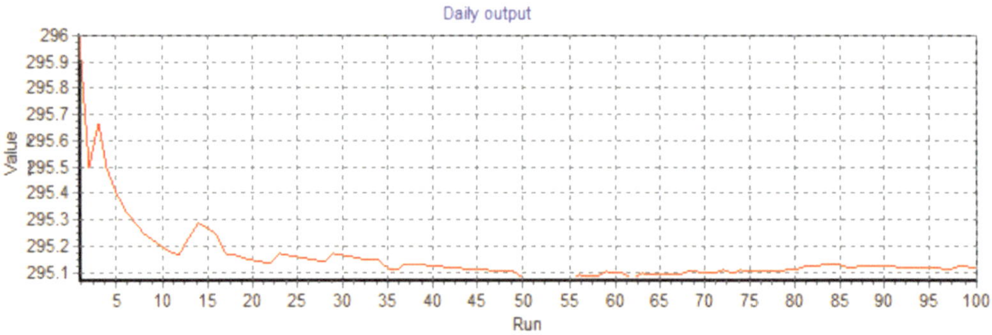

Fig. 7. Reference layout model: Average daily output for 100 observations (5 months)

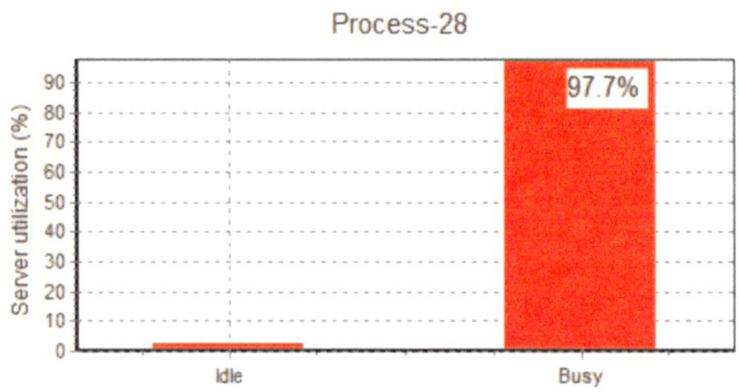

Fig. 8. Overview of the different statuses of process 28 as a percentage of total time

3.2 Results based on scenario 1

With only adding extra one machine with an operator to the reference system, the average daily output of the system increased up to 312. That means if one machine with one operator is added to the system, then around 312 trousers will be able to be produced in a day. Moreover, when other performance measures were considered, it was also observed that average content of jobs in machines increased, besides average staying of jobs in queues decreased.

Table 4 summarizes results based on scenario 1, when only one extra machine with one operator added to reference model. Fig. 9 shows the average daily output of the system according to scenario 1.

Performance measures	Average	St. Deviation	L-bound (95%)	U-bound (95%)	Min.	Max.
Average content of jobs in machines	30.22	0.06	30.21	30.23	30,07	30.38
Average staying of jobs in queues	230.97	1.28	230.72	231.22	227.24	233.64
Average daily output	312.13	0.80	311.97	312.29	311.00	315.00

Table 4. Results based on scenario 1

With scenario 1, it was found that the utility percentage of server 28 decreased to 62.6% (busy). However, despite the decrease in server utility of process 28, this time new bottleneck was appeared in other machine. The server 37 was identified as second bottleneck along the line after one extra machine was added. Fig. 10 shows its usage percentage due to total working time.

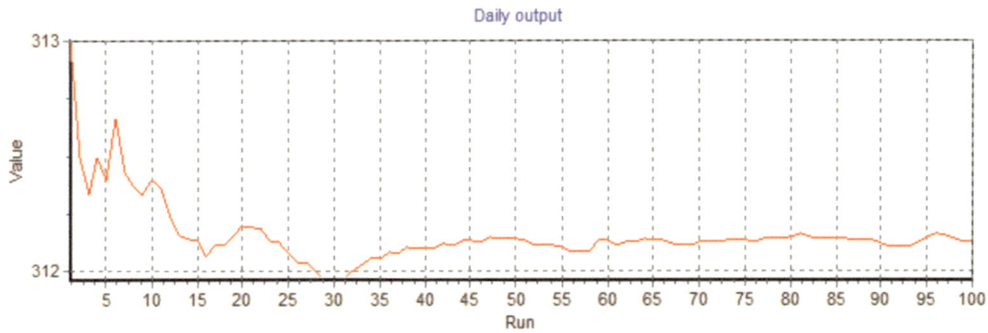

Fig. 9. Scenario 1: Average daily output for 100 observations (5 months)

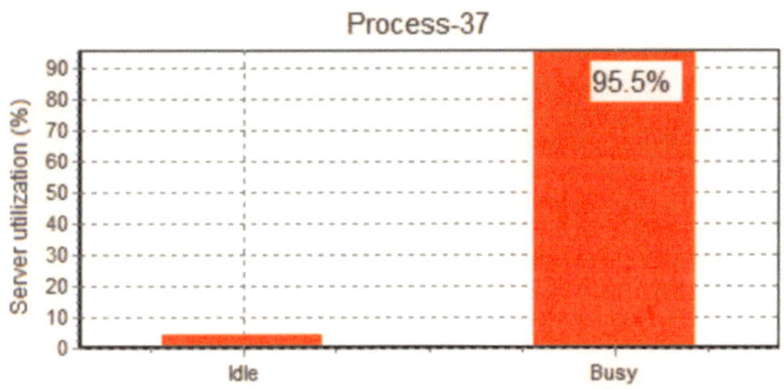

Fig. 10. Overview of the different statuses of process 37 as a percentage of total time

Therefore, as a second scenario it was decided to add one more extra machine with one more operator to the system to overcome server 37's work load. Indeed, the aim of adding new machine is to increase efficiency of sewing line.

3.3 Results based on scenario 2

By adding one more extra machine with one more operator to the system, the bottleneck problem in server 37 was also solved. By this way, the average daily output of the system increased from 312 (according to scenario 1) up to 322 as seen in Fig. 11. Moreover, average staying of jobs in queues decreased from 230 to 221 as seen in Table 5 and the work load of server 37 decreased from 95.5% to 60.30%.

Performance measures	Average	St. Deviation	L-bound (95%)	U-bound (95%)	Min.	Max.
Average content of jobs in machines	30.31	0.06	30.30	30.32	30.17	30.46
Average staying of jobs in queues	221.29	0.83	221.12	221.45	219.48	224.36
Average daily output	322.25	1.20	322.01	322.49	319.00	325.00

Table 5. Results based on scenario 2

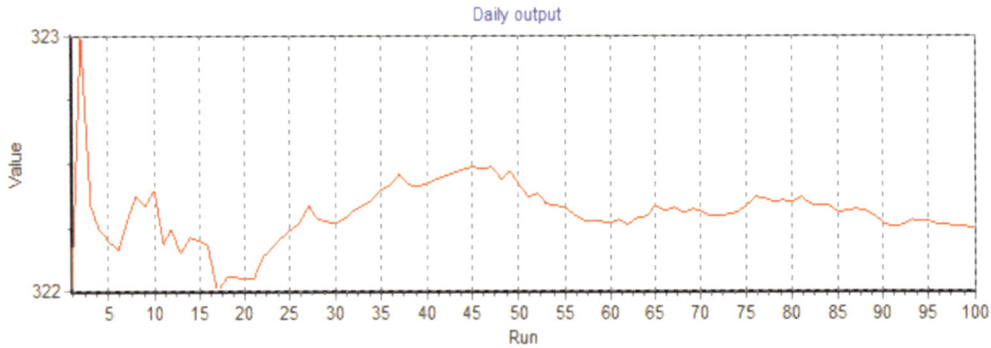

Fig. 11. Scenario 2: Average daily output for 100 observations (5 months)

However, despite the better performance measures, adding additional machine to the system brought new bottlenecks. This time, bottlenecks occurred in server 1, server 2 (see Fig. 12) and server 16.

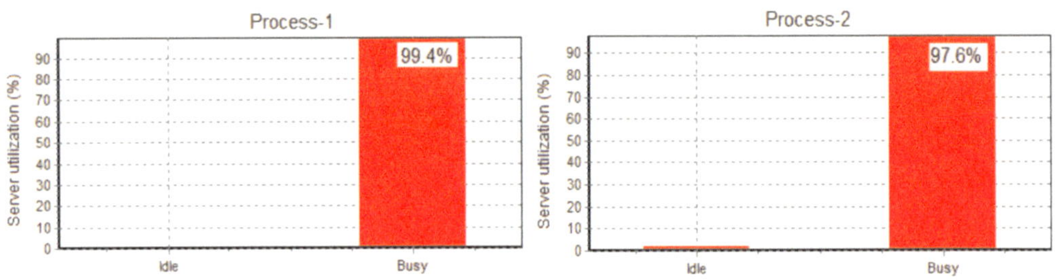

Fig. 12. Overview of the different statuses of process 1 and process 2 as a percentage of total time

Therefore, as a new strategy three machines with three operators were added to the system in order to decrease server 1, server 2 and server 16's workloads and increase the efficiency of the line as well.

3.4 Results based on scenario 3

As seen in Table 6, the daily production of trousers is increased to 340 with scenario 3. Also, with the same scenario the average content of jobs in machines was higher than the scenario 2, but the average staying times of jobs in queues was found to be slightly higher (Fig. 13) than the reference layout. Moreover with scenario 3, the workloads of server 1 and server 2 decreased to 65.7% and 64.8%, respectively. As far as performance measures are concerned, the first important thing for production is the daily output, which is directly related to the line efficiency; therefore results such as the average staying of jobs in queues can be ignored for this reason, but only when they are within acceptable limits.

However, the balance of sewing line can be increased by adding new servers to the system. Therefore, as a final scenario to increase the line efficiency more, four additional servers were added to system for recently overloaded servers; server 30 (96.8% busy), server 25 (97.9%busy), server 20 (98.8%busy), and server 14 (61.1%busy and 37% distributing).

Performance measures	Average	St. Deviation	L-bound (95%)	U-bound (95%)	Min.	Max.
Average content of jobs in machines	36.26	0.05	36.25	36.27	36.13	36.38
Average staying of jobs in queues	291.35	1,59	291.04	291.66	286.52	295.27
Average daily output	339.95	1,27	339.70	340.20	337.00	343.00

Table 6. Results based on scenario 3

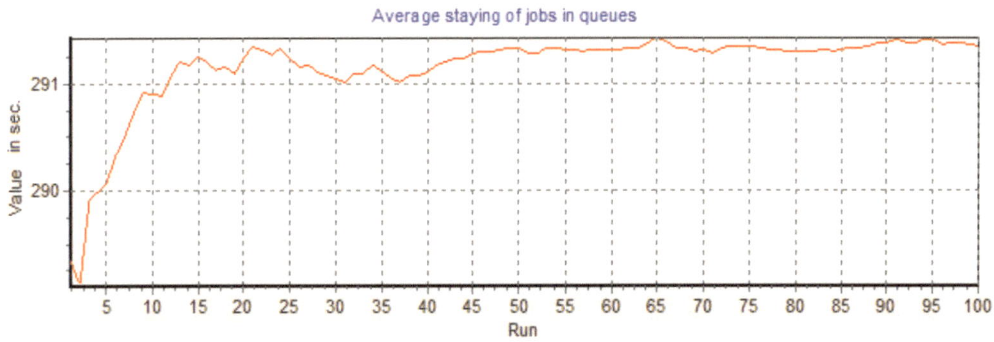

Fig. 13. Scenario 3: Average staying of jobs in queues for 100 observations (5 months)

3.5 Results based on scenario 4

With final scenario, the best performance results were obtained as summarized in Table 7. The average daily output of the system increased from 295 (according to reference layout) up to 419. Average staying of jobs in queues decreased from 260 (according to reference layout) to 186. Additionally, average content of jobs in machines increased from 28 (according to reference layout) up to 39.56. With reference to this scenario, it can be said that the balancing of sewing line seems appropriate for all performance measures.

Performance measures	Average	St. Deviation	L-bound (95%)	U-bound (95%)	Min.	Max.
Average content of jobs in machines	39.56	0.11	39.54	39.58	39.32	39.78
Average staying of jobs in queues	186.82	2.39	186.35	187.29	181.40	191.63
Average daily output	419.21	1.30	418.96	419.46	415.00	422.00

Table 7. Results based on scenario 3

Furthermore, the results of scenario 4 shows that the system became nearly balanced after 30 working days by running it for 8 hours at a 95% confidence level (Fig. 14).
Moreover, workloads of the server 30, server 25, server 20 and server 14 decreased to 62.5%, 63.9 %, 68.6 % (Fig.15), and 55.1%, respectively. As it can be understood from above also when the workloads of the all servers became around 60 % (busy), it was observed that system got nearly balanced.
As a summary, considering precedence constraints four scenarios were developed according to determined bottlenecks in the models. Table 8 summarizes the total changes that were suggested in each scenario. Scenario 1 includes one new operator with one

machine, scenario 2 includes two new operators with two machines. Scenario 3 consists of five new operators with five machines and scenario 4 consists of nine operators with nine machines.

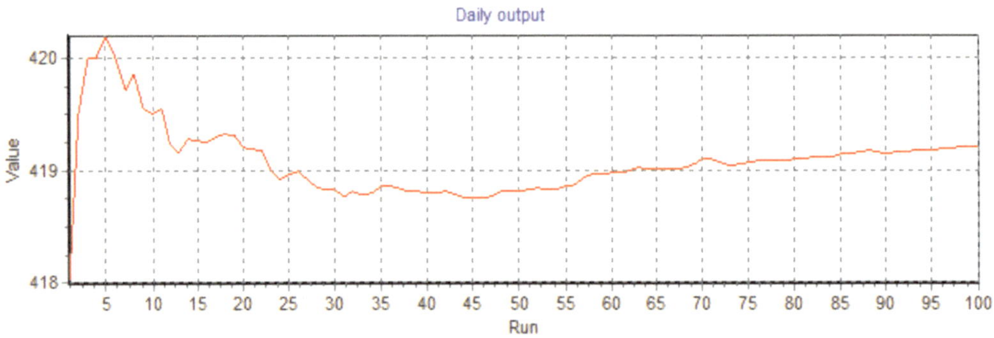

Fig. 14. Scenario 4: Average daily output for 100 observations (5 months)

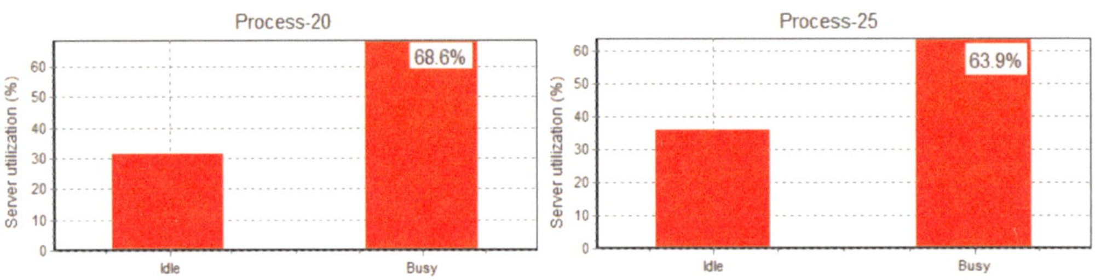

Fig. 15. Overview of the different statuses of process 20 and process 25 as a percentage of total time

	Determined bottlenecks in the model	Suggested solution for consecutive scenario	Total changes in the line according to reference layout
Reference Layout	Server 28	Add one new operator with one machine	-
Scenario 1	Server 37	Add one new operator with one machine	One new operator with one machine was added
Scenario 2	Server 1,2, 16	Add three new operators with three machines	Two new operators with two machines were added
Scenario 3	Server 30, 25, 20, 14	Add four new operators with four machines	Five new operators with five machines were added
Scenario 4	-	-	Nine new operators with nine machines were added

Table 8. Summary of suggested scenarios

As mentioned above in detail, with the suggested scenarios trousers' sewing line was tried to be balanced. It is appearent from the Fig. 16 that the best results are obtained with scenario 4. The number of averaged finished trousers per day is increased by 42% with scenario 4. The averaged content of jobs increased by 40 % whereas the average staying of jobs is decreased by 28 %. As seen from the figure; the daily output of the system is increased to 312 with scenario 1, 322 with scenario 2, 340 with scenario 3, and 419 with scenario 4. To sum up, with the suggested scenarios the efficiency of the line was increased, and the line was balanced.

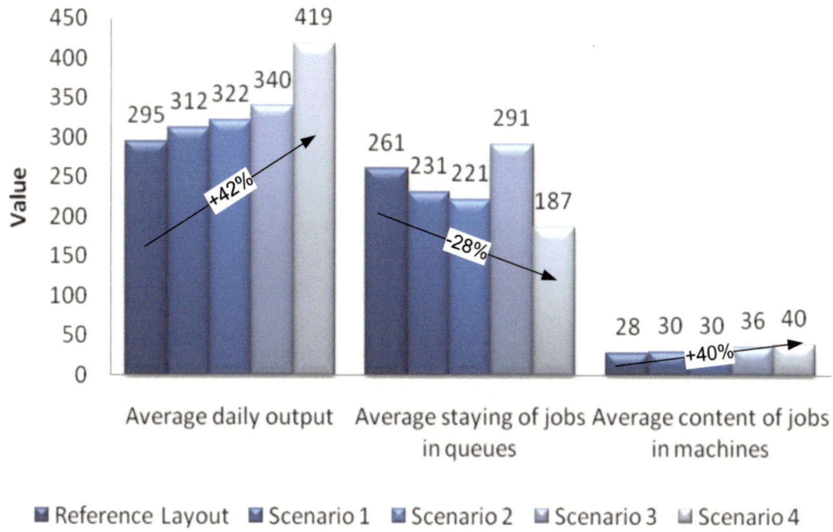

Fig. 16. Total results of suggested scenarios according to performance measures

4. Summary

In this chapter, the structure of garment assembly line was analyzed by simulation. A trousers sewing line was considered for simulation model. Firstly, the work flow of the line as well as the chronological sequence of assembly operations needed to transform raw materials into finished trousers were described in detail. Then, a detailed work and time studies were performed along the line. Secondly, real-data gathered from factory floor was tested for distribution fit, and a Kolmogrov-Smirnov test was carried out for the goodness of fit. Afterwards, the creation of model was explained. To set-up the model, all fitted data and allocation of operations to the operators with machines considering precedence constraints were transferred to simulation model. Model verification was done by comparing the results of the model with the ones of the actual system. Then, bottlenecks in the line were determined. In order to eliminate bottlenecks in the line and to balance line, the model statistics; number of current and average content in workstations in the system, cycle time, server utilization percentage, average staying time of jobs in queues, average output, throughput values of workstations.. etc. were taken into account. Due to model statistics, possible scenarios were formed by various what-if analyses in order to balance line as well as increase its efficiency. These scenarios can provide investment decision alternatives to company administrators. Moreover, in order to present more comprehensive decision alternatives, study can be enhanced by a cost analysis of the possible scenarios.

To conclude, this chapter has demonstrated the use of simulation technique to solve assembly line balancing problem in a garment production.

5. References

Brunk H. D. (1960). *An Introduction to Mathematical Statistics*, Ginn

Chan, K.C.C, Hui, P.C.L., Yeung, K.W., Ng.F.S.F. (1998).Handling the assembly line balancing problem in the clothing industry using a genetic algorithm, *International Journal of Clothing Science and Technology*, Vol.10, pp. 21-37

Chuter, A. J. (1988). *Introduction to Clothing Production Management*, Blackwell Science, Oxford, pp. 60-63.

Cooklin, G. (1991). *Introduction to Clothing Manufacturing*, Blackwell Science, Oxford, p. 104.

Eberle, H., Hermeling, H., Hornberger, M., Kilgus, R. , Menzer, D., Ring, W., (2004). *Clothing Technology*, Beuth-Verlag GmbH, Berlin

Fozzard, G., Spragg, J., & Tyler, D., (1996). Simulation of flow lines in clothing manufacture: Part 1: model construction, *International Journal of Clothing Science and Technology*, Vol. 8, pp. 17-27

Glock, R. E. & Kunz, G. I. (1995). *Apparel Manufacturing-Sewn Product Analysis*, Prentice Hall, New Jersey, p:4

Hui, C. & Ng, S. (1999). A study of the effect of time variations for assembly line balancing in the clothing industry *International Journal of Clothing Science and Technology*, Vol. 11, pp. 181-188

Incontrol Simulation Software B. V (2003). *Enterprise Dynamics Tutorial*, Netherland

Kursun, S. & Kalaoglu, F. (2009). Simulation of Production Line Balancing in Apparel Manufacturing, *FIBRES & TEXTILES in Eastern Europe* Vol. 17, No. 4 (75), pp.68-71

Law, A. W. & Kelton, D. (2000). *Simulation Modeling & Analysis* , McGraw-Hill Press (3rd Ed.)

Niebel B. (1976). *Motion and time study*, III. R. D. Irwin, Homewood

Tyler, D. J. (1991). *Materials Management In Clothing Production*, BSP Professional Books Press, London

Part 2

Optimization in Assembly Lines

6

Tackling the Industrial Car Sequencing Problem Using GISMOO Algorithm

Arnaud Zinflou[1] and Caroline Gagné[2]
[1]Institut de recherche d'Hydro-Québec (IREQ)
[2]Université du Québec à Chicoutimi
Canada

1. Introduction

In many industrial sectors, managers are confronted with problems of an ever-growing complexity. The problem could be bus route optimization for a public transporter, production cost minimization, decision-making support, electronic circuit performance enhancement, or computer system process scheduling. In many cases the situation can be expressed as a combinatorial optimization problem. Solving an optimization problem consists of determining the best solution(s) validating a set of user-defined constraints and goals. To determine if one solution is better than another, the problem must include at least one performance evaluation metric that allows solutions to be compared. The best (or *optimal*) solution, is thus the one with the best evaluation, with respect to the defined goal. When only one goal is specified (*e.g.* total distance minimization), the optimal solution is clearly defined (the one with the smallest distance).

However, in many situations there are several contradictory goals that have to be satisfied simultaneously. In fact, real-world optimization problems rarely have a single goal. This is the case for the Industrial Car Sequencing Problem (ICSP) on an automobile assembly line. The ICSP consists of determining the order in which automobiles should be produced, taking into account the various model options, assembly line constraints, and production environment goals. In this context, the optimal solution is not a single point, but rather a set of *compromise solutions* called the *Pareto-optimal front*. We can thus define two main goals in multi-objective optimization: (i) Find a set of compromise solutions whose evaluation is as close as possible to the Pareto-optimal front; and (ii) Find a set of compromise solutions as diverse as possible. Attaining these two goals in realistic time is an important challenge for any multi-objective algorithm.

However, in the literature, the ICSP, despite its multi-objective character, has been treated as a problem with a single goal or with several goals lexicographically ordered (Benoist, 2008; Briant *et al.*, 2008; Cordeau *et al.*, 2008; Estellon *et al.*, 2008; Ribeiro *et al.*, 2008). To our knowledge, the only references that treat the ICSP from a purely multi-objective viewpoint are those of Zinflou *et al.* (2009) and of de Oliveira dos Reis (2007); the latter only examines small instances (fewer than 60 automobiles).

Most of the algorithms proposed recently for multi-objective problems are Evolutionary Algorithms (EA) (Deb, 2000; Knowles & Corne, 2000a; Knowles & Corne, 2000b; Zitzler *et al.*, 2001). This is so, doubtlessly because EA's can traverse a large search space to generate

an approximation of the Pareto-optimal front in a single optimization step (Francisci, 2002). One of these algorithms, proposed by Zinflou *et al.* (2011), is a hybrid between a Genetic Algorithm (GA) and an artificial immune system. This approach, called GISMOO, has a small number of parameters to calibrate, is easy to implement, and has been shown to be efficient in solving classical benchmarks in both discrete and continuous optimization.
The goal of this chapter is to deepen the understanding of the ICSP from the Pareto viewpoint, by adapting the GISMOO algorithm to solve this problem. On one hand, we mean to show that this is an interesting approach in the solution of the ICSP from a multi-objective viewpoint. On the other hand, because only a few workers have treated this problem from a Pareto viewpoint, we wish to compare the performance of GISMOO with that of other known algorithms, on a real-world industrial problem.
This chapter is organized in the following manner. Section 2 briefly describes the industrial car sequencing problem. We subsequently describe, in Section 3, the GISMOO algorithm in order to solve the problem in a Pareto sense. Section 4 exposes the details of the numerical experiments carried out in this chapter. Sections 5 and 6 present the experimental results of GISMOO in comparison to those of two other evolutionary algorithms well known in the literature: NSGAII and PMS^{MO}. Finally, the last section offers some concluding remarks.

2. Industrial car sequencing problem (ICSP)

The ICSP analysed in this chapter was proposed by the automobile manufacturer Renault for the ROADEF Challenge 2005 (Nguyen & Cung, 2005). Each production day, the clients' orders are transmitted to the assembly factory in real time. The daily task of the factory planners consists of: (1) assigning an assembly day to each car ordered, subject to assembly line capacity and clients' delivery time constraints; and (2) scheduling cars within each production day, subject to the constraints of the three shops on the assembly line, as illustrated in Figure 1.
Renault specified that the factory technology was such that the body shop constraints could be neglected. Therefore, the ICSP consists of determining the production sequence of a set of cars (*Nb_cars*) subject to the paint and car assembly shop constraints. Since the scheduling goals are thereby achieved, the sequence determined is applied to the whole assembly line.

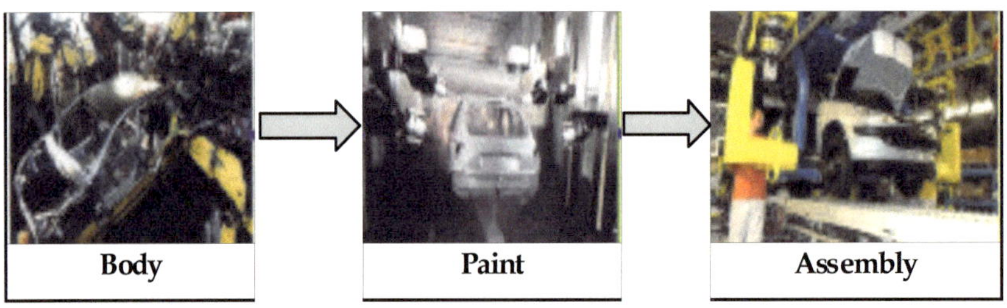

Fig. 1. The three workshops of the assembly line (Nguyen & Cung, 2005)

In the paint shop, we want to keep cars of the same colour together in consecutive order, so that the number of paint gun purges is minimized. To preserve quality, in fact, the paint guns have to be cleaned with solvents, between two cars of different colours and after a certain number (rl_{max}) of cars of the same colour. So the first goal to optimize in the ICSP is to minimize the number of colour changes (COLOR). There is also the constraint that each

purge necessarily means a colour change; therefore, any solution with a number of consecutive cars of the same colour greater than rl_{max} is not feasible.
In the assembly shop, several pieces are added to the painted body to complete the car assembly. This is characterized by a set O of options for which the corresponding workstations are designed to accommodate up to a certain percentage of the cars on the line. These capacity constraints are expressed in the form of a ratio r_o/s_o, meaning that for any subsequence of s_o cars in the assembly line, no more than r_o cars can have the option o. Cars will thus be positioned so that the workload will be well distributed at each place in the line. If it is impossible to respect the capacity constraint for an option o in a subsequence of s cars, then the number of cars that exceeds r defines what is called a *conflict*. The ICSP subdivides the assembly shop capacity constraints into two priority types in order to minimize the two following objectives: the number of conflicts for High Priority Options (HPO) and the number for Low Priority Options (LPO).
The proposed implementation puts identical cars into *classes*: cars with the same HPO's and LPO's are included in V classes for which we know the number of cars to produce (c_v). In fact these quantities represent the problem's production constraints. Table 1(a) presents an example instance with 25 (*Nb_cars*) cars, 5 options (O) generating 6 classes (V), and a possibility of 4 different colours across each class. A production sequence for this ICSP is defined by two vectors representing the car class (*Classes*) and the respective colour code (*Colors*), as presented in Table 1(b). In the rest of this chapter, the solution sequence will be noted as Y = {*Classes*/*Colors*}, and the element at position i of the sequence will be noted as $Y(i)$ = *Classes*(i)/*Colors*(i).

			Classes #					
o	*r*	*s*	**1**	**2**	**3**	**4**	**5**	**6**
1	1	2	0	1	1	0	0	0
2	2	5	1	0	1	0	1	1
3	1	3	0	1	0	0	0	0
4	3	5	0	0	0	1	0	1
5	2	3	0	1	1	0	1	0
c_v			5	5	4	4	3	4
Colours #		1	2	1	1	2	1	1
		2	1	1	0	2	1	1
		3	1	3	2	0	0	2
		4	1	0	1	0	1	0

(a)

Y	1	2	3	4	5	6	.....	21	22	23	24	15
Classes	3	5	5	4	6	4		3	1	4	5	1
Colors	4	4	2	2	2	2		3	3	1	1	1

(b)

Table 1. Example of an ICSP and its solution

Another ICSP characteristic is the link between the current production day (J), the previous day (J-1) and the following day (J+1). Any solution for day J has to take into account the

solution for *J-1* and an extrapolation of the minimum number of conflicts for day *J+1*. Moreover, we add a colour change if the colour of the first car on day *J* is different from the colour of the last car on *J-1*.
To evaluate the number of option conflicts, a binary matrix *S*, of dimension *O* by *Nb_cars*, is generated by the information contained in the solution vector *Y* = {*Classes*/*Colors*} : $S_{oi} = 1$ if the car class *Classes*(*i*) in position *i* has the option *o*, and $S_{oi} = 0$ otherwise. The decomposition of *Classes* in Table 1 to form *S* is presented in Table 2. We can see that Table 2(a) carries forward the last part of the previous day's solution, in order to compute the number of conflicts between the previous and current days. Moreover, the solution in Table 2(b) is computed partly from the following day, assuming there are no options.
For the current day *J*, options 1, 3 and 4 have no conflict: there are no cases of more than *r* cars in any subsequence of *s* cars that have the same option. However, for option 2 there are two conflicts, because there are 4 consecutive cars with option 2, in positions 1 to 5. There is also a conflict in positions 2 to 6, and two more in positions 21 to 25, because the 2/5 capacity constraint is not respected. For option 5, there is a conflict because the 2/3 constraint is violated in positions 1 to 3. Between days *J* and *J*-1, there is a conflict in positions -1 to 1 for option 1, and another in positions -1 to 2 for option 5. Between *J* and *J*+1, there is a conflict in positions 22 to 26 for option 2.
If the first 3 options are of high priority and the other two are of low priority, then there are, in this solution *Y*, 6 HPO conflicts and 2 LPO conflicts. The only step remaining in the computation of *Y* is to count the colour changes (COLOR).

		Previous day (*J*-1)					Current day (*J*)						
Positions		-5	-4	-3	-2	-1	1	2	3	4	5	6	………
Classes		4	1	4	4	2	3	5	5	4	6	4	
OPTION	1/2	0	0	0	0	**1**	**1**	0	0	0	0	0	
	2/5	0	1	0	0	0	**1**	**1**	**1**	**0**	**1**	**0**	
	1/3	0	0	0	0	1	0	0	0	0	0	0	
	3/5	1	0	1	1	0	0	0	0	1	1	1	
	2/3	0	0	0	0	**1**	**1**	**1**	**1**	0	0	0	

(a)

		Current day (*J*)						Following day (*J*+1)				
Positions		….	21	22	23	24	25	26	27	28	29	30
Classes			3	1	4	5	1					
OPTION	1/2		1	0	0	0	0	0	0	0	0	0
	2/5		**1**	**1**	**0**	**1**	**1**	**0**	0	0	0	0
	1/3		0	0	0	0	0	0	0	0	0	0
	3/5		0	0	1	0	0	0	0	0	0	0
	2/3		1	0	0	1	0	0	0	0	0	0

(b)

Table 2. Computation of the solution shown in Table 1

The ROADEF Challenge 2005 uses a weighted sum to evaluate the solution Y:

$$F(Y)=w_1*obj_1+w_2*obj_2+w_3*obj_3 \tag{1}$$

where obj_1 , obj_2 and obj_3 correspond respectively to the value of Y on each objective, for the given priority ordering. The weights w_1, w_2 and w_3 are respectively 1000000, 1000 and 1 (Nguyen & Cung, 2005). Renault's factory configuration gives three possible priority orderings: HPO-COLOR-LPO, HPO-LPO-COLOR and COLOR-HPO-LPO. For a more detailed description of this problem, the reader is referred to (Nguyen & Cung, 2005) and (Solnon *et al.*, 2008).

3. Looking for compromise solutions for the ICSP using GISMOO

The algorithm used in this chapter to solve the ICSP from the Pareto domination viewpoint is GISMOO. This algorithm was introduced by Zinflou *et al.* (2011) to solve the classical benchmarks in combinatory optimization with discrete and continuous variables. GISMOO has an original iterative process in two phases: a Genetic phase and an Immune phase. The new solutions (also called descendants) are obtained by offspring creation from the classical genetic operators and by clone creation by the principle of clonal selection in artificial immune systems. Figure 2 presents the general two-phase operation of GISMOO. Even though this is a generic multi-objective algorithm, the description presented in this chapter is specific to the ICSP.

```
Create an initial Parent population POP_0 of size N
Initialize t to 1
While no stopping rules is invoked
   While | Q_t | < N/2
      Select and recombine P_1 and P_2 ∈ POP_t to obtain E_1 and E_2
      Mutate E_1 and E_2 according to mutation probability p_m
      Evaluate E_1 and E_2 and add the best offspring found in Q_t
   End While                                             Genetic Phase
   For each non-dominated solution x∈POP_t Do
      Calculate nb_clones_x to produce for x according to Equation (5)
      cpt =0
      While cpt < nb_clones_x
         Create 2 clones C_1^clo and C_2^clo using x
         Create C_1^hypα (and C_2^hypβ) by hyper-mutation α (β) on C_1 (C_2)
         Evaluate C_1^hypα and C_2^hypβ and add the best of the two in Q_t
         cpt = cpt +1
      End While
   End For                                               Immune Phase
   Copy the N first solutions of POP_t∪Q_t into POP_t+1
   t = t +1
End While
```

Fig. 2. Outline of GISMOO procedure

The main loop of GISMOO (lines 3-21) begins with the Genetic phase (lines 4-8) and generates $N/2$ offspring. This phase has the classical GA operations: selection, crossover and mutation. Notice that the selection procedure used in GISMOO is a binary tournament selection. In addition, even if two offspring are created during the recombination, only the best of the two is added to the Descendant population Q. It is important to mention that no crossover probability is needed in the Genetic phase of GISMOO, as the number of offspring to generate is related to the Parent population (*POP*) size. However, a mutation probability (p_m) is used to determine whether the generated offspring will mutate or not (line 6). Thereafter, the Immune phase (lines 9-18) adds $N/2$ solutions to the Descendant population Q. The number of clones produced from a non-dominated solution is proportional to an isolation factor defined in Section 3.1. In this way, a more isolated solution will generate a greater number of clones. After the Genetic and Immune phases, an elitist population replacement is made, in order to keep the N best solutions of the combined Parent and Descendant populations (line 19).

3.1 Performance assignment

Before making a formal presentation of the components of GISMOO that solves the ICSP, it is important to carefully explain how solution performance is assigned. One of the main difficulties in solving multi-objective optimization problems by a Pareto EA is finding a quality metric ordering the solutions in a population. In fact, the quality of a solution depends on the evaluation of several contradictory and often incommensurable objectives. One of the mechanisms most often used by Pareto EA's is expressing the quality of the solution as a function of two factors: a *dominance factor* and an *isolation factor*. The first factor measures the solution's degree of dominance in Pareto sense, and the second evaluates the density of solutions around a given solution. Even if the EA's share this performance assignment mechanism, each one has its own definition.

GISMOO's dominance factor, similar to that in PMS^{MO} (Zinflou *et al.*, 2008b), is calculated in two steps. First, a force $S(x)$, the number of solutions dominated by x, is assigned to each solution x in the Parent population (*POP*), combined with the Descendent population Q. A feasible solution x *dominates* a feasible solution y iff there is an objective $i \in Z$ for which $f_i(x) < f_i(y)$ and such that for all other objectives $j \in Z$ $(j \neq i)$, $f_j(x) \leq f_j(y)$, where f_i is the objective function for i and Z is the number of objectives to minimize. According to the force S, the dominance factor of an individual x, noted $R^+(x)$, is determined using the following equation.

$$R^+(x) = \begin{cases} \dfrac{S(x)}{1+2*S(x)} & si \displaystyle\sum_{y \in POP_t \cup Q_t, y \succ x} S(y) = 0 \\ \displaystyle\sum_{y \in POP_t \cup Q_t, y \succ x} S(y) & \end{cases} \tag{2}$$

The above calculation thus establishes a dominance relation between solutions based on actual objectives as well as dominated solutions in the search space. Even though GISMOO's calculation is similar to that in PMS^{MO}, the considered populations in the two algorithms are not the same.

GISMOO's isolation factor is inspired by the spacing metric *sp* introduced by Schott (1995), which evaluates the distance $Dist(x)$ between an individual x and its closest neighbour y (with $y \neq x$), as indicated in Equation (3).

$$Dist(x) = \min_{y \in POP \cup Q}\left[\sqrt{\left(f_1(x) - f_1(y)\right)^2 + \ldots + \left(f_Z(x) - f_Z(y)\right)^2}\right] \tag{3}$$

We note that GISMOO does not add the dominance and isolation factors. Rather, it assigns a better performance to the solutions with a high dominance factor, and in case of tie, the equality is broken using the isolation factor.

3.2 Representation of the chromosome

Instead of a classical bit string representation that seems poorly adapted to this type of problem, our representation of a chromosome is composed of two vectors of length *Nb_cars*, corresponding to the cars' option class and colour, as already indicated in Table 1(b).

3.3 Creating of the initial population

In our proposed implementation, the initial population of solutions is generated in two ways: 70% randomly and 30% with a greedy heuristic based on the notion of interest. Two greedy heuristics are used here: *greedy_color* and *greedy_ratio*. These two heuristics were first presented by Zinflou *et al.* (2008a), to solve the ICSP lexicographically. Schematically, *greedy_color* minimizes the number of colour changes, whereas *greedy_ratio* minimizes the number of capacity conflicts for HPO's. For more detailed information, see the above reference. Note that at each iteration, there is a random choice between the two heuristics.

3.4 Genetic phase

3.4.1 Crossover operator

To solve the ICSP, the recombination operator we use is the $NCPX^{MO}$ introduced in (Zinflou *et al.*, 2008a). Schematically, this operator tries to minimize the number of car classes to reposition in the sequence, by using the information linked to the conflictual positions in the parent sequence, based on the notion of *Total Weighted Interest* (*TWI*), which calculates whether a car of class v and colour *Colors* should be in position i in the sequence, according to the following equation:

$$TWI_{v,Colors,i} = I_{v,i,HPO} * w_{HPO} + I_{v,i,COLOR} * w_{COLOR} + I_{v,i,LPO} * w_{LPO} \tag{4}$$

where w_{HPO}, w_{COLOR} and w_{LPO} correspond to the weights given to each objective (1000000, 1000 and 1 according to the objective priorities), and $I_{v,i,HPO}$, $I_{v,i,COLOR}$ and $I_{v,i,LPO}$ correspond to the "interest" in placing the car class v in position i for each objective. However, contrary to the original $NCPX^{MO}$, we do not use the TWI notion, but rather the concept of Pareto Interest (*PI*). Instead of weighting and adding the interest values for each objective, we compare them from a Pareto viewpoint. In this way, the PI allows us to find out if it is good to put a car of class v and colour *Colors* in position i, according to how it dominates (or not) the other candidate classes. A tie is broken by using an extension of a strategy introduced in (Gottlieb *et al.*, 2003), which favours a better distribution of car classes in the sequence under construction. For more details on these subjects, see (Zinflou *et al.*, 2008a).

3.4.2 Selection

Several selection strategies are possible for an EA solving the ICSP in the Pareto sense. We chose the binary tournament selection because its implementation and execution costs are low, and because it has already been shown to work for the theoretical car sequencing problem (Zinflou *et al.*, 2007) and the ICSP from the lexicographic viewpoint (Zinflou *et al.*, 2008a). The tournament games are decided by the dominance factor of the participants, and a tie is broken by using the isolation factor.

3.4.3 Mutation operators

The two mutation operators used by GISMOO to solve the ICSP are the swap operator and random exchange. These operators are often used in the literature to explore the neighbourhood of a solution in a local search method to solve the ICSP.

3.5 Immune phase

In GISMOO, the Descendant population (Q) is subdivided into two parts of equal size $N/2$. The subpopulation of offspring is generated with the selection, recombination and mutation operators, and the subpopulation of clones is generated according to the clonal selection principle introduced by De Castro and Timmis (De Castro & Timmis, 2002). To create the clones, the first step is to sort the current Parent population (POP_t) into fronts, using the same principle as the NSGAII (Deb, 2000). The non-dominated individuals of POP_t, those located in the first front, are then selected as the antibody population to be cloned. In GISMOO, the number of clones produced for each antibody is not constant, but rather proportional to its isolation factor: the individuals farther from their neighbours generate more clones. With this technique, we seek to identify and emphasize non-dominated solutions in isolated regions. The number of clones, *nb_clones(x)*, produced for each antibody x is given by Equation (5).

$$nb\_clones(x) = round\left[\left(Dist(x) * \frac{N}{2}\right) \Big/ \sum_{x=1}^{|Front_1|} Dist(x)\right] \tag{5}$$

where $|Front_1|$ is the number of non-dominated individuals of the current population and *Dist(x)* is the isolation factor of x defined in Equation (3). The function *round* rounds off the argument to the nearest integer.

Once the number of clones is determined for an individual x, two clones (copies of x) are produced and then hyper-mutated by the respective application of the α and β mutation operators (C_1^{clo} , C_2^{clo}). In the ICSP context, the operator α corresponds to a swap mutation, and β to a random exchange. No new parameters, therefore, need to be created. After the evaluation, the two mutated clones are compared, and the dominant one (in the Pareto sense) is retained. If no dominance relation can be established, one of the clones is chosen at random. In each case, the selected clone is added to the current population Q. For each individual x in $Front_1$, this process is repeated until the number of clones produced attains *nb_clones(x)*.

3.6 Replacement

GISMOO is an elitist algorithm: to conserve the best individuals in the current population, its replacement strategy is deterministic of type (λ+μ) where λ is the Parent population size and μ the Descendant population size. In our approach, $\lambda = \mu = N$.

3.7 Managing elitism

Like any elitist algorithm, GISMOO has mechanisms that retain non-dominated individuals while it is searching for solutions. It uses an archive *A* to conserve non-dominated solutions during its iterations. However, the individuals in *A* do not participate in the selection process; only the non-dominated individuals currently in the population do so. It is important to note that the size of *A* is not fixed or limited by some maximum. Finally, to ensure that the best individuals in the population are conserved in the current population, GISMOO uses the elitist replacement procedure described in the previous paragraph.

4. Numerical experiments

Until now, the ICSP has generally been approached lexicographically or by lumping together the different goals. To our knowledge, the only work that treats the ICSP from a purely multi-objective viewpoint is that of Zinflou *et al.* (2009) and of de Oliveira dos Reis (2007). However, the latter work is limited to instances of fewer than 60 cars. The experiments of Zinflou *et al.* compare the performances of PMS^{MO} to that of NSGAII, using the ROADEF 2005 instances. We use the same two algorithms to analyze GISMOO's performance from the Pareto viewpoint. The version of GISMOO solving the ICSP and used in this chapter was implemented in C++ with Visual Studio .net 2008. The computer used for the numerical experiments is a Dell model with a Pentium Xeon 3.6 GHz processor with 1GB of RAM, operating under Windows XP.

In our implementation, the main data structures are shared by all the algorithms. We thus obtain a common basis for comparing and equitably evaluating the performances of the different algorithms. In all the numerical experiments of this section, each instance of the ICSP was solved 30 times by each algorithm on the same computer. We used the following parameters:

- The size (N) of the populations was empirically fixed at 100 individuals for each instance and each algorithm.
- The mutation probability (p_m) was fixed at 0.06 for the three algorithms. Note that GISMOO does not need a crossover probability. For the other two algorithms, the crossover probability was 1.
- In the context of the ROADEF Challenge 2005, a time limit of 600 seconds on a PC Pentium4/1.6GHz/win2000/1GB RAM was fixed. Because of the computer used in our experiments, as well as the experimental conditions of the Challenge, the time limit for the algorithm was fixed at 350 seconds.
- The maximum size of the local archive for PMS^{MO} was fixed at 100 individuals.

The value of the different parameters discussed here was fixed in accordance with the numerical experiments in (Zinflou *et al.*, 2009). We also mention that the goal of this chapter is not to directly compare the performance of NSGAII to that of PMS^{MO}, but rather to analyze the performances of GISMOO in solving industrial problems. For a direct comparison between NSGAII and PMS^{MO} the reader can consult (Zinflou *et al.*, 2009).

4.1 Test problems

The numerical experiments presented in this chapter were carried out on the three samples furnished by the Renault company for the ROADEF Challenge 2005. The first sample (Set A) had 8 data sets for the scheduling of 334 to 1314 cars with 6 to 22 options, making 36 to 287 car classes with 11 to 24 different colours. This sample was used to evaluate the teams in the qualifying phase and select the 18 teams for the following round. The second sample (Set B) had 15 instances, each composed of 65 to 1270 cars with 4 to 25 options, 11 to 339 classes and 4 to 20 colours. This sample was used to select the 12 finalist teams in the Challenge. The last sample (Set X) had 19 instances with 65 to 1319 cars, 5 to 26 options, 10 to 328 classes and 5 to 20 colours. This sample was used in the final evaluation of the 12 finalists, in order to select the winning team.

4.2 Performance metrics

Evaluating the performance of a multi-objective algorithm is often an arduous task. Several metrics evaluating such algorithms are mentioned in the literature. The goal of this series of numerical experiments is to compare the performance of GISMOO with that of NSGAII and PMS^{MO} in solving the ICSP while respecting the usual standards in multi-objective optimization. To attain this goal, we use the three metrics introduced by Zitzler & Thiele (1999): the hyper-volume $\boldsymbol{H}$, the coverage of two sets metric $\boldsymbol{C}$ and the covering difference $\boldsymbol{D}$.

Schematically, the hyper-volume metric $\boldsymbol{H}$ estimates the "volume under" the "surface" formed by the points of a given solution set. In particular, when the problem has two objectives, this metric corresponds to calculating an area, and for three objectives, a volume is calculated. Formally, the size of the dominated space $\boldsymbol{H}$ is defined in the following way. Let $X = \{x_1, x_2, \ldots, x_s\}$ be a set of S feasible solutions. The function $\boldsymbol{H}$ gives the volume enclosed by the union of the polytopes $p_1, p_2, \ldots, p_s$ where each p_i is formed by the intersections of the following hyper-planes arising out of x_i, along with the axes: for each axis in the objective space, there exists a hyper-plane perpendicular to the axis and passing through the point $(f_1(x_i), f_2(x_i), \ldots, f_z(x_i))$. In the two-dimensional case, each p_i represents a rectangle defined by the points (0, 0) and $(f_1(x_i), f_2(x_i))$.

The $\boldsymbol{C}$ metric is used to represent the relative spread of solutions between two sets of solution vectors, U and V. The value $\boldsymbol{C}(U,V)$ corresponds to the percentage of solutions of V that are weakly dominated by at least one solution of U. A point x *weakly dominates* a point y iff $f_z(x) \leq f_z(y)$ for all z in Z. The C metric is not symmetric: $\boldsymbol{C}(U,V) \neq 1 - \boldsymbol{C}(V,U)$, so it is necessary to consider both values, $\boldsymbol{C}(U,V)$ and $\boldsymbol{C}(V,U)$, to obtain a reliable measure of the two compromise surfaces.

Finally, the $\boldsymbol{D}$ metric is defined in the following way. Let $U,V \subseteq X$ be two sets of decision vectors. The function $\boldsymbol{D}$ is defined by $\boldsymbol{D}(U,V) = \boldsymbol{H}(U+V) - \boldsymbol{H}(V)$ and gives the size of the space weakly dominated by U but not weakly dominated by V in the objective space.

5. Results and discussion

Tables 3 to 5 present the results obtained by the three algorithms, for the three ICSP test data sets, as measured by the $\boldsymbol{H}$ metric. Each row of these tables gives the name of the instance, the number of cars to schedule (*Nb_cars*), and the average results obtained with $\boldsymbol{H}$ for each algorithm. The best results are shaded in grey. We have indicated by (*) the instances for which the three algorithms found exactly the same set of solutions in each execution.

From the results of Tables 3 to 5, we observe that GISMOO has the best results for all the instances of all the data sets. That is, the size of the space dominated by GISMOO is always greater than or equal to that of the other two algorithms. In fact, when we compare GISMOO with NSGAII, we find that GISMOO is superior in 38 instances and equal in 4 instances. GISMOO has the same score with PMS^{MO}. Between NSGAII and PMS^{MO}, the results are divided and the scores are often very close.

Instance	*Nb_cars*	NSGAII	PMS^{MO}	GISMOO
022_3_4	485	1.16 E+8	1.17 E+8	1.22 E+8
024_38_3	1260	6.05 E+6	6.47 E+6	8.7 E+6
024_38_5	1315	1.58 E+7	1.58 E+7	1.93 E+7
025_38_1	1004	1.41 E+8	1.65 E+8	1.76 E+8
039_38_4_ch1	954	7.41 E+6	7.40 E+6	8.30 E+6
048_39_1	600	5.27 E+7	5.7 E+7	6.16 E+7
064_38_2_ch1	875	3.46 E+7	3.48 E+7	3.94 E+7
064_38_2_ch2	335	1.45 E+7	1.52 E+7	1.65 E+7

Table 3. Average hyper-volumes ***H*** of NSGAII, PMS^{MO} and GISMOO for the instances of Set A of the ICSP

Instance	*Nb_cars*	NSGAII	PMS^{MO}	GISMOO
022_S22_J1	526	1.54 E+6	1.86 E+6	2.03 E+6
023_S23_J3	1110	6.46 E+6	6.98 E+5	7.26 E+6
024_V2_S22_J1	1270	7.6 E+7	8.71 E+7	1.04 E+8
025_S22_J3	1161	8.04 E+6	9.44 E+6	1.02 E+7
028_ch1_S22_J2	365	3.41 E+6	3.35 E+6	5.43 E+6
028_ch2_S23_J3	65	8.56 E+3	9.13 E+3	9.14 E+3
029_ S21_J6	730	2.93 E+7	2.89 E+7	3.56 E+7
035_ch1_S22_J3	128	8.64 E+7	8.79 E+7	9.14 E+7
035_ch2_S22_J3	269	1.39 E+8	1.45 E+8	1.56 E+8
039_ch1_S22_J4	1231	1.24 E+8	1.37 E+8	1.42 E+8
039_ch3_ S22_J4	1000	5.86 E+6	1.06 E+7	2.01 E+7
048_ch1_ S22_J3	591	6.61 E+7	7.2 E+7	7.41 E+7
048_ch2_ S22_J3	546	9.94 E+7	1.01 E+8	1.05 E+8
064_ch1_ S22_J3	825	4.24 E+7	4.14 E+7	5.12 E+7
064_ch2_ S22_J4	412	8.04 E+7	8.11 E+7	9.15 E+7

Table 4. Average hyper-volumes ***H*** of NSGAII, PMS^{MO} and GISMOO for the instances of Set B of the ICSP

Although the hyper-volume, by definition, gives a good idea of the size of the space dominated by a solution set, this metric does not let us compare two solution sets with each other. To do that, we can use the set coverage metric ***C***. Figure 3 presents the results of the three algorithms according to the ***C*** metric. In this figure, each box corresponds to the comparison of the solutions of algorithm *U* (row) with algorithm *V* (column). The value ***C****(U,V)* indicates the percentage of elements of *V* dominated by at least one element of *U*. In each box, the scale goes from 0 at the bottom to 1 at the top. Each box has 8 graphs corresponding to the 8 instances of Set A. Each boldface horizontal bar indicates the mean of the ***C*** measures for the 30 executions, and the vertical bars indicate the spread between the maxima and minima of the executions.

Instance	*Nb_cars*	NSGAII	PMS^{MO}	GISMOO
022_S49_J2	704	1.17 E+8	1.18 E+8	1.2 E+8
023_S49_J2	1260	5.46 E+7	5.53 E+7	6.59 E+7
024_S49_J2	1319	1.02 E+8	1.02 E+8	1.1 E+8
025_ S49_J1	996	2.30 E+8	2.36 E+8	2.53 E+8
028_CH1_S50_J4	325	1.51 E+8	1.55 E+8	1.66 E+8
028_CH2_S51_J1*	65	1.00 E+3	1.00 E+3	1.00 E+3
029_S49_J5	780	1.54 E+8	1.34 E+8	1.72 E+8
034_VP_S51_J1_J2_J3	921	1.45 E+8	1.5 E+8	1.54 E+8
034_VU_S51_J1_J2_J3	231	9.05 E+7	9.11 E+7	1.00 E+8
035_CH1_S50_J4*	90	9.85 E+6	9.85 E+6	9.85 E+6
035_CH2_S50_J4	376	4.36 E+5	5.00 E+5	6.09 E+5
039_CH1_S49_J1*	1247	2.5 E+8	2.5 E+8	2.5 E+8
039_CH3_S49_J1	1037	1.59 E+8	1.67 E+8	1.81 E+8
048_CH1_S50_J4	519	7.32 E+7	7.36 E+7	8.71 E+7
048_CH2_S49_J5	459	8.5 E+7	9.03 E+7	9.62 E+7
064_CH1_S49_J1	875	3.53 E+7	3.53 E+7	4.19 E+7
064_CH2_S49_J4	273	9.76 E+7	9.76 E+7	9.77 E+7
655_CH1_S51_J2_J3_J4*	264	1.00 E+9	1.00 E+9	1.00 E+9
655_CH2_S52_J1_J2_S01_J1	219	2.08 E+6	2.17 E+6	2.37 E+6

Table 5. Average of the hyper-volumes ***H*** of NSGAII, PMS^{MO} and GISMOO for the instances of Set X of the ICSP

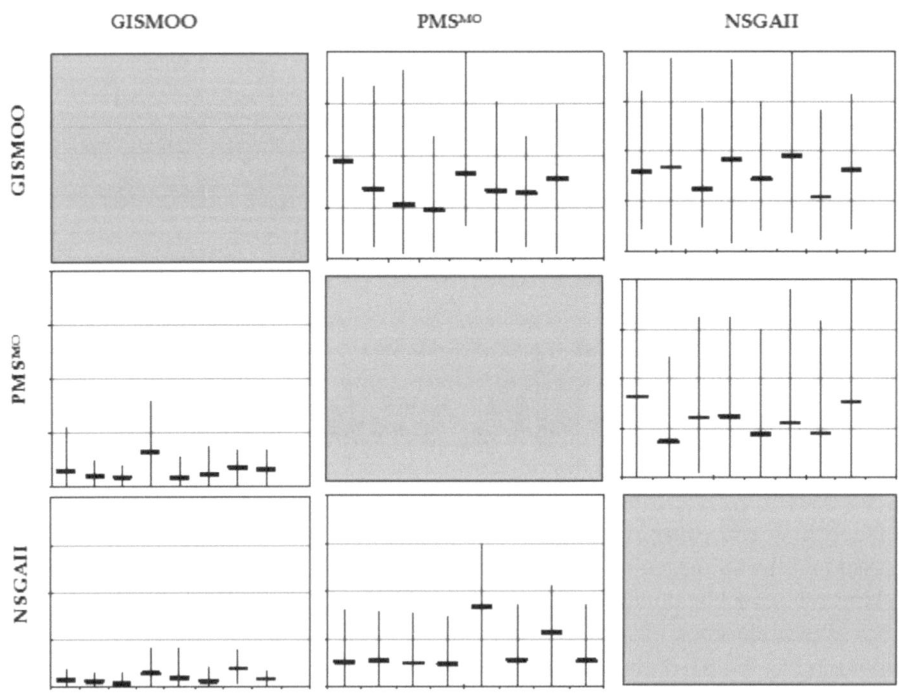

Fig. 3. Average coverage ***C*** of GISMOO, PMS^{MO} and NSGAII for the instances of Set A

In a way similar to that of Figure 3, Figure 4 presents the results of the three algorithms according to the **C** metric, for the 15 instances of Set B. Each box has 15 corresponding graphs. The boldface horizontal bar indicates the mean of the **C** measures for the 30 executions, and the vertical bars indicate the spread between the maxima and minima of the executions.

Finally, Figure 5 presents the results of the three algorithms according to the **C** metric, for 15 of the 19 instances of Set X. The results for the 4 other instances (028_CH2_S51_J1, 035_CH1_S50_J4, 039_CH1_S49_J1, 655_CH1_S51_J2_J3_J4) are exactly the same for the three algorithms, so that they are not presented here, in order to make the figure easier to read.

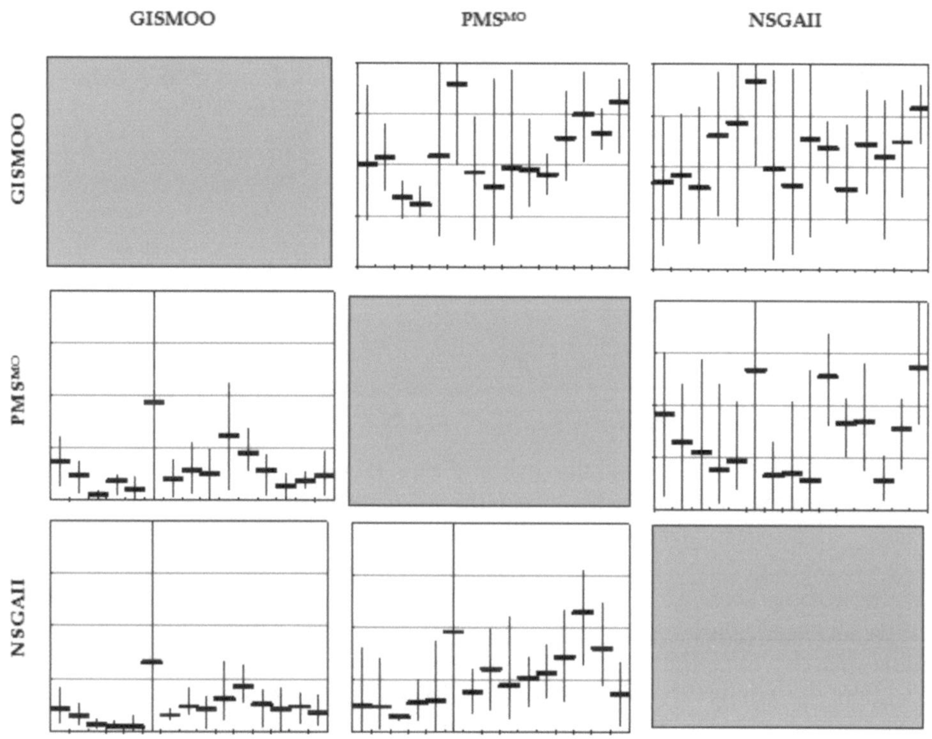

Fig. 4. Average coverage **C** of GISMOO, PMS^MO and NSGAII for the instances of Set B

The results shown are, once again, clearly in favour of GISMOO for all the instances. Indeed, the values of **C**(NSGAII, GISMOO) are always less than those of **C**(GISMOO, NSGAII). Note that for many instances, the ratio between the results for the two algorithms is greater than 10; this means that almost all the solutions found by GISMOO are non-dominated or dominate those found by NSGAII. If GISMOO is compared with PMS^MO, it is seen that the values of **C**(PMS^MO, GISMOO) are always less than those of **C**(GISMOO, PMS^MO), with a ratio greater than 10 for most instances, making almost all the solutions found by GISMOO non-dominated or dominating those found by PMS^MO. The results found by the **C** metric confirm, therefore, the results found by the hyper-volume.

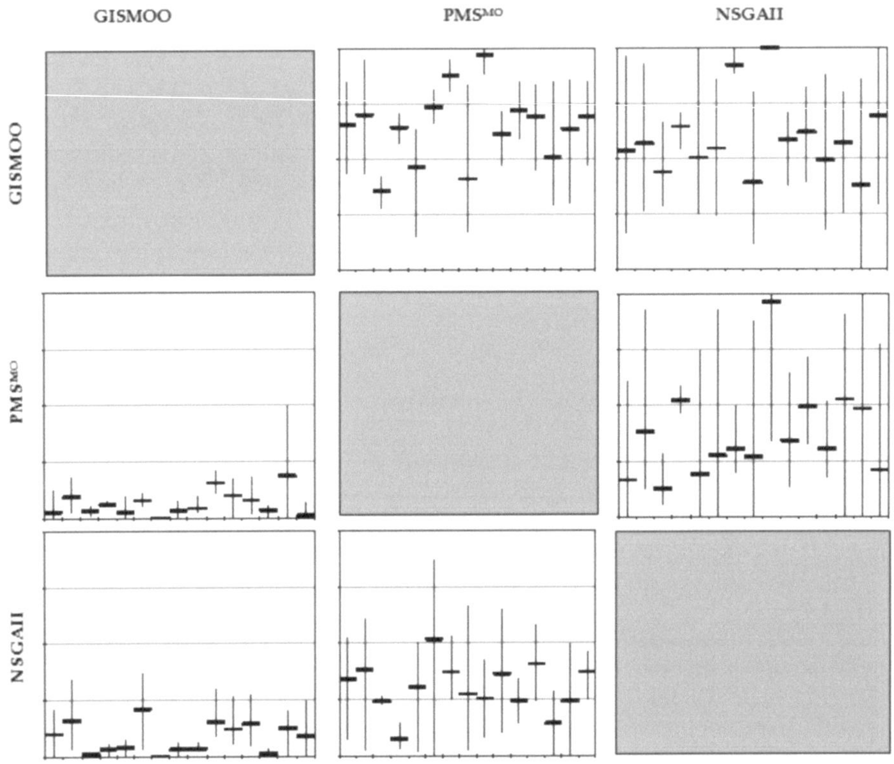

Fig. 5. Average coverage ***C*** of GISMOO, PMS^{MO} and NSGAII for the instances of Set X

Along with the hyper-volume and the covering, we also compared the three algorithms with the ***D*** metric that measures the difference in the covering by the solutions of two algorithms at a time. This metric gives an idea of the size of the solution space dominated by the solutions of one algorithm but not the other. Tables 6 to 11 resume respectively the results this metric obtained between GISMOO and NSGAII and between GISMOO and PMS^{MO}, for each data set tested. Each row of these tables gives the instance name, the number of cars to schedule and, for each algorithm, the mean results obtained with the ***D*** metric after 30 independent executions. The best results are indicated by grey shading.

Instance	**Nb_cars**	**NSGAII**	**GISMOO**
022_3_4	485	3.09 E+6	9.10 E+6
024_38_3	1260	1.16 E+5	1.19 E+5
024_38_5	1315	1.06 E+8	1.09 E+8
025_38_1	1004	8.24 E+8	8.59 E+8
039_38_4_ch1	954	4.2 E+6	5.09 E+6
048_39_1	600	1.88 E+8	1.97 E+8
064_38_2_ch1	875	2.11 E+6	2.15 E+6
064_38_2_ch2	335	8.4 E+6	1.05 E+7

Table 6. Average coverage difference ***D*** between NSGAII and GISMOO for the instances of Set A

Instance	Nb_cars	PMS^{MO}	GISMOO
022_3_4	485	3.09 E+6	7.88 E+6
024_38_3	1260	1.16 E+5	1.19 E+5
024_38_5	1315	1.06 E+8	1.09 E+8
025_38_1	1004	8.25 E+8	8.35 E+8
039_38_4_ch1	954	4.2 E+6	5.1 E+6
048_39_1	600	1.88 E+8	1.93 E+8
064_38_2_ch1	875	2.11 E+6	2.15 +6
064_38_2_ch2	335	8.5 E+6	9.82 E+6

Table 7. Average coverage difference ***D*** between PMS^{MO} and GISMOO for the instances of Set A

Instance	Nb_cars	NSGAII	GISMOO
022_S22_J1	526	1.97 E+6	2.46 E+6
023_S23_J3	1110	2.77 E+6	2.85 E+6
024_V2_S22_J1	1270	3 E+6	3.05 E+6
025_S22_J3	1161	2.4 E+6	2.77 E+6
028_ch1_S22_J2	365	1.96 E+6	2.16 E+6
028_ch2_S23_J3	65	8.66 E+2	1.44 E+3
029_ S21_J6	730	5.89 E+6	5.96 E+6
035_ch1_S22_J3	128	3.36 E+7	3.86 E+7
035_ch2_S22_J3	269	3.44 E+7	3.61 E+7
039_ch1_S22_J4	1231	1.08 E+8	1.26 E+8
039_ch3_ S22_J4	1000	1.05 E+5	1.19 E+7
048_ch1_ S22_J3	591	1 E+6	1.09 E+6
048_ch2_ S22_J3	546	1.95 E+7	2.99 +7
064_ch1_ S22_J3	825	7.38 E+6	8.26 E+6
064_ch2_ S22_J4	412	3.35 E+7	4.46 E+7

Table 8. Average coverage difference ***D*** between NSGAII and GISMOO for the instances of Set B

An examination of the six tables shows, once again, that the results are clearly in favour of GISMOO. The comparison with NSGAII (Tables 6, 8 and 10) shows that GISMOO has a larger covering difference for all instances except for four in Set X where the two algorithms have identical results. The comparison with PMS^{MO} (Tables 7, 9 and 11) shows the same advantage for GISMOO. These results confirm those obtained by the hyper-volume, which can be explained by the fact that the ***D*** metric calculations are based on those of the hyper-volume. With the hyper-volume results, the ***D*** metric results lead one to conclude that the solutions found by GISMOO have a better distribution than those found by NSGAII and PMS^{MO}.

Instance	Nb_cars	PMS^MO	GISMOO
022_S22_J1	526	1.98 E+6	2.14 E+6
023_S23_J3	1110	2.77 E+6	2.80 E+6
024_V2_S22_J1	1270	3.02 E+6	3.05 E+6
025_S22_J3	1161	2.6 E+6	2.76 E+6
028_ch1_S22_J2	365	1.96 E+6	2.17 E+6
028_ch2_S23_J3	65	8.7 E+2	8.9 E+2
029_ S21_J6	730	2.88 E+6	5.97 E+6
035_ch1_S22_J3	128	3.37 E+7	3.71 E+7
035_ch2_S22_J3	269	3.44 E+7	3.55 E+7
039_ch1_S22_J4	1231	1.09 E+8	1.13 E+8
039_ch3_ S22_J4	1000	1.04 E+6	1.14 E+7
048_ch1_ S22_J3	591	1.01 E +6	1.03 E +6
048_ch2_ S22_J3	546	1.95 E+7	2.99 E+7
064_ch1_ S22_J3	825	7.36 E+6	8.36 E+6
064_ch2_ S22_J4	412	3.38 E+7	4.39 E+7

Table 9. Average coverage difference ***D*** between PMS^MO and GISMOO for the instances of Set B

Instance	Nb_cars	NSGAII	GISMOO
022_S49_J2	704	4.8 E+6	7.82 E+6
023_S49_J2	1260	5.97 E+6	7.04 E+6
024_S49_J2	1319	1.4 E+7	1.48 E+7
025_ S49_J1	996	5.97 E+7	6.2 E+7
028_CH1_S50_J4	325	2.09 E+7	2.24 E+7
028_CH2_S51_J1*	65	0	0
029_S49_J5	780	7.84 E+7	9.62 E+7
034_VP_S51_J1_J2_J3	921	1.96 E+7	2.05 E+7
034_VU_S51_J1_J2_J3	231	2.51 E+7	3.45 E+7
035_CH1_S50_J4*	90	0	0
035_CH2_S50_J4	376	3.91 E+5	5.64 E+5
039_CH1_S49_J1*	1247	0	0
039_CH3_S49_J1	1037	1.94 E+7	2.16 E+7
048_CH1_S50_J4	519	2.88 E+7	3.02 E+7
048_CH2_S49_J5	459	1.54 E+7	1.65 E+7
064_CH1_S49_J1	875	8.31 E+6	8.97 E+6
064_CH2_S49_J4	273	2.34 E+6	2.37 E+6
655_CH1_S51_J2_J3_J4*	264	0	0
655_CH2_S52_J1_J2_S01_J1	219	2.63 E+5	2.92 E+5

Table 10. Average coverage difference ***D*** between NSGAII and GISMOO for the instances of Set X

Instance	Nb_cars	PMS^{MO}	GISMOO
022_S49_J2	704	4.82 E+6	7.21 E+6
023_S49_J2	1260	5.91 E+6	6.97 E+6
024_S49_J2	1319	1.4 E+7	1.48 E+7
025_ S49_J1	996	5.97 E+7	6.14 E+7
028_CH1_S50_J4	325	2.1 E+7	2.2 E+7
028_CH2_S51_J1*	65	0	0
029_S49_J5	780	7.85 E+7	1.16 E+8
034_VP_S51_J1_J2_J3	921	1.96 E+7	2 E+7
034_VU_S51_J1_J2_J3	231	2.5 E+7	3.39 E+7
035_CH1_S50_J4*	90	0	0
035_CH2_S50_J4	376	3.91 E+5	5.00 E+5
039_CH1_S49_J1*	1247	0	0
039_CH3_S49_J1	1037	1.94 E+7	2.08 E+7
048_CH1_S50_J4	519	2.88 E+7	3.01 E+7
048_CH2_S49_J5	459	1.54 E+7	1.6 E+7
064_CH1_S49_J1	875	8.31 E+6	8.97 E+6
064_CH2_S49_J4	273	2.34 E+6	2.37 E+6
655_CH1_S51_J2_J3_J4*	264	0	0
655_CH2_S52_J1_J2_S01_J1	219	2.63 E+5	2.83 E+5

Table 11. Average coverage difference ***D*** between PMS^{MO} and GISMOO for the instances of Set X

6. Comparison from a decision-making viewpoint

Besides the performance metrics used above, we also compare the three algorithms from a decision-making viewpoint. To graphically illustrate the results obtained, the comparison is done only with the HPO and COLOR objectives of the ICSP. We compare GISMOO with NSGAII and PMS^{MO}, and also with the results of the best team in the ROADEF Challenge 2005 (BEST_ROADEF). Remember that these latter results were obtained with a lexicographic ordering of the objectives. Moreover, the BEST_ROADEF results were obtained by optimizing the objectives in the order HPO-COLOR-LPO or COLOR-HPO-LPO. Finding these extreme solutions requires two distinct executions of the algorithm and thus requires a global execution time which is double the calculation time allocated to the three other algorithms.

Figure 6 presents a visual comparison of GISMOO, NSGAII, PMS^{MO} and BEST_ROADEF for executions of instance 022_3_4 (in Set A), for which the HPO constraints are "easy" to satisfy, according to Renault. This graphical representation confirms the results for the metrics in Section 5. It is clear that GISMOO's Pareto set, for this instance, dominates those of NSGAII and PMS^{MO}. Indeed, the curve for GISMOO is definitely lower than those of NSGAII and PMS^{MO}. We recall here that the ICSP is a minimization problem. When GISMOO's results are compared with BEST_ROADEF's, we note that a single execution of GISMOO allows us to obtain the same solutions as two distinct executions of the algorithm used by the challenge's winning team. As well, we see that GISMOO shows us several alternative solutions ignored by the lexicographic treatment imposed by the Challenge rules. By giving too much importance to one objective, the lexicographic approach used by the ROADEF 2005 teams makes their algorithm converge towards an overly restricted zone of the search space.

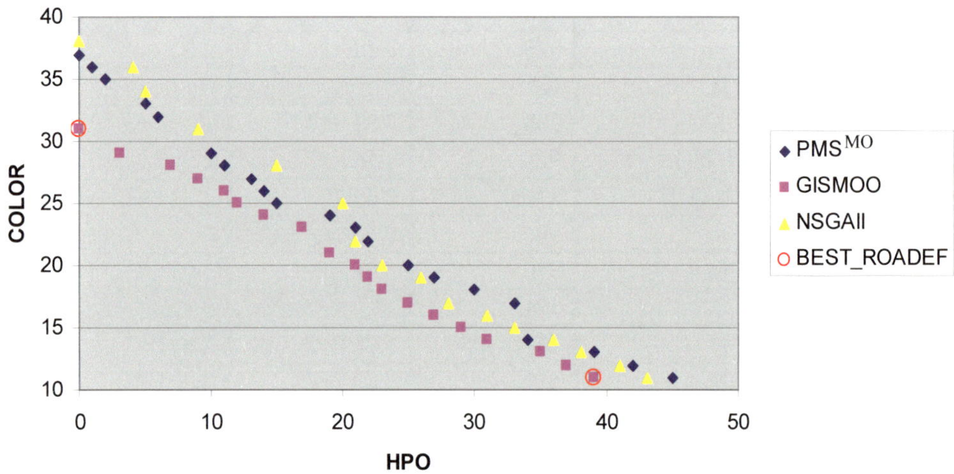

Fig. 6. Graphical performance comparison of GISMOO, PMS^{MO}, NSGAII and BEST_ROADEF for the 011_3_4 instance

From a decision-making viewpoint, GISMOO's solution set offers greater latitude to a manager by presenting him with 19 alternative solutions, instead of the two extreme solutions proposed by BEST_ROADEF. For example, a "COLOR-oriented" manager could slightly lessen the importance of that objective and save two or even four HPO conflicts. In this case, the number of colour changes would increase from 11 to 13, but the number of HPO conflicts would decrease from 39 to 35. On the other hand, a manager oriented towards HPO conflict minimization would probably be interested to see the effect of lessening the importance of that objective on the number of purges. Thanks to the various solutions presented by GISMOO, he would see that the number would diminish roughly in the same proportion. In fact, by permitting three more HPO conflicts (3) than the extreme solution (0), he would save three COLOR purges (28 instead of 31). GISMOO's solution set would allow still another manager with no preference between HPO and COLOR, to choose a balanced solution with about as many HPO conflicts (21) as colour changes (20).

It is important to note, however, that compromise solution sets cannot always be generated. Some of the ICSP instances proposed by Renault are such that all four algorithms give a unique solution optimizing all the objectives at the same time. This is the case for the instance 655_CH1_S51_J2_J3_J4 in Set X, as Table 12 shows. Each row of the table indicates, for each algorithm, the number of HPO conflicts, the number of colour changes and the number of LPO conflicts. The analysis of these results shows that the four algorithms, in all of their executions, give exactly the same solution.

Figure 7 presents a visual comparison of GISMOO, NSGAII, PMS^{MO} and BEST_ROADEF for executions on the 035_ch2_S22_J3 (from Set B), which gives a problem for which the HPO constraints are "hard" to satisfy, according to Renault. As in the example presented in Figure 6, the Pareto set proposed by GISMOO clearly dominates those proposed by NSGAII and

PMS^{MO}. However, we note that the deviation between GISMOO and the two other Pareto algorithms is not as great as that observed by the 022_3_4 instance. This situation can be explained by the fact that the 035_ch2_S22_J3 instance has only 269 cars to schedule, whereas the 022_3_4 instance has 485. Nevertheless, we note that neither NSGAII nor PMS^{MO} can obtain the extreme HPO solution obtained by BEST_ROADEF. The best solutions found by NSGAII and PMS^{MO}, with HPO as the major goal, give 448 and 438 HPO conflicts, while GISMOO's best solution gives only 385 HPO conflicts. We can suppose that the difference between GISMOO and the other two Pareto algorithms would be even larger for a larger instance with HPO constraints that are "hard" to satisfy. Along with the two extreme solutions, GISMOO offers 70 other compromise solutions to a manager. However, contrary to what was observed for instance 022_3_4, we note that there is a large difference between the extreme HPO solution and the other solutions offered by GISMOO. It is possible to explain this difference by the difficulty in satisfying the HPO constraints for this instance.

GISMOO			PMS^{MO}			NSGAII			BEST_ROADEF		
HPO	COLOR	LPO	HPO	COLOR	LPO	HPO	COLOR	LPO	HPO	COLOR	LPO
0	30	0	0	30	0	0	30	0	0	30	0

Table 12. Solution of GISMOO, PMS^{MO}, NSGAII and BEST_ROADEF for the 655_CH1_S51_J2_J3_J4 instance

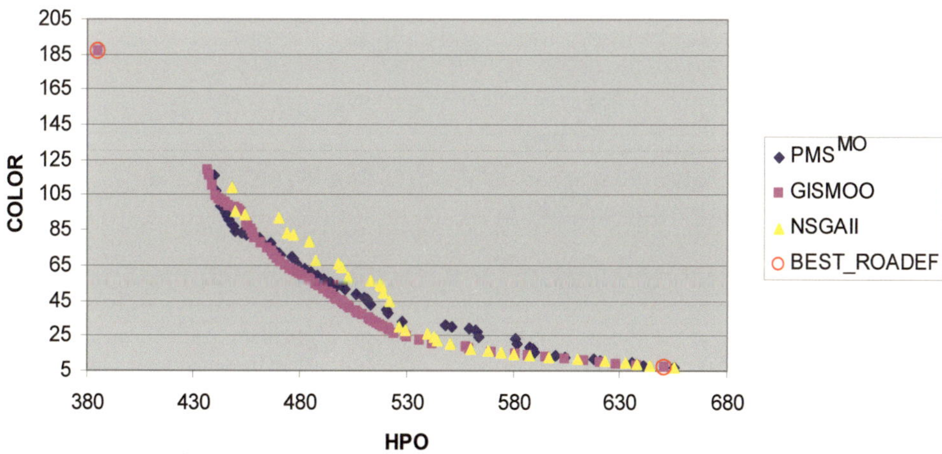

Fig. 7. Graphical performance comparison of GISMOO, PMS^{MO}, NSGAII and BEST_ROADEF for instance 035_ch2_S22_J3

Figure 8 presents a visual comparison of GISMOO, NSGAII, PMS^{MO} and BEST_ROADEF for another instance (048_ch2_S49_J5) for which the HPO constraints are "hard" to satisfy, but which have 546 cars to schedule. This graphical representation confirms the suppositions made for the preceding figure. We observe that GISMOO's solution set clearly dominates NSGAII's and PMS^{MO}'s solution sets: the difference between the curves is clearly larger for this instance than for instance 035_ch2_S22_J3 which has only about half as many cars. Figure 8 also shows that BEST_ROADEF's two solutions dominate GISMOO's solutions for instance

048_ch2_S49_J5. However, a closer look at the solutions reveals exactly the same value on the main objective. In fact, there are 3 HPO conflicts and 93 colour changes for BEST_ROADEF with HPO as the main objective, versus 3 HPO conflicts and 135 colour changes for GISMOO. With COLOR as the main objective, BEST_ROADEF obtains a solution with 58 colour changes and 282 HPO conflicts, versus 58 colour changes and 420 HPO conflicts for GISMOO. We recall that GISMOO's execution time is about half that of BEST_ROADEF. If we give GISMOO and ROADEF the same time, the performance difference is considerably lessened, as Figure 9 shows. The extreme solutions of GISMOO are almost identical to those of BEST_ROADEF: 3 HPO conflicts and 94 colour changes for BEST_ROADEF with HPO as the main objective, versus 3 conflicts and 93 changes for GISMOO, and with COLOR as the main objective, 58 changes and 284 conflicts for GISMOO versus 58 changes and 283 conflicts for BEST_ROADEF. In addition, GISMOO offers 7 compromise solutions to a decision maker.

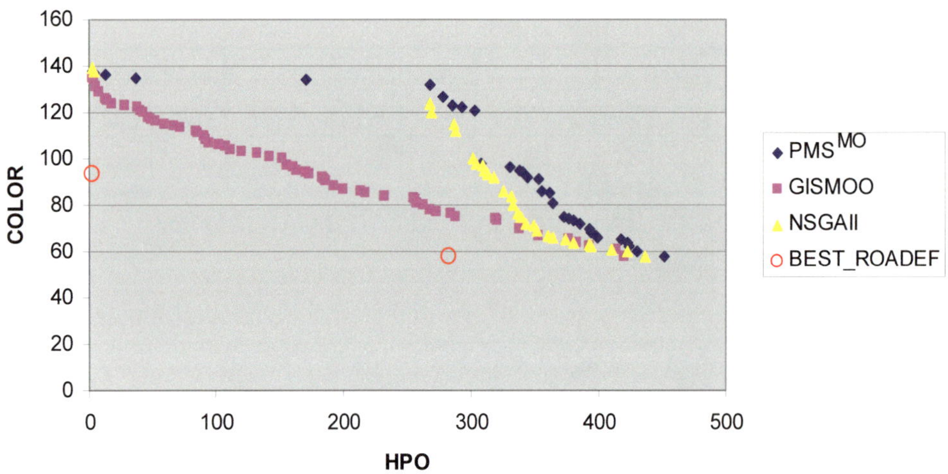

Fig. 8. Graphical performance comparison of GISMOO, PMS^{MO}, NSGAII and BEST_ROADEF for instance 048_ch2_S49_J5

7. Conclusion

In this chapter we have presented an evolutionary Pareto algorithm, GISMOO, to solve the multi-objective industrial car sequencing problem. The biggest difference between GISMOO and other multi-objective EAs mainly lies in the way in which the immune metaphor is used in a Pareto EA to identify and emphasize the solutions located in less crowded regions of the search space. Even if EAs are known to be well suited for multi-objective optimization in Pareto sense, few researchers and industrials decided to use this category of algorithms to solve the ICSP. This situation may be explained by the fact that the ICSP is generally considered as a problem with several goals lexicographically ordered and not from the Pareto viewpoint. However, the lexicographical treatment of the objectives is such that it can eliminate several "interesting" solutions for the manufacturer. Indeed, the relaxation of the importance granted to the main objective can highlight other attractive solutions.

One original effect of this use of GISMOO is that a Pareto algorithm considers the objectives of the problem integrally, without ordering them or giving them weights. To our knowledge, this problem has generally been considered from a one-goal viewpoint, or by

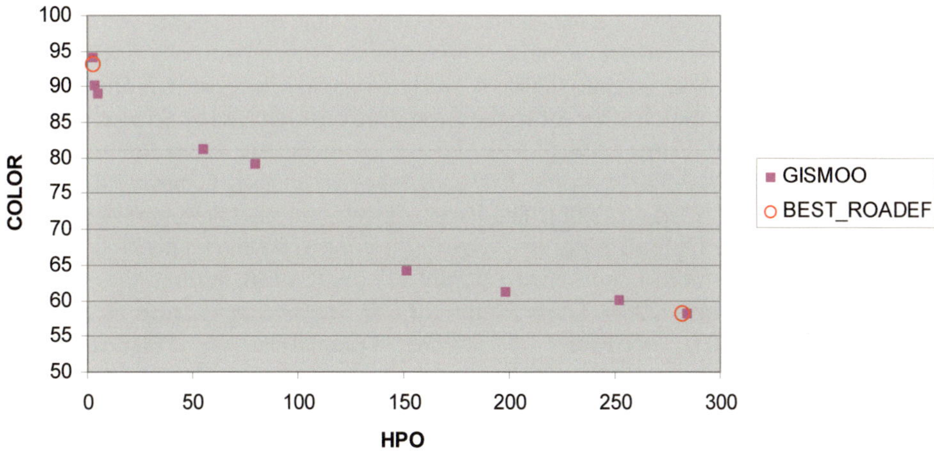

Fig. 9. Graphical performance comparison of GISMOO and BEST_ROADEF for instance 048_ch2_S49_J5, when the same execution time is allocated to the two algorithms

ordering the goals lexicographically, or by testing only small-size instances. The results obtained by GISMOO for a variety of test cases of the ICSP have showcased compromise solutions that were not obtainable by a lexicographic treatment. Instead of a single solution in function of an *a priori* ordering of goals, GISMOO offers the decision maker a panoply of solutions from which he can choose according to his own preferences. Another positive effect of using GISMOO is to bridge the gap between theoretical approaches and practical situations for this type of industrial problem. Indeed, the experimental results confirm the excellent performance of our algorithm in coping with a real-life situation.

In future work, we will seek to extend the application field of GISMOO to others multiple-objective problems such as other industrial problems or continuous test problems with or without constraints.

8. References

Benoist, T. (2008). Soft car sequencing with colors: Lower bounds and optimality proofs. *European Journal of Operational Research*, Vol. 191, No. 3, pp. 957-971.

Briant, O., D. Naddef & G. Mounié (2008). Greedy approach and multi-criteria simulated annealing for the car sequencing problem. *European Journal of Operational Research*, Vol. 191, No. 3, pp. 993-1003.

Cordeau, J. F., G. Laporte & F. Pasin (2008). Iterated Tabu Search for the Car Sequencing Problem. *European Journal of Operational Research*, Vol. 191, No.3, pp.945-956.

De Castro, L. N. and J. Timmis (2002). *Artificial Immune Systems: A New Computational Intelligence Approach*. London.

de Oliveira dos Reis, R. J. (2007). *Solving the Car Sequencing Problem from a Multiobjective* Perspective Lisbonne, Universidade Técnica de Lisboa. Master's thesis.

Deb, K. (2000). A Fast Elitist Non-Dominated Sorting Genetic Algorithm for Multiobjective Optimization : NSGA II. *Parallel problem Solving form Nature – PPSN VI*, M. Schoenauer et al. (Eds), Springer Lecture Notes in Computer Science, p. 849-858.

Estellon, B., F. Gardi & K. Nouioua (2008). Two local search approaches for solving real-life car sequencing problem. *European Journal of Operational Research*, Vol. 191, No. 3, pp.928-944.

Francisci, D. (2002). *Algorithmes évolutionnaires et optimisation multi-objectifs en data mining,* Laboratoire informatique, signaux et systèmes de Sophia Antipolis UMR 6070.

Gottlieb, J., M. Puchta & C. Solnon (2003). A study of greedy, local search and ant colony optimization approches for car sequencing problems. *Computers Science*: pp.246-257.

Knowles, J. D. & D. W. Corne (2000a). M-PAES : A Memetic Algorithm for Multiobjective Optimization. In *Proceedings of the 2000 Congress on Evolutionary Computation, p. 325-332.*

Knowles, J. D. & D. W. Corne (2000b). The Pareto-Envelope based Selection Algorithm for Multiobjective Optimization. In *Proceedings of the Sixth International Conference on Parallel Problem Solving from Nature* (PPSN VI), pp. 839-848, Berlin.

Nguyen, A. & V.-D. Cung (2005). Le problème du Car Sequencing Renault et le challenge ROADEF' 2005, *Proceedings of Journées Francophones de Programmation par Contraintes,* 3-10.

Ribeiro, C. C., D. Aloise, T.F. Noronha, C. Rocha & S. Urrutia (2008). An efficient implementation of a VNS/ILS heuristic for a real-life car sequencing problem. *European Journal of Operational Research,* Vol. 191 No. 3, pp. 596-611.

Schott (1995). *Fault tolerant design using single and multicriteria genetic algorithm optimization.* Department of Aeronautics and Astronautics, Massachusetts Institute of Technology. Master's thesis.

Solnon, C., V.-d. Cung, A. Nguyen & C. Artigues (2008). The car sequencing problem: overview of state-of-the-art methods and industrial case-study of the ROADEF'2005 challenge problem. *European Journal of Operational Research,* Vol 191, No. 3, pp.912-927.

Zinflou, A., C. Gagné & M. gravel (2007). Crossover operators for the car-sequencing problem. *Seventh European Conference on Evolutionary Computation in Combinatorial Optimisation (EvoCOP 2007),* LNCS 4446, C. Cotta and J. van Hemert (Eds.), Springer-Verlag Berlin Heidelberg, pp. 229-239.

Zinflou, A., C. Gagné & M. Gravel (2008a). Design of an efficient genetic algorithm to solve the industrial car sequencing problem. *Advances in Evolutionary Algorithms,* Advances in Evolutionary Algorithms, I-Tech Education and Publishing, Vienna, Austria, Witold Kosinski Editor, Chapter 19, pp. 377-400, November 2008, ISBN 953761911-7.

Zinflou, A., C. Gagné, M. Gravel & W.L. Price (2008b). Pareto memetic algorithm for multiple objective optimization with an industrial application. *Journal of Heuristics* Vol. 14, No. 4, pp. 313-333.

Zinflou, A., C. Gagné & M. Gravel (2009). Solving the industrial car sequencing problem in a Pareto sense. Proceedings of *the 12th International Workshop on Nature Inspired Distributed Computing,* Rome, Italy.

Zinflou, A., C. Gagné & M. Gravel (2011). GISMOO: A New Hybrid Genetic/Immune Strategy for Multiple Objective Optimization. (submitted).

Zitzler, E., M. Laumanns & L. Thiele (2001). *SPEA2 : Improving the Strength Pareto Evolutionary Algorithm,* Technical Report 103, Computer Engineering and Communication Networks Laboratory (TIK), Swiss Federal Institute of Technology (ETH) Zurich, Gloriastrasse 35, CH-8092 Zurich, Switzerland.

Zitzler, E. & L. Thiele (1999). Multiobjective evolutionary algorithms: a comparative case study and the strength Pareto approach. *IEEE Transactions on Evolutionary Computation,* Vol.3, pp. 257-271.

A Review: Practice and Theory in Line-Cell Conversion

Ikou Kaku[1], Jun Gong[2], Jiafu Tang[3] and Yong Yin[4]
[1]*Akita Prefectural University*
[2,3]*Northeasten University*
[4]*Yamagata University*
[1,4]*Japan*
[2,3]*China*

1. Introduction

Since the 1990s, Japanese manufacturers had been faced with a dynamic production environment of decreased market demands and increased product variations. To survive in such an extremely tough business environment, the traditional high-volume conveyor assembly lines were no longer fulfilled. Speedy adjustments were needed to handle transitions in product models and demands. A company's competitiveness was becoming dependent on whether or not it can respond to these transitions. Instead traditional conveyer assembly line, several flexible manufacturing methods were developed to handle the outside changing factors like as varying product types, smaller batch sizes, varying task sizes, and inside changing factors like as flexible layout and planning, cross-training of workers. These challenges and innovations of manufacturing methods have resulted a remarkable improvement in productivity, reduction on capital investment, shortened lead times, saving of manufacturing work space, improvement in product quality, decreasing work in process and parts storage, and so on.

Traditionally, such innovations have been called totally *Seru Seisan* (cellular manufacturing or shorten it by CM) in Japan, because the system was adopted usually a U-shaped layout in which one (or multiple) worker carries out all of the operations of a job. However, even it is not a problem in Japan because the special characteristic of Japanese, the naming may confuse the understanding of the innovations against traditional cellular manufacturing. In fact, we refer the new innovations now in Japan include the traditional cellular manufacturing. For example, Tanaka (2005) reported that there are seven manufacturing methods have been used to correspond the new manufacturing in RICOH UNITECHNO Inc., which is a middle scale Japanese company to manufacture facsimiles/copy machines/printers. Those methods are as follows: (1) One worker-One machine method (the product will be assembled by only one worker, he should do all of the assembly operations); (2) Two workers-One machine method (there are too operational works to assemble for a larger machine that can not complete by one worker, in such case two workers should be assigned to do this assembly operation); (3) Cart pulling method (instead conveyor line a cart is used as transport tool, which is pulled among several workers to complete the assembly operations); (4) Relay method (the form of assembly line is existed but the workers assigned in the line do not only those operations for themselves but also the operations not assigned for them, by their operation ability); (5) Conveyor assembly line (traditional assembly method is also remained for those large lot size products); (6) One worker cell (only one worker does all of the assembly operations of products usually in a U-shaped layout. The difference with method (1) is that the worker in the cell can do all of assembly operations of several products, that means he has a higher operation ability.); (7) Division cell (several workers are assigned in a cell, who may do the assembly operations using the methods of (3), (4) or (6)). Those methods and their combinations are used to correspond flexibly different kinds (over 400 kinds of products) and different quantities (70% of products are under 100 units/month) of products, and successful performances were gained. Therefore, to distinguish the circumstance of converting traditional conveyor assembly line (shorten by CAL) to new production system (including CM) in Japanese manufacturer's real life from previous situation of changing traditional function layout to CM, the circumstance of converting CAL to CM is called line-cell conversion in this chapter. It should be pointed that line-cell conversion is based on the reflection of mass conveyor manufacturing and is for searching more effective production systems, so that converting old conveyor assembly line to new production systems is a considerable strategy in order to increase the productivity of companies. However, how to complete this conversion is a very complicate decision making problem for those companies who wanted to do such conversion. That means when a company faces a changing production environment and wants to convert their conveyor assembly line to a new production system, then the company must decide that how many cells should be formatted, how many workers should be assigned in each cell and how many workers should be rested in the shortened conveyor line. Moreover, how to evaluate the performance improvement through the line-cell conversion is also an important decision making factor. We define such technical and decision making problems in the line-cell conversion as line-cell conversion problems.

In this chapter, a review of line-cell conversion based on a serial research developed by Kaku, et al. (2008a,2008b,2009b,2009c) is proposed. Firstly a content analysis is examined by reviewing a technical journal (*Factory Management 1995-2006)* published in Japan, in which total 24 cases of line-cell conversion were reported. It can be understood why the line-cell conversion is porpular in Japan and what are the insights of such innovation through the content analysis. Secondly a mathematical model has been built to analyze quantitatively the system performances of line-cell conversion problems. By using the model it can be clarified that how are the system performances influenced by which operating factor of inside and outside. Also a linear weighted method is used to solve the multi objective optimization problem, in which two objectives of total throughput time and total labor power are optimized together. Thirdly the influences of operating factors are analyzed by a L27 arrays experiments. Here four factors are designed to represent the complex production environment, which are multiple types of product, different batches and batch sizes, number of stations, and be very important in the decisions of line-cell conversion.

The remainder of this chapter is organized in the following way. An overview of line-cell conversion is presented in the second section. Then the mathematical model is proposed in third section. A linear weighted method is illustrated by using a simple numerical example in fourth section. A L27 arrays is designed and executed in fifth section, also the result analysis are presented. Finally conclusions are given in the sixth section.

2. Overview of line-cell conversion in Japanese industry

An early document of line-cell conversion was reported by Tsuru (1998), which is based on a questionnaire of 13 factories and one consulting company. These anonymous factories converged in electronic and automobile industries. The main standpoint of the document claimed that line-cell conversion can be recognized as a form of the knowledge of Toyota Production System which has been historically transferred to other industries. At the same period, other large-scale investigation on Japanese manufacturing firms (Economic Research Institute 1997) was reported that 48.2% of the respondents had adopted or were planning to implement line-cell conversion. A tremendous achievement of line-cell conversion was brought from CANON Inc., a famous Japanese electronic company. Takahashi, Tamiya and Tahoku (2003) reported that by introducing line-cell conversion into their factories in CANON, since 1995 there were over 20,000 meters of belt conveyor had been withdrawn and 720,000 square meters of working space from 54 related factories were emptied. The total cost rate was decreased from 62% to 50% during past eight years. Since then the line-cell conversion was coming into fashion in Japanese manufacturers.

2.1 Limitations of conveyor line

It is no necessary to describe the greatness of Fordist paradigm afresh. Its economies of scale attained by mass production and shorter throughput time brought material civilization and modern industrial innovation, and lead the production revolution in past 20th century. In fact, further developments and adaptations brought by variant systems such as automated transfer-lines, mixed-model production lines, and robotized flexible assembly lines which were better suited to new business and competition circumstances, were based on the recognition and consideration of the reflection of conveyer production line. Even the famous Toyota Production System is also not exceptional.

However in recent years, after many companies shifted their production organizations to out of Japan, those manufacturers left behind in Japan have been changing their production systems remarkably. Several manufacturing methods have been developed for strengthening their competitive power. In addition, instead conveyor mass production, only the products what suited the needs of customers (the kinds of products are changing dynamically) should be manufactured flexibly when they were needed (the production quantities are also variable). This changing of production system is constructed with the limitations of classical conveyer line, which have been discussed by a series of Japanese field studies (see Shinohara 1997, Tsuru 1998, Isa and Tsuru 1999, Sakazume 2005, Miyake 2006). Briefly two inefficiencies of conveyer line have been investigated. One is that conveyor line presents a series of detrimental aspects for productivity which may be epitomized by the following seven categories of waste: (1) underutilization of workforce due to the face that line cycle time is bounded by slowest worker; (2) waste of time in reaching work-piece on conveyor and returning it onto conveyor after task completion; (3) waste of inventory due to the holding of work-in-process (WIP) between successive stations; (4) waste due to defective parts and rework; (5) waste of resource capacity during product model changeover; (6) waste due to difficulty in promoting mutual support; (7) waste of waiting time by workers operating partially automated short cycle process that does not allow handling of multiple machines. The other is that conveyer line lacks manufacturing flexibilities in product model changes; introduction of new products; changeover of jigs and devices costly and time-consuming, and layout reconfiguration extremely difficult.

2.2 Motivations for the development of line-cell conversion

Japanese manufacturers have been under strong pressure to devise a more effective and agile production system in face of the limitations of their former systems. The line-cell conversion problems had arisen under this challenging context as a promising and competitive production system alternative. Yagyu (2003) speculated that large manufacturers in Japanese electronics industries were the pioneers to embark on the experimentation of line-cell conversion by the first half of 1990s. Shinohara (1997) contributed to increase the awareness on this matter surveying the initiatives taken by samples of manufacturers that had implemented production systems based on this emerging organization pattern.

Different goals and motivations are driving Japanese manufacturers to embrace line-cell conversion. Among the primary motivations, Yagyu (2003) proposed that (1) the flexibility constraints of production systems organized around conveyor lines and dedicated automated machines to cope with high-mix small-lot production and its fluctuating nature became increasingly evident; (2) the wastes and deficiencies that are intrinsic of conveyor lines have become critical restrictions in the increasingly complex market environment; and (3) an opportunity has been perceived in this shift to reinvigorate the workers' morale and motivation by refreshing production organization practices and establishing more autonomous settings.

The above viewpoints indicate that Japanese manufacturers have identified striking potentials in the line-cell conversion as an alternative that may make up for the incapacity of the conveyor line system to coping with the new issues imposed by the current market and labor environments. Thus, the line-cell conversion can be admitted as an outcome that emerged from the amalgamation of the efforts towards the development of an alternative production system which were driven by these motivations.

2.3 Case studies of line-cell conversion

In this chapter we invested 24 cases of line-cell conversion reported in *Factory Management (1995-2006)* (Appendix 1), which is a common Japanese technical journal facing to factory managers. The cases covered mostly electric appliances and information equipment such as digital camera, printer, CD&DVD player and facsimile machines (92% of the cases). Several manufacturers in the cases like Canon and Fujifilm are playing the leading role in their manufacturing field. These cases represented that line-cell conversion can be adapted to various types of processes, including assembly, finishing, testing, packing, forming, casting, heat treating and so on.

Two kinds of implemental changes in line-cell conversion were classified. One is division method of labor power by setting up U-shaped layout or multiple cells (40% of the cases) and by improving the worker's level of skill to do all of the operations in an assembly process (44% of cases). The other is changing of production layout and equipment by removing of conveyors and expensive automated equipment (44% of the cases) and by composing a line with simple equipment (20% of the cases).The first change shows a change in the division of labor in the line system. This is a series of changes that reorganizes the line so that one or a few workers can do all of the operations, by reducing the number of workers who are involved in the division of labor per line, and expanding the extent of operations per worker. The second change indicates a change of production equipment in the related lines. This is part of a series of changes to the conventional production line carried out by removing the automated equipment such as conveyors and robots, implementing simpler equipments and jigs and tools, and formulating a line with a simple workbench and part boxes.

The system performance improvements achieved by line-cell conversion are popularity with those measures related to productivity, parts storage, work-space, lead time, operators, work-in-process and so on. For example in our cases, by introducing line-cell conversion into their factories, SANYO TOKYO manufacturing, IKEDA Electric, YAMADA Metal, CANON can easily adapt multi-item small-sized products and production volume changes. SONY Mexico also can manufacture 15 models of television and increased their productivity 29%. CANON declared that there are 400 Kaizen activities arranged by workers in 9 months and ULVAN COATING declared that the work defectiveness decreased 50%. TOKIN decreased lead time from one month to a week. SHOWAD DENKI shortened the cycle time from 2 minutes 26 seconds to 1 minute 56 seconds. However, it should be considered that line-cell conversion is a very complex and difficult operation. Sengupta and Jacobs (1998) found environments where the conventional assembly line outperformed assembly cells in a plant that assembles television sets. These environments occurred when conversion also results in an increase in task time or other loss of efficiency in the assembly cells. Shmizu (1997) reported that the performance of the assembly cells used in Volvo was inferior to the more traditional assembly line. There are also several researches reported the advantages and disadvantages of line-cell conversion (see Tsuru 1998, Isa and Tsuru 1999, Sakazume 2005). Combining those researches and the content analysis, we simply illustrate the advantages and disadvantages of line-cell conversion in table 1.

<Advantages>

- Increased adaptability to the market demand fluctuations

(1) Easily adaptable to product volume changes
(2) Easily adaptable to frequent model changes
(3) Easily adaptable to multi-item small-sized products

- Improvement in Q.C.D. competitiveness

(1) improvement in product quality and productivity (13 cases)
(2) decrease the parts storage (5 cases)
(3) decrease the work-space (9 cases)
(4) decrease the lead-time (7 cases)
(5) decrease the operators (8 cases)
(6) decrease the work-in-process inventory (3 cases)

<Disadvantages>

(1) Operators are required to have higher skills
(2) It takes time to acquire the required skills
(3) Operators are required to have a stronger sense of responsibility
(4) Increase the input of machines

Table 1. The advantages and disadvantages of line-cell conversion

As shown in Table 1, the interrelations among 9 advantages were classified that three items concerned the adaptability to the market demand fluctuations, while six items concerned the improvement of Q.C.D (quality, cost, delivery). These advantages suggest that line-cell conversion can lead an effective production system for companies pursuing flexibility in production with a stable Q.C.D competitiveness, especially under market conditions in which 1) there are drastic fluctuations in demand; 2) frequent model changes are necessary; 3) the company has been obliged to adopt small-lot multi-kind production. Several disadvantages of line-cell conversion were also classified. The first three of disadvantages were related with cross-training of workers and the last is related with possible increasing capital investment.

Based on these findings, the product and process conditions of successful line-cell conversion can be summarized as follows. First, there are three product conditions: 1) low total assembly man-hours while production volume is high; 2) small number of assembly components; 3) small products and components. Secondly, there are five process conditions: 1) high possibility of securing multi-skilled workers because of the low production volume; 2) few difficult operations requiring a high level of proficiency; 3) no need for expensive equipment; 4) high possibility of sharing equipment; and 5) small equipment.

As mentioned above, line-cell conversion is often used to increase manufacturer's competitive ability and may has different multiple objectives. In order to increase the productivity manufacturer may for example either shorten their assembly time per product (per lot) or reduce operators or take both decision polices. It makes the line-cell conversion problems become difficult to analyze and evaluate. That is how could an operating factor influence what objective still be not clear through the content analysis and clarifying such relationship is a very key issue in successful line-cell conversion.

3. Modelling of line-cell conversion problem

3.1 Problem description and modeling

Following mathematical model of line-cell conversion is built by Kaku, et al. (2009) and cited below. We consider a real problem of not only assembly but also assembly manufacturing: there exists a traditional belt conveyor assembly line with multiple assembly stations. Workers were assigned at each station according to a traditional job design method but they have had abilities to do more tasks than that were assigned to them. We assume that the worker's abilities are different with stations and products. Multiple products will be assembled in the conveyor line, each product is able to have different batch sizes but with a known distribution of demand. Products should be assembled by a given scheduling rule like as First Come First Service (FCFS) but with a full batch (i.e., we do not consider batch splitting). When the products are assembled in the conveyor line, the stations and workers used to complete the assembly jobs are active. Because workers have different abilities to do those jobs (which belong to stations and products) how long the batch will be finished is dependent on the worker who has the slowest speed to do the jobs. That means the abilities of the other workers were not useful sufficiently, which may lead to decreasing the motivation of workers. On the other hand, all of the products should be assembled at the same conveyor line with a fixed order; there may be some waiting times in the assembly processing so that we can not response flexibly to the customer's variant demand. Assume that the workers will do all of jobs that they can do even that are not assigned for them, there are several KAIZEN methods be able to implement the assembly conveyor line. For example, workers who have higher abilities could help other workers in the conveyor line; or converting the conveyor line to some assembly cells; or converting part of assembly line to cells in which the frequent flexibilities exist and assigning workers who have higher abilities, and remain the part of conveyor line for workers who have lower abilities otherwise.

Here we consider three types of assembly systems including a pure cell system, a pure assembly line and a hybrid type of cells + assembly line. Because it does not influence the system performance either the cells are set to front or behind of assembly line (Van der Zee and Gaalman 2006). For simplicity and without lose of generality, we assume assembly line is formatted behind assembly cells in the hybrid assembly system as shown in Fig. 1. We propose a two step approach to design the assembly system from Fig. 1. First step is a cell formation approach: if there were only cells formatted in the system (pure cells), we assign all of workers to cells according to their abilities which are different with products and stations (jobs); if there were part of assembly line should be converted to cells, we assign the workers who have higher abilities to cells and remain the workers who have lower abilities into assembly line. As a special example, the case of workers can help each other in the assembly conveyor line just should be considered like as a pure cell in which all workers are assigned in a cell. Finally, pure assembly line is the traditional belt conveyor line.

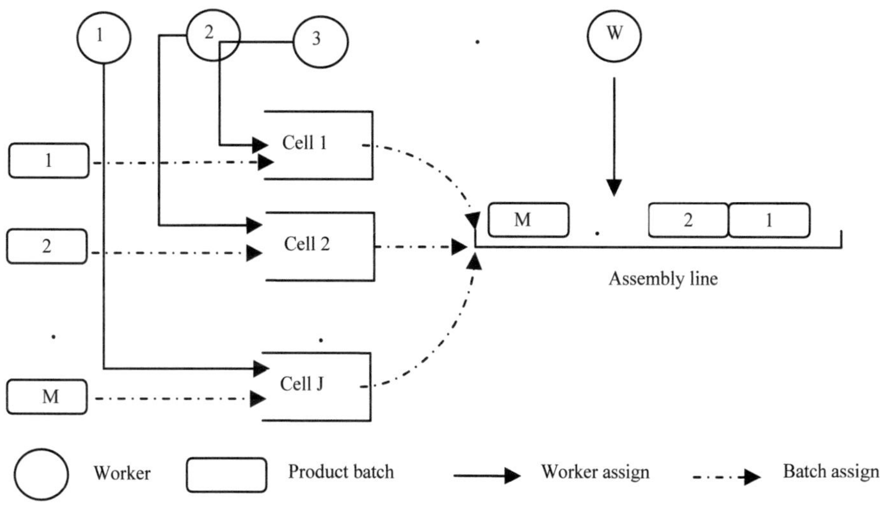

Fig. 1. A hybrid assembly system

The second step is a scheduling approach: We use the FCFS rule to assign assembly product batches to cells or line. In the case of pure assembly line the product batches are just scheduled according to the order of their coming; in the case of pure cells the product batches are scheduled according to not only the order of their coming but also the ability of workers (that means that product should be assigned prior to the worker (cell) who has higher ability to assembly the product). In the case of hybrid system, the product batches are firstly assigned to cells with the FCFS rule, then assigned to assembly line with the order calculated by the earliest finish time rule. Fig. 2 shows an example of the hybrid system with four batches and three cells, where the length of rectangle chart in Fig. 2 states the flow time of that assembly product batch.

For evaluating the system performance two criteria are considered. Firstly we define total throughput time to represent the system productivity that is the time of all of product batches had been assembled. That is to say, for given assembly product mix instead assembly line the new production system should have a shorter total throughput time. Secondly we define total labor power (hours) to represent the work efficiency that is the cumulative working time of all of workers assigned in the system. Therefore, our problem is to determine the number of cells and number of workers in each cell to minimize the total throughput time and total labor power.

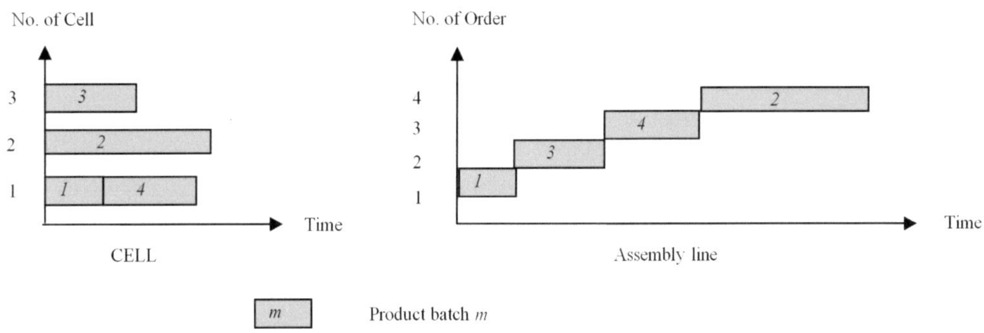

Fig. 2. An example of scheduling in the joint cells + assembly line system

3.2 Problem features and assumption

Following assumptions are considered in this chapter to construct the model:

1. Multiple products are planned to assembly with a product mix.
2. The products are assembled with different batches and different batch sizes.
3. The types and batches of products are known and constant.
4. The number of assembly tasks is the same to all of product types (if the tasks of products were different then assume the task time to do the different tasks was zero so that we can treat the products with different assembly tasks).
5. If the assembly system is a conveyor line, just one conveyor line is considered.
6. The number of workers is same with the number of tasks on assembly line.
7. A worker only does one assembly task in assembly line.
8. The number of workers in each cell may be different but limited.
9. The number of tasks assigned to each cell is the same.
10. The number of tasks assigned to each cell is at least greater than a constant (that means the workers assigned in the cell should do more tasks than in assembly line).
11. A worker assigned in a cell can operate all the tasks assigned in a cell.
12. An assembly product batch is just processed in a cell.
13. Setup time is considered when two different types of product have been assigned into a cell, but the setup time between two batches with the same product type is zero.

3.3 Notations

We define the following terms:

- Indices

i : Index set of workers $(i = 1,2,...,W)$.
j : Index set of cells $(j = 1,2,...,J)$.
n : Index set of product types $(n = 1,2,...,N)$.
m : Index set of product batches $(m = 1,2,...,M)$.

k : Index set of sequences of product batches in a cell ($k = 1,2,...,M$).
r : Index set of orders of product batches in the assembly line ($r = 1,2,...,M$).

- Parameters

$W_{\max}$: Maximum number of workers in one cell.
$S_{\min}$: Minimum number of stations in one cell.
V_{mn} : A 0-1 binary variable where $V_{mn} = 1$, if product type of product batch m is as same as product type n ; otherwise 0.
B_m : Size of product batch m .
T_n : Balanced cycle time of product type n in the assembly line.
SL_n : Setup time of product type n in the assembly line.
SC_n : Setup time of product type n in a cell.
ε_i : Coefficient of influencing level of skill to multiple stations for worker i .
η_i: Upper bound on the number of tasks for worker i in a cell, if a number of tasks assigned to a worker is over than it, the task time will become longer than ever.
β_{ni} : Level of skill of worker i for one task of product type n .

- Decision variables

$X_{ij} = 1$, if worker i is assigned to cell j , otherwise 0.
$Y_i = 1$, if worker i is assigned to line, otherwise 0.
$Z_{mjk} = 1$, if product batch m is assigned to cell j in sequence k , otherwise 0.if $k = 0$, $Z_{mjk} = 0$.
$O_{mr} = 1$, if product batch m is assembled by order r in the assembly line, otherwise 0. Moreover, if $r = 0$, then $O_{mr} = 0$.

- Variables

C_i : Coefficient of variation of assembly task time of worker i in each cell accounting for the effect of multiple stations.
SC_m : Setup time of product batch m in a cell.
TC_m : Assembly task time of product batch m per station in a cell.
FC_m : Flow time of product batch m in a cell
FCB_m : Begin time of product batch m in a cell.
SL_m : Setup time of product batch m in the assembly line.
TL_m : Assembly task time of product batch m per station in the assembly line.
FL_m : Flow time of product batch m in the assembly line.
FLB_m : Begin time of product batch m in the assembly line.
$P = 1$, if the assembly line exists in the hybrid assembly system, otherwise 0.

3.4 Problem formulation

Here we consider the assembly planning problem which is based on a fixed assembly product mix with M product batches and N product types. W workers are assigned to the system which may be pure assembly cells or pure assembly line or a joint type system. Given the upper bound $W_{\max}$ on the number of workers and the lower bound $S_{\min}$ on the number of stations in a cell, the objective is to determine the number of cells and workers in each cell to minimize the total throughput time and the total labor hours.

3.4.1 Scheduling of assembly batches in cells

For defining the total throughput time of the assembly batch assignments in cells, the assembly plan will be scheduled with a given scheduling rule under the worker assignments to cells. Firstly, a worker's level of skill is able to vary with the number of tasks. If the number of tasks is over an upper bound η_i, the task time will become longer. This can be represented as below:

$$C_i = 1 + \varepsilon_i \max((W - \sum_{i'=1}^{W} Y_{i'} - \eta_i), 0) \quad \forall i \tag{1}$$

Secondly, the task time of a product is also able to vary with workers. Consequently, the task time of a product is calculated by mean task time of all workers in the same cell. Actually, the task time of product batch m per station in a cell is represented via following equation:

$$TC_m = \frac{\sum_{n=1}^{N}\sum_{i=1}^{W}\sum_{j=1}^{J}\sum_{k=1}^{M} V_{mn} T_n \beta_{ni} C_i X_{ij} Z_{mjk}}{\sum_{i=1}^{W}\sum_{j=1}^{J}\sum_{k=1}^{M} X_{ij} Z_{mjk}} \tag{2}$$

Then, using the FCFS rule, the setup time SC_m, the flow time FC_m and the begin time FCB_m of product batch m are represented as below.

$$SC_m = \sum_{n=1}^{N} SC_n V_{mn} (1 - \sum_{m'=1}^{M} V_{m'n} Z_{m'j(k-1)}), \quad (j,k) \big| Z_{mjk} = 1, \forall j,k \tag{3}$$

$$FC_m = \frac{B_m TC_m (W - \sum_{i=1}^{W} Y_i)}{\sum_{i=1}^{W}\sum_{j=1}^{J}\sum_{k=1}^{M} X_{ij} Z_{mjk}} \tag{4}$$

$$FCB_m = \sum_{s=1}^{M}\sum_{j=1}^{J}\sum_{k=1}^{M}\sum_{k'=0}^{k-1} (FC_s + SC_s) Z_{mjk} Z_{sjk'} \tag{5}$$

Where, equation (3) states the setup time of product batch m. If there are more than two batches assigned in a cell and the type of those batches is same then the setup time will be set to be zero. Equation (4) states the flow time of product batch m. Equation (5) states the begin time of each product batch. There is no waiting time between two product batches so that the begin time of one product batch is the aggregation of flow time and setup time of all the prior product batches which are in the same cell.

3.4.2 Scheduling of batches production in the assembly line

For defining the total throughput time of the assembly batch assignments in the assembly line, the assembly plan will be scheduled with a given scheduling rule under the worker assignments. Therefore, if all workers are assigned to the assembly line, then that is a traditional assembly line system; otherwise, that is a hybrid assembly system. Here, the task time is calculated by the longest task time among the workers in the assembly line. Actually, the task time of product batch m is represented via the following equation:

$$TL_m = \sum_{n=1}^{N} \max(V_{mn} T_n \beta_{ni} Y_i) \qquad \forall i \tag{6}$$

Then, using the FCFS rule, the setup time SL_m , the flow time FL_m and begin time FLB_m of product batch m are presented as below.

$$SL_m = \sum_{n=1}^{N} SL_n V_{mn} (1 - \sum_{m'=1}^{M} V_{m'n} O_{m'(k-1)}) \quad O_{mk} = 1, \forall k \tag{7}$$

$$FL_m = \sum_{n=1}^{N} \sum_{i=1}^{W} V_{mn} T_n \beta_{ni} Y_i + TL_m (B_m - 1) \tag{8}$$

$$FLB_m = \begin{cases} \max(FCB_m + FC_m + SC_m, FLB_{m'} + FL_{m'} + SL_{m'}) \\ \qquad \left\{ m' \middle| O_{mk} = 1, O_{m'(k-1)} = 1 \quad k = 2,3,...,M \right\} \\ FCB_m + FC_m + SC_m \qquad O_{m1} = 1 \end{cases} \tag{9}$$

Where, equation (7) states the setup time of product batch m. If the product type of coming product batch is as same as the preceded product batch, the setup time of this product batch will be set to be zero. Equation (8) states the flow time of product batch. Equation (9) states the begin time of each product batch. If the production system is the hybrid one, the waiting time between two product batches will be considered, otherwise no consideration for waiting time. In the hybrid model, the begin time of product batch m is the maximum value between the end time of the prior product batch in assembly line and the end time of product batch m in cells. In the assembly line model, the begin time of product m is the end time of the prior product batch which is assigned by the FCFS rule.

3.4.3 The comprehensive mathematical model

The comprehensive mathematical model is given in equation (10)-(23) as below.
Objective functions:

$$TTPT = Min \left\{ \underset{m}{Max} \left[(1-P)(FCB_m + FC_m + SC_m) + P(FLB_m + FL_m + SL_m) \right] \right\} \tag{10}$$

$$TLH = Min \sum_{m=1}^{M} \sum_{i=1}^{W} (\sum_{j=1}^{J} \sum_{k=1}^{M} FC_m X_{ij} Z_{mjk} + FL_m Y_i) \tag{11}$$

Subject to

$$\sum_{i=1}^{W} X_{ij} \le W_{\max} \qquad \forall j \tag{12}$$

$$\sum_{i=1}^{W} X_{ij} \le \sum_{i=1}^{W} X_{ij'} \qquad \forall j > j', (j = 1,2,..J) \tag{13}$$

$$\sum_{i=1}^{W} Y_i \leq W - S_{\min} \tag{14}$$

$$\sum_{j=1}^{J} X_{ij} + Y_i \leq 1 \qquad \forall i \tag{15}$$

$$\sum_{j=1}^{J} \sum_{k=1}^{M} Z_{mjk} = 1 \qquad \forall m \tag{16}$$

$$Z_{mjk} \leq Z_{m'j(k-1)} \qquad \forall m, m' = 1,2,...,M, m' \neq m, k = 2,3...,M \tag{17}$$

$$\sum_{m=1}^{M} \sum_{k=1}^{M} Z_{mjk} = 0 \qquad \{j \,|\, \sum_{i=1}^{W} X_{ij} = 0, \forall j\} \tag{18}$$

$$FCB_m \leq FCB_{m+1} \qquad \forall m \tag{19}$$

$$\sum_{m=1}^{M} O_{mr} = 1 \qquad \forall r \tag{20}$$

$$\sum_{r=1}^{M} O_{mr} = 1 \qquad \forall m \tag{21}$$

$$O_{mr} FLB_m \geq O_{m'(r-1)} (FLB_{m'} + TL_{m'} + SL_{m'}) \quad \forall m, m' = 1,2,...,M \tag{22}$$

$$P = \begin{cases} 1 & \sum_{i=1}^{W} Y_i \geq 1 \\ 0 & \sum_{i=1}^{W} Y_i = 0 \end{cases} \tag{23}$$

Where, equation (10) states the objective to minimize the total throughput time (TTPT) of the total product batches assignments. The total throughput time is the due time of the last completed product batch. The first part is the throughput time in cells. The second part is the throughput time in the assembly line. Equation (11) states the objective to minimize the total labor hours (TLH) of the product batches assignments. The total labor hours is the time of all workers assembly the total product batches. The first part is the labor hours in cells. The second part is the labor hours in the assembly line. Equation (12) is a cell size constraint because the space of a cell is limited. The value of the maximum number of workers in one cell will be a function of plant size, design and process technology. Equation (13) is the rule of cell formation ensures that the number of workers in prior cell is greater than that in next cell. Equation (14) is a minimum number of tasks in each cell which means if there were no tasks assigned to cells the production system will become traditional assembly line. Equation (15) is the rule of worker assignment ensures that each worker should be at most assigned to one cell or the line. The sign of inequality means that the worker who has the worse ability is discarded possibly. Equation (16) is the assignment rule in which a product batch is only assigned to a cell. Equation (17) is the assignment rule in which product batches must be assigned sequentially. Equation (18) are the rules of assigning constraints, that means a product must be assigned to a cell in which a worker is assigned at least. Equation (19) is the FCFS rule which means the prior product batch must be assembled before the next product batch. Equation (20) ensures that a product batch must be assigned by a fixed scheduling order. Equation (21) ensures that an order also must be assigned to fix the product batches. Equation (22) ensures that the begin time of a product batch must be late the end time of the prior product batch. Equation (23) is a flag variable shown whether the assembly line exists in the system. This rule can lead a smaller search space of feasible solutions but guarantee the optimality of solutions.

4. A linear weighted method to solve the multi-objective model

4.1 The consideration

By using formula (10)-(23), the line-cell conversion can be described completely that whether the conveyer assembly line should be converted to cell(s) and how to do such conversion. In the above model there are two objective functions of total throughput time and total labor hours, which are most important evaluation factors in line-cell conversion. Usually, increasing manufacturer's productivity can be executed by shortening total through put time or decreasing number of workers or some other efforts. However, shorten total throughput time may lead to increasing demand of workers which offend against the other objective function of total labor hours, and vice versa. So that the objective functions should be solved simultaneously. In this chapter, we just use a linear weighted method to construct the total throughput time and total labor hours into a new utility function in which objective functions are related with a linear weight (usually the linear weighted method should be used in a convex space of solutions, later we show the convex property of solution space with an example but not theoretical proof).

Based on the consideration, the two objectives of total throughput time (*TTPT*) and total labor hours (*TLH*) can be construted together with the linear factors α_1, α_2 as below.

$$U(X) = \alpha_1 \cdot TTPT + \alpha_2 \cdot TLH$$

where $U(X)$ is the utility function when the solution of the model is X, and α_1, α_2 are the linear weights. For determining the α_1, α_2, we consider following simultaneous equations:

$$\alpha_1 \cdot TTPT^*(X_{TTPT}) + \alpha_2 \cdot TLH^0(X_{TTPT}) = c$$
$$\alpha_1 \cdot TTPT^0(X_{TLH}) + \alpha_2 \cdot TLH^*(X_{TLH}) = c$$

Where, c is a constant (c≠0) and

$$TTPT^*(X_{TTPT}) = \min_X TTPT(X)$$
$$TLH^*(X_{TLH}) = \min_X TLH(X)$$
$$TTPT^0(X_{TLH}) = TTPT(X_{TLH})$$
$$TLH^0(X_{TTPT)}) = TLH(X_{TTPT})$$

Solving the simultaneous equations and set $\alpha_1 + \alpha_2 = 1$, then α_1, α_2 can be shown as bellows.

$$\alpha_1 = \frac{TLH^0(X_{TTPT}) - TLH^*(X_{TLH})}{TLH^0(X_{TTPT}) - TLH^*(X_{TLH}) + TTPT^0(X_{TLH}) - TTPT^*(X_{TTPT})}$$

$$\alpha_2 = \frac{TTPT^0(X_{TLH}) - TTPT^*(X_{TTPT})}{TLH^0(X_{TTPT}) - TLH^*(X_{TLH}) + TTPT^0(X_{TLH}) - TTPT^*(X_{TTPT})}$$

Hence, linear weights α_1, α_2 show the percentages of the objectives. Following we give a simple numerical example to illustrate how to calculate α_1, α_2.

4.2 A simple example

In the hybrid line-cell conversion model, workers assigned in the conveyer line can be considered not only re-assign into a cell but also may remain in the shortened line because they have not enough ability to do those operations in cell. For a given number of workers (*X* + *Y*), the objective functions are not linear but bounded. Hence, we must conduct an exhaustive search over *X* + *Y*. For simplifying the calculation but not lose generality, following example only considers the case where all of workers in line be reassigned to cells. Table 1, Table 2 and Table 3 show the parameters, level of skill of workers and coefficient of influencing level of skill to multiple stations for workers, respectively.

Factor	Number	Parameter
Stations	5	$W = 5$
Workers	5	$W = 5$
Lot sizes	10	$B_m = 10$
Batches	5	$M = 5$
Product Types	3	$N = 3$

Table 1. The parameters of line-cell conversion example

Product/Worker	1	2	3	4	5
1 (A)	0.97	0.93	1.19	1.17	1.11
2 (B)	0.96	0.9	1.28	1.26	1.17
3 (C)	0.94	0.87	1.38	1.34	1.23

Table 2. level of skill of workers (β_{nw})

Worker	1	2	3	4	5
ε_i	0.18	0.16	0.29	0.28	0.25

Table 3. Coefficient of influencing level of skill to multiple stations for workers (ε_i)

From Table1, Table 2 and Table3, it can be observed that the conveyer line has five stations in which 3 types of product are manufactured with 5 batches and the lot size of each batch is

10. Five workers are assigned into the line and have different skill level to do those operations of products. The ability of workers is also different with stations. When we are going to convert the line to a cellular manufacturing system, consider that cell may be constructed into several form like one worker cells or multi workers cells, there are total 52 feasible solutions in above case (see Appendix 2). By using the model we can calculate TTPT and TLH for each solution. Therefore,

$$TTPT^*(X_{TTPT}) = \min_X TTPT(X) = 166.13$$

$$TLH^*(X_{TLH}) = \min_X TLH(X) = 766.65$$

$$TTPT^0(X_{TLH}) = TTPT(X_{TLH}) = 201.84$$

$$TLH^0(X_{TTPT)}) = TLH(X_{TTPT}) = 811.14$$

Then the linear weights can be calculated as bellows.

$$\alpha_1 = \frac{TLH^0(X_{TTPT}) - TLH^*(X_{TLH})}{TLH^0(X_{TTPT}) - TLH^*(X_{TLH}) + TTPT^0(X_{TLH}) - TTPT^*(X_{TTPT})} = \frac{811.14 - 766.65}{811.14 - 766.65 + 201.84 - 166.13} = 0.55$$

$$\alpha_2 = \frac{TTPT^0(X_{TLH}) - TTPT^*(X_{TTPT})}{TLH^0(X_{TTPT}) - TLH^*(X_{TLH}) + TTPT^0(X_{TLH}) - TTPT^*(X_{TTPT})} = \frac{201.84 - 166.13}{811.14 - 766.65 + 201.84 - 166.13} = 0.45$$

Finally, the utility function is as bellow.

$$U(X) = 0.55 \cdot TTPT + 0.45 \cdot TLH$$

By using this utility function, the values of *U(X)* can be calculated, which are also shown in Appendix 2. According to the calculating results, optimal solution is two cells {(12), (345)} (() represents one cell and {}represents a solution). if we only consider TTPT, whereas is four cells {(45), (1), (2), (3)} if we only consider TLH. However the total optimal solution is 2 cells {(34), (125)} when the multi objectives should be considered.

Moreover as shown in Figure 3, all of the feasible solutions of the defined line-cell conversion problem are plotted in a two demention space, in which vertical axle is TLH and horizontal axle is TTPT. Then *U(X)* is a straight line contacted the solution of {(12), (345)} and {(45), (1), (2), (3)} and the sloop of dotted line is parallel to α_1/α_2. Moving the line parallely the total optimal solution {(34), (125)} is obtained at the point which contacted with the line and solution space. However, it does not mean that the space of solutions is convex so in general case the linear weighted method may is not appropreate. Here a very theoretical work to make a proof of convexity should be done in future.

Because the linear weights are very important for solving the multi objective line-cell conversion problem, its sensitivity is calculated in Table 4.

As shown in Table 4, there are three intervals in the solution space. In interval of $\alpha_1 \le 0.4$ the optimal solution is {(45),(1),(2),(3)} which means TLH (hours productivity) is considered preferentially; in interval of $0.4 \le \alpha_1 \le 0.8$ the optimal solution is {(34),(125)} which means TTPT and TLH are considered simultaneously; in interval of $\alpha_1 \ge 0.9$ the optimal solution is {(3),(12),(45)} which means TTPT (product productivity) is considered preferentially. Consider that how to increase labor productivity is a main theme of line-cell conversion, it is noticeable that cellular form has an advantage over almost interval of α_1 in above example.

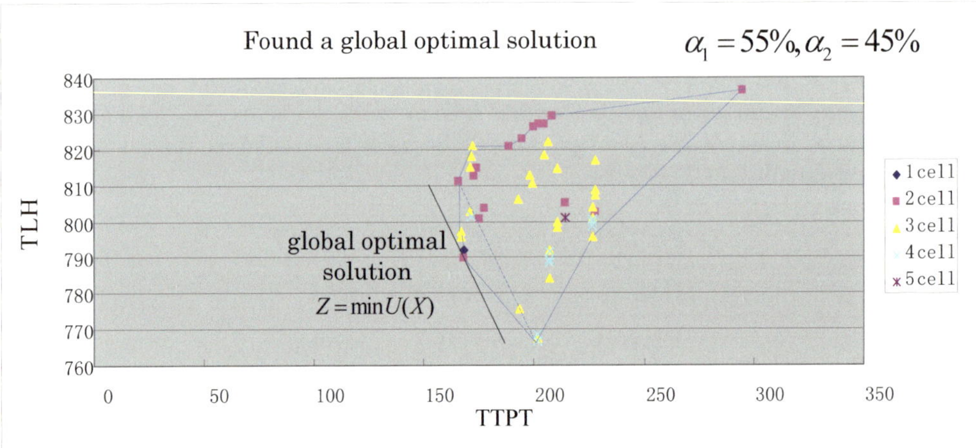

Fig. 3. Geometrical illustration of optimal solution

No	α_1	U(X)	Optimal solution
1	0.1	710.17	(45) (1) (2) (3)
2	0.2	653.69	(45) (1) (2) (3)
3	0.3	624.82	(45) (1) (2) (3)
4	0.4	540.73	(45) (1) (2) (3)
5	0.5	478.86	(34) (125)
6	0.6	416.71	(34) (125)
7	0.7	354.56	(34) (125)
8	0.8	292.41	(34) (125)
9	0.9	229.86	(3) (12) (45)

Table 4. Sensitivity analysis of linear weights

5. Stochastic analysis for operating factors with L27 array

5.1 The L27 experiment design

Therefore, the line-cell conversion problem in a special production environment can be solved completely where how to convert the line to cell and how to assign workers to each cell are optimally determined. However, line-cell conversion in real world is usually considered with changing production environment and the decision-making is depended on those factors influence their production environment. Hence, which factor is how to influence the environment should be defined. In this chapter, we design a L27 experiment to determine which factors most affect the system performance improvement of line-cell conversion with a minimum amount of experimentation thus saving time and resources. The system performance of line-cell conversion is represented by using a multi objective function constructed with total throughput time and total labor hours. Table 5 shows that there are four factors are organized to represent the complex production environment, which are multiple types of product, different batches and batch sizes, number of stations (workers). The first three factors are representing outside influence and the last factor is representing the inside influence. Each factor has three varied levels.

Factors	Level 1	Level 2	Level 3
Station A	4	8	12
Product type B	1	10	20
Lot size C	10	30	50
Batch number D	5	10	15
Worker (A)	4	8	12

Table 5. Experiment factors design

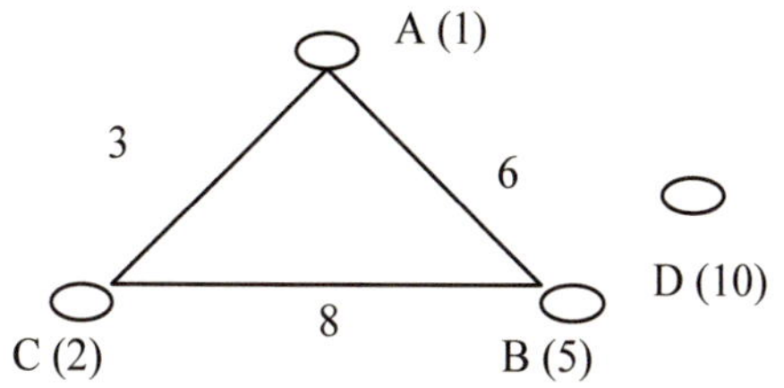

Also three specific 2-factors interactions are investigated through the experiment. Above graph shows the factors (A, B, C and D) and their specific 2-factors (A×B, A×C and B × C) interactions and which column they will be in Appendix 3.

5.2 Analysis and discussion

According to the above design, we do numerical experiments to simulate the effects of factors influenced on the performance improvement of line-cell conversion and show the computational results of the L27 experiment in Appendix 3. Generally, we define an index P which represents the ascendancy of line-cell conversion. i.e., the positive value of P shows an ascendancy of cellular form over line form, and the negative value of P shows the reverse. Figure 4 shows all 27 results in where 10 cases show cellular form is at an advantage over line form, and 17 cases show line form is at an advantage over cellular form. That means cellular form can be used to improve the system performance well when the system (operations) is comparatively smaller. Against, line form is appropriate when there are many stations (operations) needed to assembly a product. However, it does not mean that line form should be used but effort of shortening the line into several cellular forms should be done to improve their manufacturing performance. This is the key strategy for successful line-cell conversion which has been executed by Japanese industries.

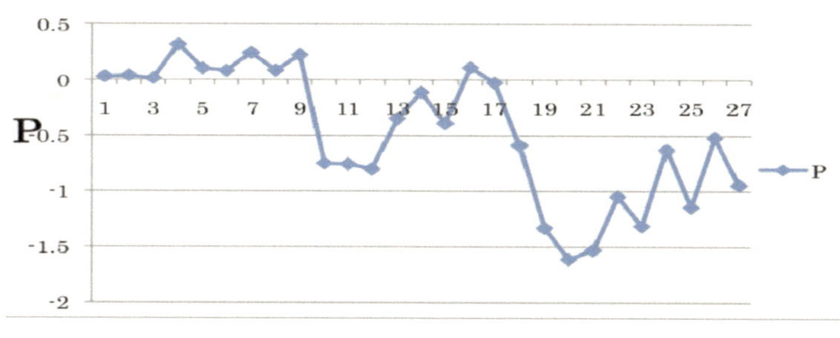

Fig. 4. The index P in 27 cases

For analyzing the effects of each factor and the specific 2-facotrs interactions which may influence the performance of line-cell conversion, detail calculations were made as below. Table 6 shows the calculation results. In Table 6, each column shows the factors, S1, S2, S3 mean the sum of data in all level, m1, m2, m3 show the average value of the data in all level, R1 shows the error and Rank shows the ranking of the factors. From Table 6 it can be observed that workstations (workers), product types, and their interaction are strong, however lot size (which has been considered as a barrier of line-cell conversion) is not almost influencing the performance improvement of line-cell conversion. It can be understood that workstation may give a negative influence on line-cell conversion because the longer of the line, the worse the performance improvement of line-cell conversion, and the product type may give a positive influence vice versa. However, it seems that either line or cell can treat the problem of large lot size their own production form. It is a fact that even by using cell form large lot size production can be executed with same or small throughput time of line with several like lot splitting techniques. Moreover, batches of product show more completed behaviors. For clarifying the tendency of influenced factors, Figure 5 shows the calculated results of each factor in different level respectively.

	A	B	A×B	C	A×C	B×C	D
S1	1.126	-6.705	-3.415	-3.922	-4.636	-3.453	-5.486
S2	-3.669	-3.343	-4.411	-4.116	-3.872	-4.723	-2.704
S3	-10.064	-2.559	-4.781	-4.569	-4.099	-4.431	-4.417
m1	0.125111	-0.745	-0.37944	-0.43578	-0.51511	-0.38367	-0.60956
m2	-0.40767	-0.37144	-0.49011	-0.45733	-0.43022	-0.52478	-0.30044
m3	-1.11822	-0.28433	-0.53122	-0.50767	-0.45544	-0.49233	-0.49078
R1	1.243333	0.460667	0.151778	0.071889	0.084889	0.141111	0.309111
Rank	1	2	4	7	6	5	3

Table 6. The computational results of L27

From Figure5, it can be clearly observed that the system performance improvement was increasing with product types (B) and decreasing with workstations (A) and lot sizes (C). That means varying product types is promoting companies to convert their line to cell, and how to reduce the negative effect of stations and lot sizes is a key issue for a successful conversion. In fact, flexible layout and lot splitting technologies are useful in such KAIZEN activities. However, the system performance improvement is increasing when product batch is changing in a smaller interval but decreasing when product batches become larger.

Also the effects of factors and specific 2-factors interactions are estimated by using the analysis of variance (ANOVA). Table 7 shows the source of variation, degree of freedom, sum of squares, variance and the F-value, respectively. Because the critical value of F-test at 5% significance is $F_{2,6}(0.05) = 5.14$ and $F_{4,6}(0.05) = 4.53$, it can be recognized that three factors (product types, batches and stations), and two specific 2-factors interactions (A×B,

A×C) are significant at 5% level. However, the F value of the 2-factors interactions is near the critical value. It can be considered that the interactions are strongly influenced by A because the F value of A is too big. For illustrating those 2-factors interactions under the designed conditions, Figure 6 shows the specific 2-factors interactions (A×B, A×C, B×C). It can be observed from Figure 6 that the curves are not on a parallel with each other so that they will cross at some other point. That means the specific 2-factors interactions should not be ignored in some special production environment.

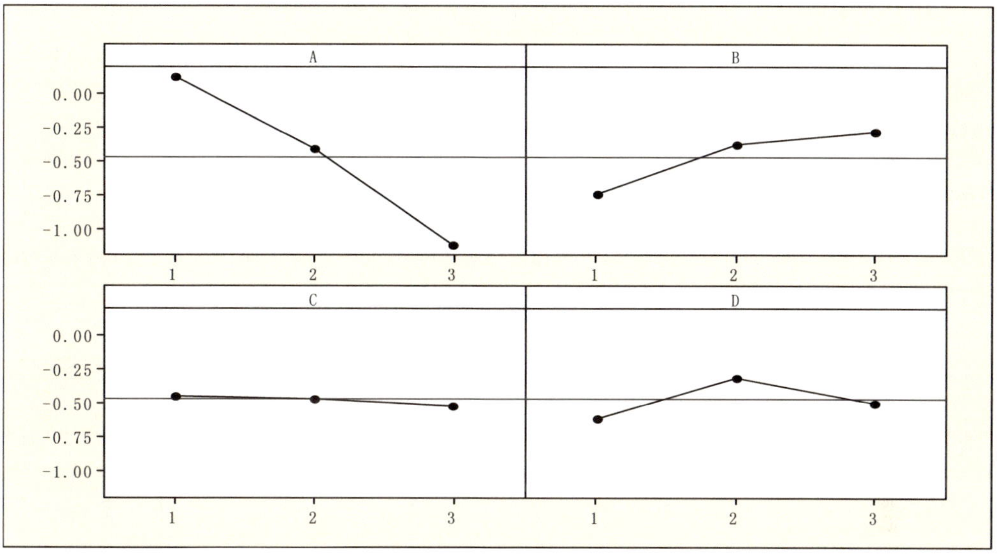

Fig. 5. The influence tendency of factors

Source of Variation	Df	SS	V	F
A	2	7.01532	3.50766	339.79
B	2	1.07815	0.53908	52.22
C	2	0.02457	0.01229	1.19
D	2	0.43788	0.21894	21.21
A×B	4	0.22636	0.05659	5.48
A×C	4	0.18888	0.04722	4.57
B×C	4	0.13871	0.03468	3.36
Error	6	0.06194	0.01032	
Total	26	9.17181		

Table 7. ANOVA results

General speaking, the numbers of station in a belt conveyer assembly line is the largest barrier in line-cell conversion indisputably. For example, a worker could not assembly an automobile only by himself. How many operations (stations) should be assigned to a worker is appropriate depends on many other factors include not only the outside and inside discussed above but also like cross training of workers, complexity of products, learning effect and so on. However, it may start up partially for converting an assembly line to cell not needs complete cross trained worker ability.

6. Conclusions

In this chapter we totally studied line-cell conversion problem. Several contributions are proposed. Firstly an overview of line-cell conversion was proposed by investigating 24 Japanese manufacturing cases. Content and mechanisms analysis on the cases generated the insights of line-cell conversion problems. Secondly we constructed a mathematical model of line-cell conversion with multi objective functions. Thirdly we applied a linear weight method to solve the multi objective problem. Fourthly we investigated some operating factors stochastically by using a L27 array. Through the simulation experiments several operating factors of line-cell conversion were clarified their contributions.

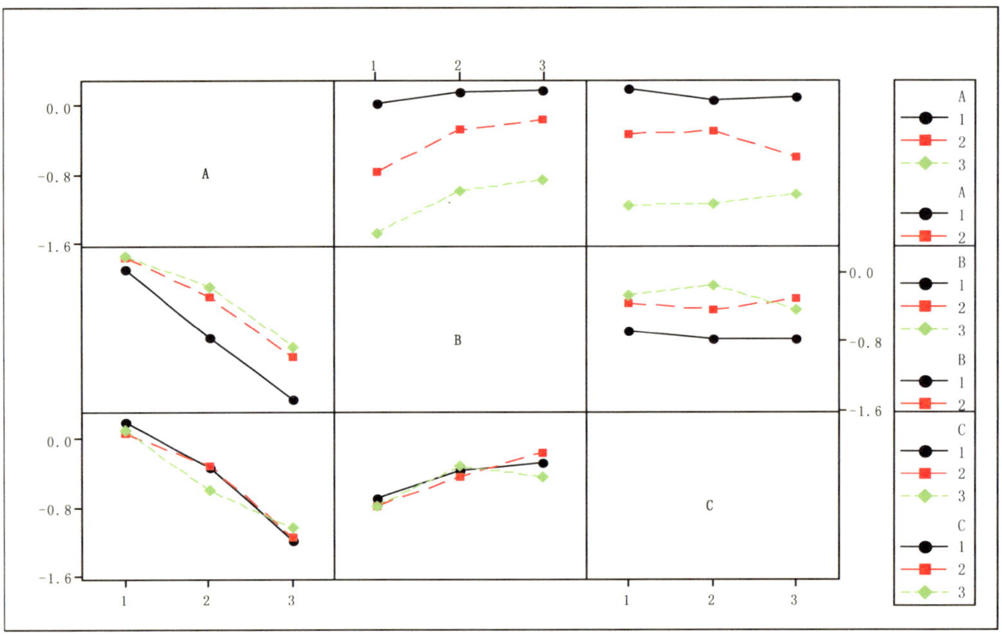

Fig. 6. The 2-factors interactions

7. Acknowledgements

This work is the result of our cooperating research group. I appreciate Dr. Jun Gong, Prof. Jiafu Tang and Prof. Yong Yin for their contributions. I also thank Ms. Lei Yu and Mr. Yang Liu (who were my graduate students) for their helps.

8. Appendix

Firm or factory name	Location	Introducti on period	Product
Fuji film Corporation, Yoshida-Minami	Yoshida-Minami Mie		Printer
Fuji film Corporation, Yoshida-Minami	Yoshida-Minami Mie		Digital camera
FUKUTOME Meat Packers LTD	Hiroshima	2004,3	Ham, sausage
SANYO Tokyo Manufacturing Co.,LTD	Tokyo	2002	Cryogenic power generation
SANYO Electric Co.,LTD	Saitama	2002,1	Absorption chiller
Sony Mexico Factory	Mexico	2001	Camcorder
Chinontec Industries Co.,LTD	Nagano	2001,4	Optical equipment
Ikeda Electric Co.,LTd	Osaka		Electric equipment
YAMADA Metal Co.,LTD	Sendai		Automobile installation
ULVAC COATING CORPORATION	Saitama		ULCOAT
SHOWAD DENKI Co.,LTD	Osaka	2000	Electric wires
KANOU SHOU JUAN Co.,LTD	Nagahama, shiga		Japanese-style confection
Unitilka Group Film Division	Osaka		Plastic, resin
Sony EMCS Corp. Minokamo TEC	Gifu		Digital camera
Canon	AmiPlant, Ibaraki	1999	Digital copier
NEC TOCIN Corporation	Sendai	2002	Electron element
Harmonic Drive Systems Inc.	Saitama		Smart theater
Pioneer Corporation	Kawagoe	2000	CD player
Itoki Crebio Corporation	Osaka		Comnet table
STANLEY ELECTRIC Co., LTD	Yamagata	1997	Automobile accessories
Tokin Corporation	Tokyo	2001	Battery cell
Pioneer Corporation, MEC	Kawagoe	2002	CD,VCD player
Pioneer Corporation	Shizuoka		Laser, Display
Nagahama Canon	Nagahama, Shiga	1998	Laser beam printer

Sources: adapted from following materials.
Factory management, 1999, 45 (7), 63-74; 2000, 46(8), 40-42, 57; 2001, 47(14), 5-6, 36; 2003, 49(1), 23-24; 49(2), 4-5, 19, 111-112; 2004, 50(1), 14-19, 28-29; 50(10), 19, 20-27, 31, 39, 51-52, 53-60; 2005, 51(3), 21-22, 45, 65; 51(10), 66-73.

Appendix 1. Reported cases of Japanese firms and factories

	Combinations	TTPT	TLH	U(x)	
1	(12345)	168.39	791.93	448.98	
2	(1)(2345)	227.38	802.34	486.11	
3	(2)(1345)	214.28	805	480.10	
4	(3)(1245)	172.73	812.43	460.60	
5	(4)(1235)	174.26	814.79	462.50	
6	(5)(1234)	294.5	836.34	538.33	
7	(12)(345)	166.13	811.14	456.38	TTPT*
8	(13)(245)	208.22	829.13	487.63	
9	(14)(235)	204.72	826.85	484.68	
10	(15)(234)	194.77	822.89	477.42	
11	(23)(145)	202.43	826.8	483.40	
12	(24)(135)	199.85	826.32	481.76	
13	(25)(134)	188.98	820.55	473.19	
14	(34)(125)	168.11	789.61	447.79	Z=minU(X)
15	(35)(145)	175.41	800.49	456.70	
16	(45)(123)	177.69	803.6	459.35	
17	(123)(4)(5)	172.61	820.86	464.32	
18	(124)(3)(5)	171.4	815.05	461.04	
19	(125)(3)(4)	171.52	818.08	462.47	
20	(134)(2)(5)	207.32	784.14	466.89	
21	(135)(2)(4)	193.29	775.46	455.27	
22	(145)(2)(3)	207.32	792.18	470.51	
23	(234)(1)(5)	226.2	795.67	482.46	
24	(235)(1)(4)	226.2	801.08	484.90	
25	(245)(1)(3)	226.2	803.77	486.11	
26	(345)(1)(2)	201.84	767.37	456.33	
27	(1)(23)(45)	227.38	807.24	488.32	
28	(1)(24)(35)	227.38	808.6	488.93	
29	(1)(25)(34)	227.38	816.93	492.68	
30	(2)(13)(45)	210.8	798.38	475.21	
31	(2)(14)(35)	210.8	799.75	475.83	
32	(2)(15)(34)	210.8	814.72	482.56	
33	(3)(12)(45)	166.97	795.84	449.96	
34	(3)(14)(25)	192.998	806.065	468.88	
35	(3)(24)(15)	199.64	810.47	474.51	
36	(4)(12)(35)	166.97	797.26	450.60	
37	(4)(13)(25)	192.998	806.16	468.92	
38	(4)(15)(23)	198.35	812.68	474.80	
39	(5)(12)(34)	170.99	802.67	455.25	
40	(5)(13)(24)	204.63	818.41	480.83	
41	(5)(23)(14)	206.47	822.18	483.54	
42	(12)(3)(4)(5)	171.52	801.32	454.93	
43	(13)(2)(4)(5)	207.32	788.77	468.97	
44	(14)(2)(3)(5)	207.32	791.48	470.19	
45	(15)(2)(3)(4)	207.32	789.48	469.29	

46	(23)(1)(4)(5)	226.2	799.93	484.38	
47	(24)(1)(3)(5)	226.2	798.61	483.78	
48	(25)(1)(3)(4)	226.2	800.62	484.69	
49	(34)(1)(2)(5)	201.84	768.71	456.93	
50	(35)(1)(2)(4)	201.84	766.7	456.03	
51	(45)(1)(2)(3)	201.84	766.65	456.00	TLH*
52	(1)(2)(3)(4)(5)	214.28	800.77	478.20	

Here combinations of workers show possible form of cells. For example, (12345) shows one cell in which five workers are assigned and (1) (2) (3) (4) (5) show five cells in each of them only one worker is assigned.

Appendix 2. computational results of the simple example

	Contents							Multi objective function			Only conveyer line			Index
Factors	A	B	A×B	C	A×C	B×C	D	Linear weights		Utility function	Objectives		Utility function	P=
No.	1	2	3	5	6	8	10	α_1	α_1	$U_c(X)$	TTPT	TLH	$U_l(X)$	$U_l(X)-U$ / $U_l(X$
1	1	1	1	1	1	1	1	0.203	0.797	323.080	134.59	383.00	332.475	0.028
2	1	1	1	2	2	2	2	0.481	0.519	1470.63	695.38	2300.00	1527.094	0.036
3	1	1	1	3	3	3	3	0.154	0.846	5062.35	1684.57	5751.00	5124.431	0.012
4	1	2	2	1	1	2	2	0.911	0.089	275.66	356.76	878.00	403.658	0.317
5	1	2	2	2	2	3	3	0.354	0.646	2662.08	1347.86	3858.00	2968.470	0.103
6	1	2	2	3	3	1	1	0.339	0.661	1441.91	655.58	2040.00	1570.246	0.081
7	1	3	3	1	1	3	3	(0,1)	(0,1)	(313,1205)	593.79	1410.00	(593,1410)	0.242
8	1	3	3	2	2	1	1	0.346	0.654	861.28	407.18	1224.00	941.034	0.084
9	1	3	3	3	3	2	2	0.873	0.127	1467.61	1524.60	4390.00	1890.215	0.223
10	2	1	2	1	2	1	3	0.222	0.778	3246.29	507.37	2241.0	1854.577	-0.750
11	2	1	2	2	3	2	1	0.511	0.489	3747.81	598.99	3735.0	2132.730	-0.757
12	2	1	2	3	1	3	2	0.084	0.916	12505.45	1195.78	7470.0	6940.255	-0.801
13	2	2	3	1	2	2	1	0.822	0.178	403.09	194.78	777.0	298.323	-0.351
14	2	2	3	2	3	3	2	0.992	0.008	1160.45	1012.68	4888.0	1042.309	-0.113
15	2	2	3	3	1	1	3	0.847	0.153	5260.41	2288.18	12032.0	3775.413	-0.393
16	2	3	1	1	2	3	2	0.992	0.008	389.71	428.76	1629.0	437.997	0.110
17	2	3	1	2	3	1	3	0.932	0.068	2169.64	1707.99	7670.0	2110.883	-0.027
18	2	3	1	3	1	2	1	0.818	0.182	2021.88	691.58	3885.0	1273.556	-0.587
19	3	1	3	1	3	1	2	0.591	0.409	2705.42	410.98	2245.0	1161.523	-1.329
20	3	1	3	2	1	2	3	0.229	0.771	21095.82	1257.97	10101.0	8071.205	-1.613
21	3	1	3	3	2	3	1	0.432	0.568	8759.90	634.99	5612.0	3460.054	-1.531
22	3	2	1	1	3	2	3	0.653	0.347	3551.30	731.54	3624.0	1734.084	-1.047
23	3	2	1	2	1	3	1	0.625	0.375	3730.45	479.18	3507.0	1613.588	-1.311
24	3	2	1	3	2	1	2	0.984	0.016	2995.85	1668.60	12274.0	1838.504	-0.629
25	3	3	2	1	3	3	1	0.628	0.372	1241.22	230.78	1169.0	579.413	-1.142
26	3	3	2	2	1	1	2	0.984	0.016	1800.23	1084.68	7365.0	1185.327	-0.518
27	3	3	2	3	2	2	3	0.774	0.226	12723.23	2822.19	19289.0	6542.556	-0.944

For each experiment we first calculated α_1 and α_1 to found the optimal solution $U_c(X)$ by repeating the calculations presented at section 3. Using the same parameters the utility function of line is also calculated for comparison. Then the index P is calculated.

Appendix 3. Results of L27 experiments

9. References

Economic Research Institute, (1997) Cellular manufacturing and production system innovation (in Japanese).

Isa, K. and Tsuru, T. (1999) Cell production and workplace innovation in Japan: Toward a new model for Japanese manufacturing?, *Industrial Relations*, 4(1), Issue 4, 548-578.

Johnson, D.J. (2005) Converting assembly lines to assembly cells at sheet metal products: Insights on performance improvements, *International Journal of Production Research*, 43(7), 1483–1509.

Kaku, I., Murase, Y. and Yin, Y. (2008) A study on human tasks related performances of converting conveyor Assembly line to cellular manufacturing, *European Journal of Industrial Engineering*, 2(1), pp.17-34.

Kaku, I., Gong, J., Tang, J. and Yin, Y. (2008) A mathematical model for converting conveyor cssembly line to cellular manufacturing, *International Journal of Industrial Engineering and Management Science*, 7(2), 160-170.

Kaku, I.,Gong, J., Tang, J. and Yin, Y. (2009) Modeling and numerical analysis of line-cell conversion problems, *International Journal of Production Research*, 47(Issue 8), 2055-2078.

Kaku, I., Liu, Y., Gong, J., Tang, J. and Yin, Y. (2009) Using L27 array to investigate the performance improvements in line-cell Conversion, *Proceedings of the Symposium on Group Technology and Cellular Manufacturing*.

Miyake, D. I. (2006) The shift from belt conveyor line to work-cell based assembly system to cope with increasing demand variation and fluctuation in the Japanese electronics industries, *Report paper of CIRJE-F-397*.

Sakazume Y. (2005) Is Japanese cell manufacturing a new system?: A comparative study between Japanese cell manufacturing and cellular manufacturing, *Journal of Japan Industrial Management Association*, 12, 89-94.

Sengupta, K. and Jacobs, F.R. (1998) Impact of work teams: a comparison study of assembly cells and assembly lines for a variety of operating environments, *Technical Report of Department of Management, Bucknell University*.

Shimizu, K. (1995) Humanization of the production system and work at Toyota Motor Co. and Toyota Motor Kyushu, *In Enriching Production: Perspectives on Volvo's Uddevalla Plant as an Alternative to Lean Production*, 4, 383-403.

Shinohara T. (1997) Why single-worker production is profitable?, *Plant Management*, 43(4), 17-63. (In Japanese)

Takahashi S., Tamiya H. and Tahoku H. (2003) A miracle cannon production system: It is not only Toyota, *Special report of Weekly Toyo Economics*, November 27-January 3, 123-136. (In Japanese)

Tanaka T. (2005) Birth of a cart pulling style production system, *IE Review*, 46(4), 31-37,(in Japanese).

Tsuru T. (1998) Cell manufacturing and innovation of production system, *Report of Economic Research Institute, Japan Society for the Promotion of Machine Industry*, H9-9. (in Japanese)

Van der Zee D. J. and Gaalman G. J. C. (2006) Routing flexibility by sequencing flexibility – exploiting product structure for flexible process plans, *The Proceedings of the Third International Conference on Group Technology / Cellular Manufacturing*, 195-202.

Yagyu, S. (2003) Synchronizes cellular manufacturing system, *Nikkan Kyhyo Simbu*.

8

Small World Optimization for Multiple Objects Optimization of Mixed-Model Assembly Sequencing Problem

Huang Gang, Tian Zhipeng, Shao Xinyu and Li Jinhang
State Key Laboratory of Digital Manufacturing Equipment & Technology
Huazhong University of Science & Technology
China

1. Introduction

With the continuously accelerating of economic globalization and subdividing of product demand, the production mode in manufacturing industry is always evolving. As the diversity and rapid changes of customer needs increasing, as the competition among enterprises with unprecedented speed and intensity expanding in global range, with the shorten lifecycle of product technology and market, the large-scale production with the characteristics of single variety, high-volume, continuous, and lacking of flexibility cannot meet the demands of current and future market, and the flexible production of small batch and multi-variety products, which are certain scaled and customer-oriented, have been favored in industry.

The evolution of production mode has generated the mixed assembly line. Products of different types and yields have been produced on the same assembly line by changing the organization of production without changing the existing production conditions and capacity, to meet the personalized demands of different consumers in the shortest possible time and quick response to the changes of market demands, then improve the competitiveness of enterprises.

Sequencing problem is one of the key issues of effectiveness on the mixed model assembly line, which determines the processing order of different products on the assembly line. A reasonable sort of production has important significance to improve production efficiency, reducing waste of resources and increasing competitiveness of enterprise in a market economy.

Sequencing problem is a typical combinatorial optimization problem. Exact approach and Heuristic method (Solnon et al., 2008) have been generally used to solve such problems. Exact approach generally includes Constraint Program, Integer Program and branch and bound algorithm. And Heuristic Approach involves Greedy Approach, Local Search Approach, Genetic Algorithm, Ant Colony Optimization and Particle Swarm Optimization (Chao-Tang & Ching-Jong, 2008). The computational complexity of sequencing problems has often grown exponentially, and the exact solution is difficult to deal with the traditional larger-scale problem, so in recent years Heuristic Approach has been the best approach to solve the sequencing problem.

Johnson's paper (Johnson, 1954) in 1954 is the first research on sequencing problem. Classical sequencing problem has been studied widely in last 30 years. For the production line, this kind of problem is called Flow Shop Problem (FSP). Classic FSP generally has four basic assumptions:

- **Resource type assumption:** the same machine can only process a workpiece at any time and one workpiece only can be processed on one machine at any time;
- **Determinacy assumption:** all input parameters are known in advance and fully identified;
- **Computability assumption:** all parameters can be calculated without regarding how to determine the date of delivery;
- **Single objective and regularity assumption:** Assuming that the goal of sorting is one-dimensional non-decreasing function of job completion time.

The so-called classical sequencing problem is generally difficult to use directly in production practice due to the strict assumptions. In recent years, mixed-model assembly sequencing problem in automotive industry has become a hot spot according to the research literatures and practical applications. The following sections fully described this problem in the automobile industry, proposed a model of general assembly sequencing problem with balancing the assembly load, smoothing the production flow and reducing the operation changeover as goals, and established a multi-objective car sequencing problem. Then we designed the transformation strategy for the discrete problem by the research of small world effect on a continuous problem, and proposed the small world algorithm to solve the discrete sorting problem. Finally, we solved the Car Sequencing Problem considering three objectives by Small World Optimization, found Pareto optimal solution of multi-objective car sequencing problem with actual production data. The result shown a good performance of Small World Optimization algorithm in multi-objective mixed sequencing.

2. The multi-objective mixed-model sequencing problem in automobile assembly workshop

Car Sequencing Problem was first put forward by Parrllo et al. in 1986, but Renault Company was the first one who began to solve the simple car sequencing problem by using the simulation annealing software (Solnon et al., 2008). This problem was focused a lot since it's coming out, and developed to a classical problem.

2.1 Status of the art of car sequencing problem

The researchers mainly considered two objectives in optimizing the mixed-model assembly sequencing: balancing the assembly load and smoothing the production flow. The objective of balancing the assembly load is to maximize the production capability, but without going beyond its utmost (Kenjiro & Hajime, 1979). The objective of smoothing the production flow was put forward by Toyota Company in JIT environment, and in this objective, the key to organize the mixed-model assembly for multi-product is to level the production line (Miltenburg, 1989). The core problem of heijunka is how to optimize the production order in the mixed-mode assembly line, in order to keep a balanced and steady production.

In recent years, many scholars have conducted research on the Car Sequencing Problem (CSP). To solve this problem in a precise way, Estellon (Bertrand & Karim, 2007) proposed two different local search methods to solve the real CSP. Prandtstetter Matthias (Matthias & Günter, 2008) proposed that using integer linear programming and mixed variational areas to solve the CSP. In terms of heuristic approach, Nils (Nils & Malte, 2006) and Sara (Sara et al., 2009) proposed a method based on ant colony algorithm, in which the theory of ants find the shortest path was simulated, to solve the CSP. Chul (Chul et al., 1998) used a new genetic algorithm to solve the multi-objective sequencing problem, and proposed a new gene fitness function and selecting rule. In his method, the three objectives considered are minimizing the excessive working time; keeping a uniform consumption speed of accessories and minimizing the adjusting cost. The result shows that this genetic algorithm is better than other genetic algorithm. Jianfeng (Jianfeng, 2006) proposed Pareto result by using genetic algorithm with multi-objective, which consist of the objective of smoothing the logistics and minimizing completion time. Allahverdi0(Allahverdi & Al-Anzi, 2006) put forward a model which considered the set-up time and minimum completion time, and proposed an algorithm to solve this model. This algorithm can get the best result faster than PSO and Taboo Search when the size of work-piece is larger than 50.

2.2 Three different objects of sequencing problem in assembly workshop

People in industry area have always been committed to improving the production efficiency; the way they take is improving the equipment and optimizing the production sequence. The cost of optimizing production sequence is very low, so it will get a better effect when put more attention on this way. Usually, there are many different objectives of car sequencing problem, but we only take three of them into consideration: balancing the assembly load, smoothing the production flow and reducing the operation changeover.

The components in the assembly process could be sorted in three types: the current component, i.e. the electronic adapter in the television assembly. Televisions of different model take the same adapter. The second type is the key component, i.e. the kinescope in the television assembly. All the televisions need this component, but different models need different kind of kinescope components. The third type is selected component, i.e. the stereo module. This type is selected only by certain product. For current component, their assembly process will not be influenced by the production sequencing because all the products need the same component. In this paper, we mainly talk about the key component and the selected component in the mix-mode assembly sequencing.

The objective of balancing the assembly load is to reduce the bottleneck in the production line; the objective of smoothing the production flow is to reduce the inventory of work in process.

In order to describe the assembly sequencing problem conveniently, some expressions were defined as follows:

- V : The set consist of the product' type, thus the total number of productions' type is $|V|$
- D_T : The sequence of all the products in one production zone, it's an orderly set. Thus, the total output in this production zone is $|D_T|$.
- Z_v : The set consist of v th product in D_T , thus the output of v th product is $|Z_v|$, and $\sum_{v \in V} |Z_v| = |D_T|$
- D_j : The j th product in production sequence D_T , $j = 1,2,...,|D_T|$, and $D_j \in V$
- $s_{j,v}$: The symbol that indicates whether product v was produced in station j in the production sequence.

In addition, $v \in V, s_{j,v} = \begin{cases} 0 & \textit{product } v \textit{ is not produced in position } j \\ 1 & \textit{product } v \textit{ is produced in position} \end{cases}$, $j = 1,2,...,|D_r|$. It's obvious that $\sum_{v \in V} s_{j,v} = 1$.

2.2.1 Sequencing aimed at balancing the assembly load

In the automobile's final assembly line, the automobile in process was put on a transmitting belt which moving with a fixed speed, and several working stations were distributed on both sides of the transmitting belt based on the assembly sequence. Each working station has an assembly group which must finish the stated task during the period from the automobile moving into this working station to its leaving. If the working station cannot finish the assembly in this period, production on the flow line would be influenced.

The main reason lead to the assembly unfinished is the unbalance of assembly load. For example, when there are too many selected components should be assembled continuously, the task would be hard to finish since operator's heavy working load. Therefore, in real production, the selecting frequency of each selected component should have an upper limit, which usually expressed as $H_x : N_x$, that means in the N_x products, only H_x products select the component x at most. For example, if there is a selected component A, and its selecting frequency $H_A : N_A$ is 2:3, that is to say, in any three discretional products in the production sequence only 2 products need to assemble component A. If more than 2 products select this component, the operator will not have enough time to assemble the component for the third product. In that way, the production line would be stopped.

Take one product's assembly for example, based on the requirement of the assembly load, the selecting frequency of each selected component was shown as Table 1, and the production plan was shown as Table 2. In table 2, the number in the first row stands the assembly order, and number in the second row stands the type of product. For example, 1 means type 1, 2 means type2, etc. In table 2, the number 1 in the data part means component type 1 need to be assembled, if nothing in the data part, that means no component need to be assembled.

Selected component	$H_x : N_x$
O[1]	2:3
O[2]	2:4
O[3]	3:5
O[4]	2:6

Table 1. The selecting frequency of each selected component

Production sequence		1	2	3	4	5	6	7	8	9	10	11	12	13	14
Selected component	$H_x : N_x$	1	6	3	4	5	1	2	6	1	3	4	5	6	1
O[1]	2:3		1	1	1			1	1		1	1		1	
O[2]	2:4			1	1						1	1			
O[3]	3:5					1		1					1		
O[4]	2:6	1			1		1			1		1			1

Table 2. Production sequence plan to be scheduled

In these two tables, there are 14 products be produced ($|D_T| = 14$); 6 different types ($|V| = 6$) and 4 selected components(O[1], O[2], O[3], O[4]) included. It's obvious that the production sequence in Table 2 disobey the restriction of selecting frequency in Table 1. Table 3 shows a production sequence meet the objective of balancing the assembly load:

Production sequence		1	2	3	4	5	6	7	8	9	10	11	12	13	14
Selected component	$H_x : N_x$	1	1	2	3	5	3	1	4	6	5	6	6	1	4
O[1]	2:3			1	1		1		1	1		1	1		1
O[2]	2:4				1		1		1						1
O[3]	3:5			1		1					1				
O[4]	2:6	1	1					1	1					1	1

Table 3. A production sequence meets the objective of balancing assembly load

The objective of balancing the assembly load is setting the selected component's assembling frequency reasonably. This objective offers a new idea to solve the bottleneck problem resulted from the unsmooth load.

In order to describe the model of sequencing problem about balancing the assembly load vividly and conveniently, some symbols are defined as follows：

E The set consists of selected components.

$a_{v,x}$ The notation indicates if the component be selected or not.

$$a_{v,x} = \begin{cases} 1 & \text{if product } v \text{ need selected component } x \\ 0 & \text{other} \end{cases}, x \in E.$$

$H_x : N_x$ In the N_x continuous products, there are only H_x products select the component x, $x \in E$.

According to the description above, the formula below can describe a Constraints Satisfaction Problem (Kenjiro & Hajime, 1979):

$$\sum_{v \in V} s_{j,v} = 1 \qquad j = 1,2,\ldots,|D_T| \tag{1}$$

$$\sum_{j=1}^{|D_T|} s_{j,v} = |Z_v| \qquad v \in V \tag{2}$$

$$\sum_{m=j}^{j+N_x} \sum_{v \in V} a_{v,x} s_{j,v} < |H_x| \qquad j = 1,2,\ldots,(|D_T| - |N_x| \tag{3}$$

$$s_{j,v} = \{0,1\} \qquad j = 1,2,\ldots,|D_T|, v \in V \tag{4}$$

2.2.2 Sequencing aimed at smoothing the production flow

Sequencing aimed at smoothing the production flow was rooted from the automobile industry. In order to describe the model of sequencing problem about smoothing the production flow vividly and conveniently, some symbols are defined as follows：

k The k th producing step

The moment when the former k products are finished was called the k th producing step. $k=1,2,\ldots,|D_T|$

r_i The ideal output ratio of the i th product, $i=1,2,\ldots,|V|$

That is to say, the ratio of i th product's output and the overall output. $r_i=\dfrac{d_i}{|D_T|}$

$g_{i,k}$ The i th product's output till the k th step. Obviously, $g_{i,k}=\sum_{j=1}^{k}s_{i,j}$, $d_i=g_{i,|D_T|}$

The understanding of these notations could be referenced as Table 4:

		The real production sequence				
		$k=1$	$k=2$	……	$k=\|D_T\|$	
Production's type	$i=1$					$\sum_{j=1}^{\|D_T\|}s_{1,j}=g_{1,\|D_T\|}=d_1$
	$i=2$					$\sum_{j=1}^{\|D_T\|}s_{2,j}=g_{2,\|D_T\|}=d_2$
	…		$s_{i,j}=\{0,1\}$			$\sum_{j=1}^{\|D_T\|}s_{i,j}=g_{i,\|D_T\|}=d_i$
	$i=\|V\|$					$\sum_{j=1}^{\|D_T\|}s_{\|V\|,j}=g_{\|V\|,\|D_T\|}=d_{\|V\|}$
	$\sum_{i=1}^{\|V\|}s_{i,j}$	1	1	……	1	$\sum_{i=1}^{\|V\|}\sum_{j=1}^{\|D_T\|}s_{i,j}=\|D_T\|$

Table 4. The model of sequencing problem on smoothing the production flow

According to the requirement of smoothing the production flow, the ideal numerical model of sequencing problem of smoothing the production flow is as follows:

$$\begin{aligned} &min. \quad f_l=\sum_{k=1}^{|D_T|}\sum_{i=1}^{|V|}\left|g_{i,k}-kr_i\right| \\ &\text{s.t.} \quad \sum_{i=1}^{|V|}g_{i,k}=k \qquad k=1,2,\ldots,|D_T|, g_{i,k}\ \textit{is nonnegative integer} \end{aligned} \tag{5}$$

Still taking Table as an example, it's obvious that $|D_T|=14$. Choosing step $k\le|D_T|$, such as $k=10$, the objective of smoothing the production flow is for every k, the value of object function in formula(5 could be as small as possible. That is to say, it's required that in any step, the products' output in different types could be close to the overall products' output.

2.2.3 Sequencing aimed at reducing the operation changeover

The objectives of these two models are the same according to their definition: balancing the production. The former one requires balancing the assembly line's load, and the latter one requires balancing the output of products and smoothing the production flow.

With the diversification of customer's requirement, more and more enterprises take the order-oriented production strategy, which lead the fact that more and more different types of products be produced in the same assembly line, and also lead more and more components have to be assembled in the same working station. However, according to Table 3, if the two objectives mentioned before have to be realized, it will bring a great discrepancy between two continuous products in the one working station. Since the production sequence determinate the component's assembly order, the frequent change of product's type always means the frequent switching of component assembled in the working station.

For one same working station on the assembly line, if a production sequence lead the component assembled in this working station switching less, obviously, the set-up time will decrease a lot, and the efficiency of the assembly will increase a lot. Therefore, in terms of ergonomics, it's significant to schedule the production sequence and decrease the switching component's type. This paper proposed a sequencing problem which aimed at decreasing the switching frequency of work-in-process's type and increasing the similarity of product.

Symbols are introduced to describe the sequencing problem which based on the product's similarity in the assembly line:

v_n The n th assembly procedure of v th product

$C(v_n, v_n^{'})$ The similarity value of v_n and $v_n^{'}$, if v_n and $v_n^{'}$ have the same type, $C(v_n, v_n^{'})$ is 0, otherwise, $C(v_n, v_n^{'})$ is 1, i.e. $C(v_n, v_n^{'}) = \begin{cases} 0, & if\ v_n = v_n^{'} \\ 1, & if\ v_n \neq v_n^{'} \end{cases}$

The model aimed at reducing the operation changeover could be expressed as follows:

$$\begin{aligned} min. \quad & f_p = \sum_{j=1}^{|D_T|-1} C(v_n, v_n^{'}) \\ s.t. \quad & C(v_n, v_n^{'}) = \begin{cases} 0, & if\ v_n = v_n^{'} \\ 1, & if\ v_n \neq v_n^{'} \end{cases}, n = 1,2,\ldots,|D_T|-1 \end{aligned} \tag{6}$$

Obviously, when all the products in the production sequence are the same type, f_p has its minimum value: 0.

In addition, the key component influences the efficiency and quality in the assembly process greatly. Therefore, for simplicity, the paper only considered the key component in the assembling process when definite the product's changeover.

3. Small world optimization for discrete sequencing problem

There are many proven algorithms to solve discrete sequencing problem and we commonly use the heuristic algorithms such as Ant Colony Algorithm and Genetic Algorithm. When solving the discrete sequencing problem, the Ant Colony Algorithm may be convergence to local optimum and the Genetic Algorithm has low solving efficiency as producing a lot of redundancy iteration by using insufficient feedback information. Therefore, we need an algorithm which has a strong capability of global searching and fast convergence speed. Small World Optimization just meets our requirements.

The research on small world firstly originates in 1929 by a Hungarian writer F.Karinthy. He put forward the theory "Any two people on earth can be contacted through a chain composed

by five other persons" (Braun, 2004). In 1967, S.Milgram confirmed the small world phenomenon through the famous letters delivery experiments and proposed six degrees of separation theory (Travers & Milgram, 1969). With the study on small world effect by Watts and Strogatz (Watts & Strogatz, 1998; Strogatz, 2001), small world phenomenon was gradually valued and quickly became a hot spot of complex system and complexity theory research.
This paper designed a fast search algorithm called Small World Optimization (SWO) which was inspired by the hierarchical categorization tree model and multi categories method based on small world theory. The solution space can be divided into the hierarchical categorization tree model using hamming distance. Two bijective mapping solution spaces were adopted to establish multi categories method. Small World Optimization evaluated the relationship between nodes of the short and long distance neighbors in the two spatial networks and put the envelopes to the target, and then found the optimal solution.

3.1 Designing small world optimization

3.1.1 Basic idea of small world optimization

The basic ideas of Small World Optimization: simulating Milgram's letters delivery experiments (J.Travers & S.Milgram, 1969), first deliver several envelopes from different places in the solution space and regard each envelope holder (node called in this paper) as a candidate solution; then find out each node's best node which is estimated by objective function and deliver the envelope to this node; at last, each envelope will be closer to the object node through multiple delivery in the small world network.
The extensive research on small world effect also makes Small World Optimization achieve excellent results in solving actual problem. Huang (Huang et al., 2011) made a deep research on Small World Optimization and designed a fast search algorithm for continuous problem. With some test of standard continuous problems for Small World Optimization, Small World Optimization has fast convergence ability. Inspired by the Small World Optimization on continuous problem, we study the solving strategy of Small World Optimization in discrete problem.

3.1.2 Definition of short and long neighbors

For problems of continuous type, we can easily use network address division method to solve the hierarchical structure of the solution space. The relationships between the different addresses of the same subnet network are short neighbors. But the addresses of different subnet networks have long neighbor relationships.
However, sequencing problem is a kind of discrete problem for which we can't apply Internet Protocol division rules in continuous problem because each solution is a sequence. Normally, hamming distance is used to express the distance between different sequences of discrete problem. So the two sequences between which the hamming distance is 2 are defined as short neighbor relationship. The number of short neighbors of a sequence node is C_n^2 in a sequence of length n. Such as sequence 1 2 3 4 5 and sequence 1 3 2 4 5 have short neighbor relationship. If hamming distance of two sequences is greater than 2, they have long neighbor relationship.
Acquaintance probability of any two long nodes is expressed by index function $e^{-\alpha d_{ij}}$, α is acquaintance index and d_{ij} is the hamming distance of two long nodes. From the function, we realize that the smaller the hamming distance, the greater the acquaintance probability, and vice versa.

3.1.3 Construction of multiple space conversion

The highlight of small world optimization algorithm is multiple space synchronization search, solving multiple space mapping is a key problem. According to the above discussion, Small World Optimization can find the best solution through searching in double spaces which combine two bijective mapping solution spaces with original solution spaces.

1. Discrete space similar conversion rules of gray yards

For discrete problem, sequence is usually described as real integer. Such as, for a Flow shop problem with 5 workpieces, workpieces sequence A, B, C, D, E is corresponded with 1, 2, 3, 4, 5 respectively.

Adjusting the workpieces sequence will get a new solution; we define all the solution as the first heavy solution space. The second heavy solution space is from the first heavy space through some conversion rules. According to the conversion rules of binary gray yards for continuous space, we design a similar conversion rule for discrete space, the conversion code rules are as follows:

Conversion rules. For a sequence $X = \{1\ \ 2\ \ 3\ \ 4\ \ 5\}$, its length $M = 5$, conversion procedure below:

i. According to the conversion rules of gray yards, first consider shifting the sequence coding, circularly shift N byte rightward. If $N = 2$, the sequence will be $X_1 = \{4\ \ 5\ \ 1\ \ 2\ \ 3\}$.
ii. In order to increase the differences between the original sequence and new sequence and enlarge its hamming distance length, we can reverse the original sequence, and the sequence will be $X_2 = \{3\ \ 2\ \ 1\ \ 5\ \ 4\}$.
iii. According to the two sequences which take exclusive by bitwise operation in gray yards conversion, we can do the same operation in discrete sequence. For sequence 3 2 1 5 4, if overall add m, then take more to the sequence length M. $G_i = (X_i + m - 1)\%M + 1$. If $m = 1$, the sequence will be $G = \{4\ 3\ 2\ 1\ \ 5\}$.

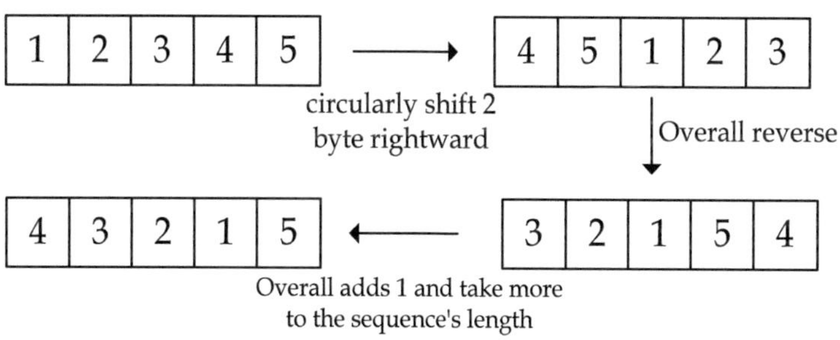

Fig. 1. The procedure of conversion rules

2. The conversion of gray yards for discrete space

In order to make discrete sequence fully meet the conversion of gray yards, first the real integer will be converted to binary coding. For a sequence 2 5 4 3 1, respectively convert the real integer sequence to binary coding of which length is 3. As Fig. 2 (a) describes.

Then we get a new binary coding sequence by conversion of gray yards to the binary coding sequence. At last, convert the binary coding to integer coding and get the final sequence as Fig. 2(b):

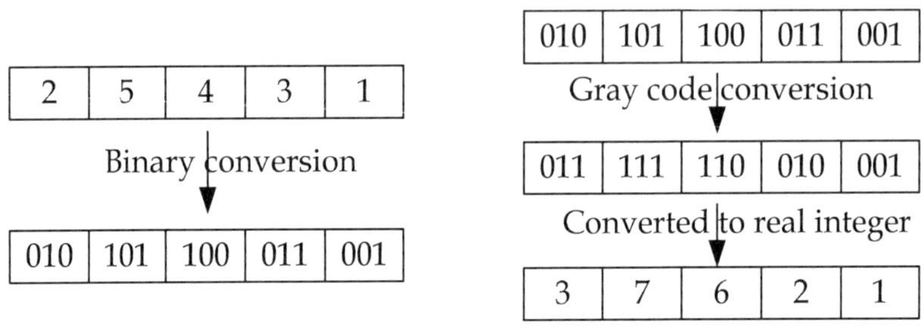

(a) (b)

Fig. 2. The binary conversion of Real integer

Usually we can correct the sequence by Smallest Order Value rule, the sequence will be: 3 5 4 2 1.In fact, we don't have to correct the final sequence, because we only get the second heavy space for finding short neighbor nodes. So, for the sequence 3 7 6 2 1, we can do nothing. The sequence 6 7 3 2 1 is a short neighbor node, which will be 4 5 2 3 1 after converting to the first heavy space.

3.1.4 Updating of the envelope node

According to above definition, for a sequence whose length is N , the number of its short neighbor node is C_N^2 and the number of its long neighbor nodes is $A_N^N - C_N^2 - 1$.
Each envelope holder will search part of its short neighbor nodes and long neighbor nodes in double space and find a better envelope by some rules to update it. General selection rules have two kinds: the first is only finding a better envelope to update; the second is searching specific number of long and short neighbor nodes in double space and finding the best envelope to update.

3.2 Summarizing the procedure of small world optimization

The process of Small World Optimization is as follows:

1. Set iterative counter $c = 0$;
2. Create initial envelope population $Q(t_0)$: build the mapped relationship between solution space and coding style according to the constraint definition, randomly select envelopes as the initial nodes.
3. Inquire the neighbor nodes: A certain number of short and long neighbor nodes n are inquired for each envelope node. The inquired neighbor nodes include the nodes coming from the original solution space and the mapping space.
4. Evaluate the neighbor nodes of each node: calculate the evaluation function for each neighbor node, and sort them.
5. Deliver envelope: select a envelope node which better than the current envelope node, then delete the current envelope node. The selection strategy have two kinds: one is calculating all the selected neighbor nodes and selecting the node with best evaluation value. The other one is calculating the evaluation value of each node from this space, and delivering the current envelope node since it's easy to find high quality nodes in the mapping space. When appearing a better node, stop calculation.
6. Determine whether meet the terminal conditions: if it meets, stop searching and print the result; else, keeping seaching.

7. Update all envelope nodes: when all envelope nodes have compete their dilivering, the nodes compose a new envelope group, and can get $Q(t+1)$.
8. $c = c+1$, return to procedure 3).

The parameters needed in Small World Optimization: the length of coding is n; the number of short neighbor nodes is mc; the acquaintance index of two long neighbor nodes is α; the number of envelope nodes is m.

Here, n is equal with the length of sequence and the problem will be more complex when n largen. The number of short neighbor nodes decides the efficiency and quality of search. α is an important parameter to select long neighbor nodes and generally the range from 0.4 to 0.6. Usually the number of envelope nodes is range from 10 to 40.

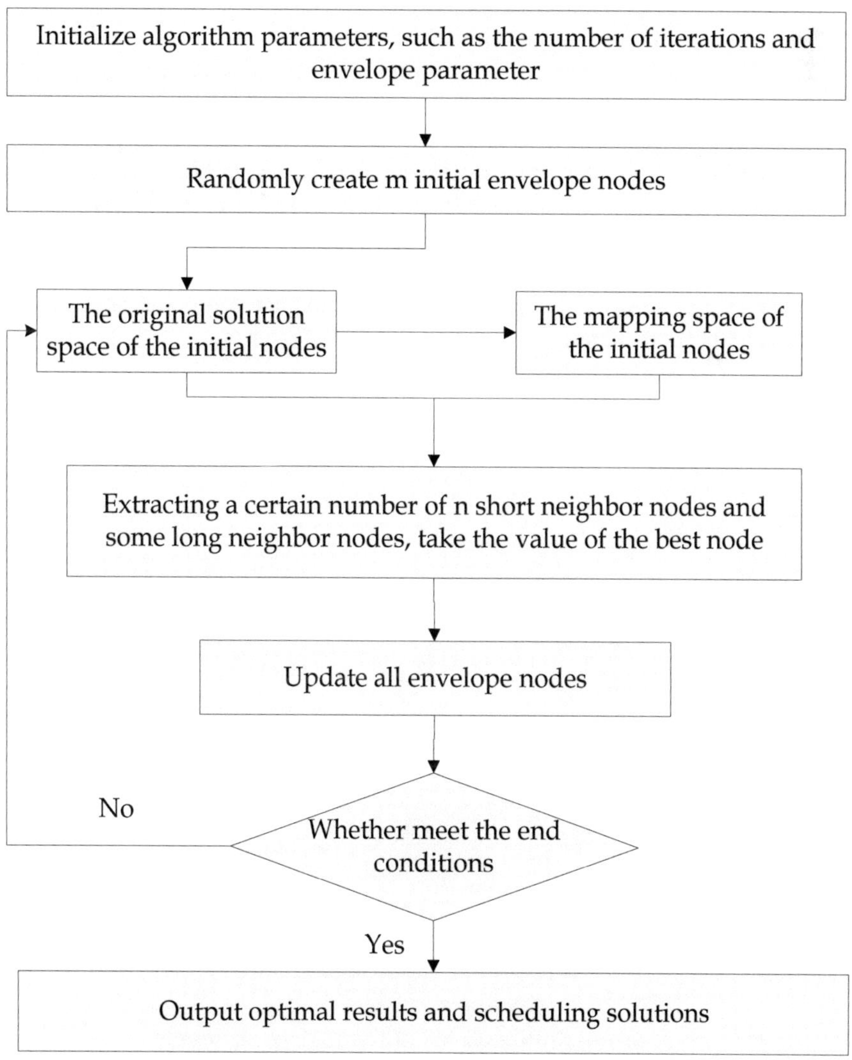

Fig. 3. The flowchart of Small World Optimization algorithm

4. Numerical experiments with small world optimization on the multi-objective car sequencing problem

By the multi-objective mixed-model sequencing problem in automobile assembly workshop Established in section 2, we use the Small World Optimization, which is introduced in section 3, to solve a practical example of automotive industry production.

4.1 Simplifying the mixed-model sequencing problem of multi-objective

In section 2.2, we analyzed three types of car sequencing problem in the current assembly workshop. The three goals are shown as below:

- f_c : The goal of balancing the assembly load
- f_l : The goal of smoothing the production flow
- f_p : The goal of reducing the operation changeover

Here, we get a Pareto solution set by considering these three objectives simultaneously to optimize the car sequencing problem. According to section 2.2.1, balancing the assembly load can describe as a Constraints Satisfaction Problem, and just needs to be satisfied. In addition, for an actual auto enterprise, the imbalance of the assembly load may lead the line stopping running. Therefore, we'll regard the goal f_c as the primary goal and put the balance of the assembly load as the constraint conditions.

After conversion, the three-objective problem changed into a two-objective problem with a constraint satisfaction. Minimize the goal of balancing the assembly load and minimize the goal of reducing operation changeover have the same importance.

$$\begin{aligned} &min. \quad f_l = \sum_{k=1}^{|D_T|}\sum_{i=1}^{|V|}\left|g_{i,k} - kr_i\right| \\ &min. \quad f_p = \sum_{j=1}^{|D_T|-1}\sum_{n=1}^{\Theta} C(v_n, v_{n+1}) \end{aligned} \tag{7}$$

$$s.t. \quad \sum_{m=j}^{j+N_x}\sum_{v\in V} a_{v,x}s_{j,v} < |H_x| \qquad j = 1,2,\ldots,(|D_T| - |N_x|) \tag{8}$$

$$\sum_{i=1}^{|V|} g_{i,k} = k \qquad k = 1,2,\ldots,|D_T|, g_{i,k} \text{ is Nonnegative integer} \tag{9}$$

$$C(v_n, v_{n+1}) = \begin{cases} 0, & if \quad v_n = v_{n+1} \\ 1, & if \quad v_n \neq v_{n+1} \end{cases}, n = 1,2,\ldots,|D_T| - 1 \tag{10}$$

In addition, formula (7)is the final two objects, formula (8)is the constraint that balancing the assembly load and the formula (10)expresses the switching times of parts use.

4.2 Preparing the actual production data for numerical experiments

A certain assembly workshop produces Multi-Purpose Vehicles. According to the different configurations can be divided into models, and each batch has 20 vehicles. The major plan as Table 5 below:

Product NO	1	2	3	4	5	6	7	8	9	10	11	12
Product model	a	b	c	d	e	f	g	h	i	j	k	l
number	1	2	1	2	1	1	3	3	1	1	2	2

Table 5. The major plan of actual production data

Model 1 concern the balance of assembly load. Here, we consider five selected options (O [1], O [2], O [3], O [4], O [5]). In discussing smoothing the production flow, in order to simplify the solution, we only consider these product parts (K [1], K [2], K [3], K [4], K [5]) which switch frequently and are more critical. The selected options and critical parts of each product model list as Table 6 below:

Product	O[1]	O[2]	O[3]	O[4]	O[5]	K[1]	K[2]	K[3]	K[4]	K[5]
a	1	0	0	1	1	1	1	1	1	1
b	1	1	0	0	0	1	2	2	2	2
c	1	0	1	1	0	1	2	3	2	2
d	0	1	0	0	1	2	2	1	1	3
e	0	0	1	0	0	3	3	2	1	3
f	0	1	0	1	1	4	2	2	2	1
g	0	1	1	0	1	3	3	3	1	2
h	0	1	0	1	0	4	1	2	2	2
i	1	0	0	0	1	1	3	2	1	1
j	0	0	1	1	0	1	2	3	1	1
k	0	0	0	0	1	4	2	2	2	2
l	0	0	0	1	1	3	3	1	1	2

Table 6. The selected options and critical parts of each product model

The selecting frequency of each selected component in Table 6 shows in Table 7 below:

The selected component in	**O[1]**	**O[2]**	**O[3]**	**O[4]**	**O[5]**
$H_x : N_x$	3:4	4:7	3:5	2:3	2:3

Table 7. The selecting frequency of each selected component

4.3 Solving the multi-objective CSP by small world optimization

The multi-objective model considers smoothing the production flow and reducing the operation changeover in this paper and we use Small World Optimization to solve it. According to the major plan, each production batch has 12 kinds of vehicles and 20 vehicles. In order to code the vehicles, use the integer to coding them, as Table 8 below:

Product type	a	b	b	c	d	d	e	f	g	g	g	h	h	h	i	j	k	k	l	l
Coding style	1	2	3	4	5	6	7	8	9	10	11	12	13	14	15	16	17	18	19	20

Table 8. The coding style of each product model

Initialize the envelope nodes. Randomly generate sequences whose coding length is 20, and then compute each target of the sequences. Select 100 envelope nodes, which meet the balance of assembly load, as initial nodes. The initial solution space is shown as Fig. 4 :

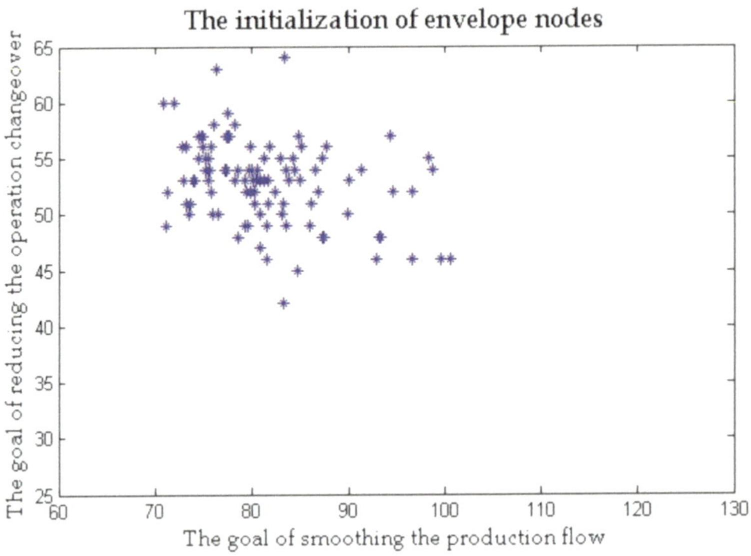

Fig. 4. The initialization of 100 envelope nodes

Consider the single object. According to the initial envelope nodes, when smoothing the production flow is the unique goal, 27 is the minimum value. The final envelope nodes after optimizing are shown as Fig. 5(a)and (124, 27) is the right margin of the Pareto solution set. When only thinking about the goal of smoothing the production flow, the minimum value is 67. The final envelope nodes after optimizing are shown as Fig. 5 (b) and (67, 54) is the left margin of the Pareto solution set.

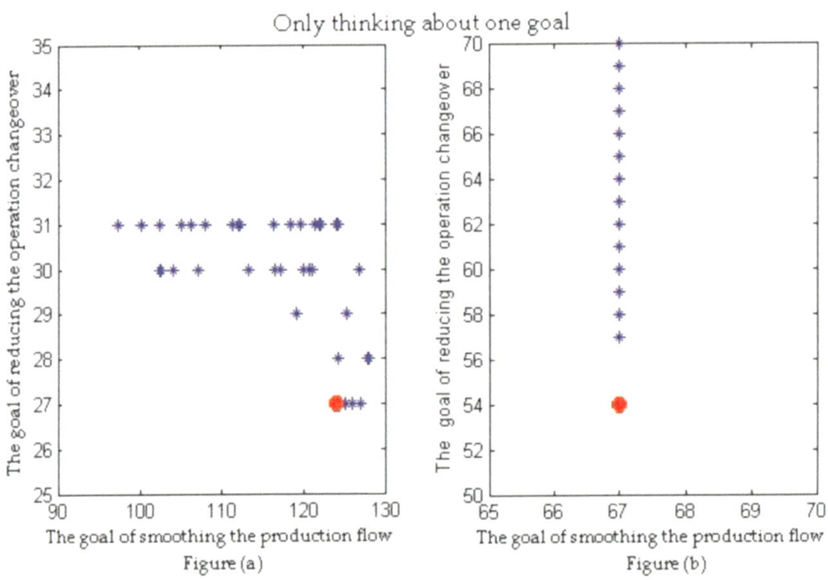

Fig. 5. Concerning the single goal

Consider two objectives simultaneously. Considering the goal of smoothing the production flow and the goal of reducing the operation changeover in the condition of that the balance of product load meets the requirements. In optimization process of Small World Optimization, for the target of each envelope nodes $Goal_i = (f_{l,i}, f_{p,i})$, only when $f_{l,i+1} < f_{l,i}$ and $f_{p,i+1} < f_{p,i}$, the envelope nodes begin updating, $Goal_{i+1} = (f_{l,i+1}, f_{p,i+1})$.
The parameters in Small World Optimization are set as follows: the number of envelope nodes is 100; each envelope node select 20 short neighbor nodes; the acquaintance index of two long neighbor nodes $\alpha = 0.4$; the iteration times is 400; and we optimize the problem in two space.
Finally, we get the Pareto solution set of multi-object Car Sequencing Problem by Small World Optimization. The all Pareto solutions are shown as Table 9. The upper and lower boundary in Table 9 is according with the result of single objective. To a certain extent, it also proves that Small World Optimization has a strong global searching capability.

Pareto solutions	1	2	3	4	5	6	7	8	9
the similarity of product goal	27	29	30	31	35	37	38	39	41
the equilibrium of logistics goal	124	118.1	102.7	85.9	80.6	77.7	77	74.2	71.6
Pareto solutions	10	11	12	13	14	15	16	17	
the similarity of product goal	42	43	44	45	46	47	51	54	
the equilibrium of logistics goal	71.4	68.7	68.4	68.2	67.8	67.7	67.2	67	

Table 9. The All Pareto solution

The optimization of Small World Optimization keeps the envelopes delivering and updating, thus the initial envelope node is always changing its position. The final envelope nodes distribute as Fig.6. The red line connects all the Pareto solutions, and the red dots indicate the Pareto solution. It can be seen that when two objectives have to be met, no other nodes are better than Pareto best solution.
The final envelope nodes distribute as Fig. 6. The final envelope nodes are close to the Pareto solution nodes which reflect that Small World Optimization algorithm has good convergence ability. And the global distribution of the envelope nodes reflects the superiority of multidimensionality space synchronous searching ability of Small World Optimization.

5. Discussion and conclusions

In this paper, we offer three main contributions to the multiple objects car sequencing problem, in the areas of modeling, solution engine, and to the quality of numerical results for real industrial problems.
On modeling: This paper mainly discusses the problem of production line in assembly workshop and establishes an all-sided model. Based on the actual production, we put forward a model which considers the goals of balancing the assembly load, smoothing the production flow and reducing the operation changeover. Thus, the problem which we discuss is a multi-object Car Sequencing Problem.
The solving model was improved and revised in order to simplify the calculation process. The goal of balancing the assembly load is the highest-level objective since the unbalance of production load will stop the production line. To simplify the solving process, it was regarded as a constraint that has to be satisfied in the multi-objective model.

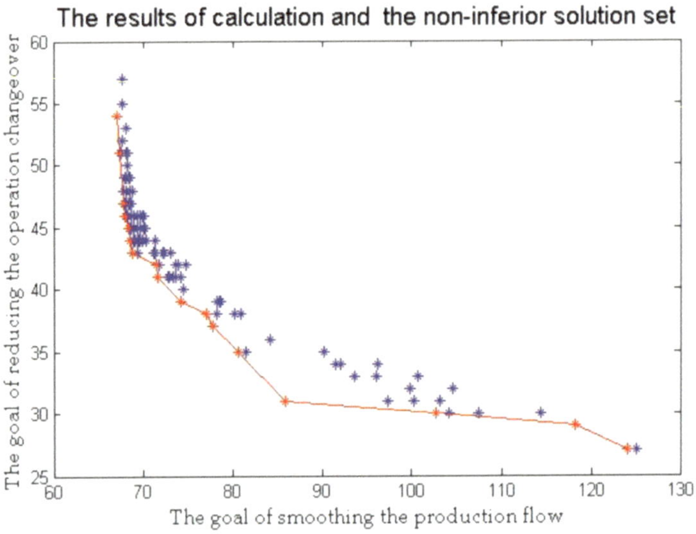

Fig. 6. The final distribution of the envelope and the Pareto solution set

On the solution engine: We present a new and powerful solution engine for the multiple objects car sequencing problem. This paper designed a fast search algorithm called Small World Optimization (SWO) which was inspired by the small world effect in nature.

According to the characteristic of "Six Degrees of Separation" in small world, we can optimize the current solutions by distributed searching in the original solution space and the mapping space synchronously. Small World Optimization fully displays its good search performance and global searching capability by solving the multi-object Car Sequencing Problem in section 4.

The result of solving multi-object Car Sequencing Problem reflects the excellent overall searching performance of Small World Optimization algorithm. Small World Optimization realizes the multi-point searching based on the envelope transmission system. Each envelope records the node with best value in the search process and each envelope has its independence. This search mechanism in solving multi-objective question has better searching performance. It can record many Pareto best solutions synchronously and get a better Pareto solution set in solving multi-object sequencing problem.

On the numerical results: For an actual industrial problem of real dimensions, we present computational results that show that the Small World Optimization approach has the ability rapid convergence and global optimization characteristic. For the multi-object Car Sequencing problem which belongs to NP-Hard problem, Small World Optimization can get a good solution by solving with its superior calculation performance.

Our work also suggests extensions that would broaden the range of the models presented, but that were not included in this paper because they would not have allowed direct comparisons to available results from current practice.

Firstly, as the production batch and the constraints of assembly load are little, there are so many solutions when the constraint of balancing the assembly load is satisfied. So we can convert the three goals optimization problem to two goals optimization problem. However, when the scale of sequencing problem enlarges and the number of constraints increases, the balance of assembly load may not be satisfied. In this case, we have to regard it as a three-goal problem.

The objective of balancing the assembly load is to minimize the times of violating the balance principle; the model was described as below:

$$min. \qquad f_c = \frac{1}{2}\sum_{i=1}^{|E|}\sum_{j=1}^{|D_T|+|N_x|} rep_{i,j}$$

$$s.t. \qquad rep_{i,j} = \begin{cases} 0, & if\ |H_x| - \sum_{k=0}^{|N_x|-1}\sum_{v\in V} a_{v,x}s_{j+k,v} \geq 0 \\ 1, & other \end{cases}, s_{j,v} = \{0,1\}, j = 1,2,\ldots,|D_T|$$

$$v \in V,\ x \in E$$

Secondly, we could also consider other requirements in the car assembly workshop, like the balancing of working time in different workstation. The equilibrium of working time of each working station can also optimize the production tact time.

Finally, for Small World Optimization algorithm, the research in this paper is not mature enough. However it also achieved an excellent effect on the dispersing problem by designing the relative algorithm, the searching method and the updating strategy. What's more, the Small World Optimization algorithm has more strength on the multi-objective problem, since the global searching of multi-node will help to find the Pareto optimal solution. Small World Optimization will have a better development when more and more scholars doing research on it.

6. Acknowledgements

The research reported in this chapter is supported by the National Natural Science Foundation of China (NSFC) under Grant No. 50775089, 50825503, 50875101.

7. References

Allahverdi, Ali & Al-Anzi, F.S (2006). Evolutionary heuristics and an algorithm for the two-stage assembly scheduling problem to minimize make span with setup times. *International Journal of Production Research*, Vol.44, No.22 (December, 2006), pp.4713-4735, ISBN 0020-7543

Bertrand, Estellon & Karim, Nouioua (2007). Two local search approaches for solving real-life car sequencing problems. *European Journal of Operational Research, Vol.*191, No.3 (December, 2008), pp. 928-944, ISBN 0377-2217

Braun T (2004) · Hungarian priority in network theory · *Science*, Vol.304, No.5678 (June, 2004), pp.1745, ISBN 0036-8075

Chao-Tang, Tseng & Ching-Jong, Liao (2007). A discrete particle swarm optimization for lot-streaming flow shop scheduling problem. *European Journal of Operational Research*, Vol.191, Issue 2, 1(December, 2008), pp.360-373, ISSN 0377-2217

Chul, J.H.; Yeongho, K. & Yeo, K.K (1998). A genetic algorithm for multiple objective sequencing problems in mixed model assembly lines. *Computers & Operations Research*, Vol.25, No.7-8(July, 1998), pp.675-690, ISBN 0305-0548

Huang Gang; Li Jinhang & Jia Yan (2011). Small World Optimization: A fast search algorithm based on small world effect. *Computer Science*, (July, 2011), ISSN 1002-137X

Jianfeng, Yu; YueHong, Yin & Zhaoneng, Chen (2006). Scheduling of an assembly line with a multi-objective genetic algorithm. *The International Journal of Advanced Manufacturing Technology*, Vol.28, No.5-6 (2006), pp. 551-555, ISBN 0268-3768

Johnson, S.M (1954). Optimal two-and three-stage production schedule with setup times included. *Naval Research Logistics Quarterly*, Vol.1, No.1 (1954), pp. 61-68

Kenjiro, Okamura & Hajime, Yamashina (1979). Heuristic algorithm for the assembly line model-mix sequencing problem to minimize the risk of stopping the conveyor. *International Journal of Production Research*, Vol.17, No.3 (1979), pp. 233-247, ISBN 0020-7543

Matthias, Prandtstetter & Günter, R. Raidl (2008). An integer linear programming approach and a hybrid variable neighborhood search for the car sequencing problem. *European Journal of Operational Research, Vol.*191, No.3 (December, 2008), pp.1004-1022, ISBN 0377-2217

Miltenburg, G.J (1989). Level schedules for mixed-model assembly lines in just-in-time production systems. *Management Science*, Vol.35 (1989), pp.192-207, ISBN 0025-1909

Nils, Boysen & Malte, Fliedner (2006). Comments on "Solving real car sequencing problems with ant colony optimization". *European Journal of Operational Research*, Vol.182, No.1 (October, 2007), pp.466-468, ISBN 0377-2217

Sara, Morin; Caroline, Cagne & Marc, Gravel (2008). Ant colony optimization with a specialized pheromone trail for the car-sequencing problem. *European Journal of Operational Research*, Vol.197, No.3 (September, 2009), pp.1185-1191, ISBN 0377-2217

Solnon, Christine; Cung, Van Dat & Nguyen, Alain (2007). The car sequencing problem: overview of state-of-the-art methods and industrial case-study of the ROADEF'2005 challenge problem. *European Journal of Operational Research*, Vol.191, No.3 (December, 2008), pp. 912-927, ISSN 0377-2217

Strogatz, S. H. (2001). Exploring complex networks. *Nature*, Vol.410 (March, 2001), pp. 268-276, ISBN 0036-8075

Travers, J. & Milgram, S. (1969). An experimental study of the small-world problem. *Journal of Information Storage and Processing systems*, Vol.32 (1969), ISBN 1099-8047

Watts, D. J. & Strogatz, S. H. (November, 1997). Collective dynamics of 'small-world' networks. *Nature*, Vol.393 (June, 1998), pp.440-442, ISBN 0036-8075

9

Optimizing Feeding Systems

Shramana Ghosh and Sarv Parteek Singh
NSIT, University of Delhi
India

1. Introduction

1.1 What are feeders?

Application of automation to assembly processes is vitally important to meet the requirements of manufacturing and production .Feeders form a critical part of automated assembly lines (Singh et al., 2009) . They are used to feed discrete parts to assembly stations or work cells or assembly cells on the production line from bulk supplies. They convert the randomness of parts into a flow in geometrical patterns such that the parts can become an integral contribution to the production process and get delivered at a pre-determined rate. Most of the time these parts are added to other parts to become the finished product at the end of an assembly line (Mitchell Jr., 2010). Feeders are sometimes also used as inspection devices (Boothroyd, 2005). Part feeders can be designed to reject certain kinds of defective parts, which when fed to the machine may result in its breakdown. Assembly process requires the presence of the correct parts, in the suitable amounts, at the appropriate places and at the right times, in the absence of which the entire production line may come to a halt. Ad-hoc setting of system parameters results in either starvation or saturation, where too less or too many parts are delivered to the work cells respectively.

This chapter describes a method of studying the behaviour of part feeding devices via a statistical analysis of the given system, carried out to formulate its empirical model. Once the model has been formulated and its validity confirmed, it can be used to suggest values of inputs and operating factors to get the desired output. The regression model can also be used to find the local optimum.

1.2 Types of feeders

Vibratory feeders are the most widely employed and versatile part-feeding devices in the industry (Boothroyd, 2005). Detailed theoretical analysis of vibratory feeders has been carried out (Redford and Boothroyd, 1967; Parmeshwaran and Ganpathy, 1979; Morrey and Mottershead, 1986, Ding and Dai, 2008 etc.).

While vibratory feeders remain the part-feeders of choice for general purpose requirements, many other designs of feeders have been developed for feeding parts having special features like headed parts or abrasive materials. Such feeders can usually be classified under:

i. Reciprocating feeders
ii. Rotary feeders
iii. Belt feeders

Some common feeders that fall under these categories have been listed in Table 1.

Type	Feeder	Description
I	Reciprocating-Tube Hopper Feeder 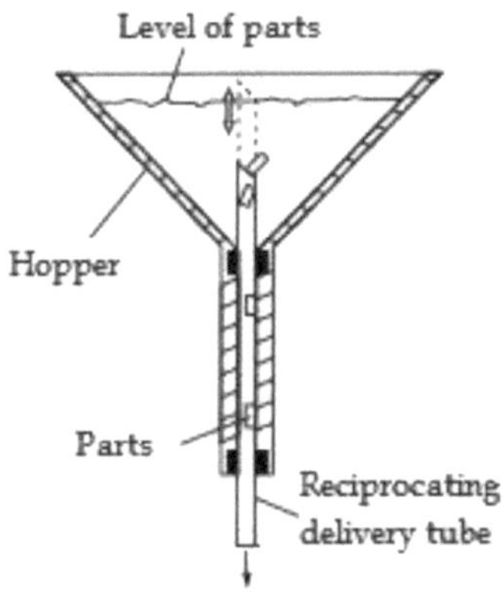	It consists of a delivery tube passing through a hole at the centre of a conical hopper where the parts are populated. Relative motion between the hopper and tube by means of reciprocation of either results in the transfer of parts from hopper to the tube when the latter falls below the level of parts. Top of the tube is machined at an angle to allow parts to fall clear into it rather than blocking the opening during the tube's upward motion. This kind of feeder is best suited for cylindrical parts
I	Reciprocating-Fork Hopper Feeder 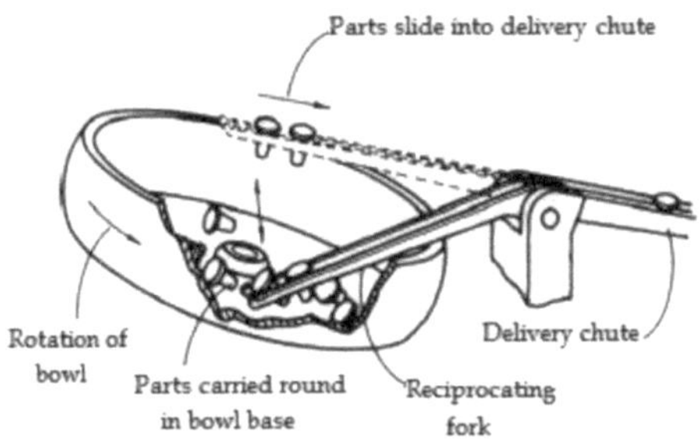	It consists of a shallow cylindrical bowl that rotates about an axis inclined at a small angle to the vertical plane. A two pronged fork reciprocates in the vertical plane. The fork dips into the rotating bowl where centrifugal force pushes the parts into it. As the fork moves upward, parts with the right orientation (see adjoining fig.) are transferred into the delivery chute due to the effect of gravity. This kind of feeder is best suited for headed parts like bolts and rivets.
I	Reciprocating-Centerboard Hopper Feeder 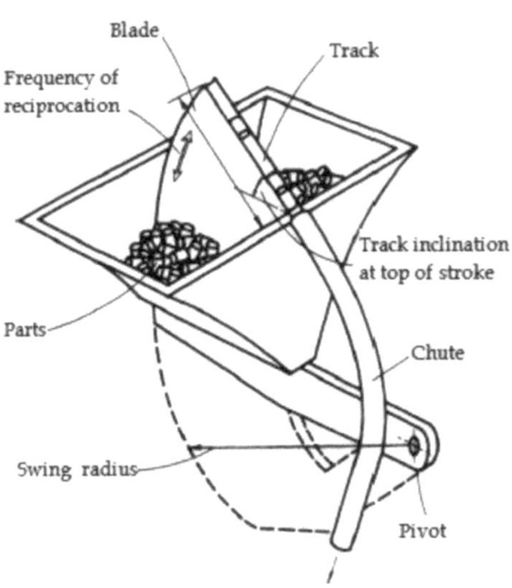	It consists of a blade with a shaped track along its upper edge. This blade undergoes a reciprocating motion in the vertical plane, being pushed into and out of a hopper populated with parts in a cyclic fashion. In its topmost position, the blade is in line with a chute. Thus, all the parts which enter into the blade during its upward motion, fall into the chute as it reaches the uppermost position. This kind of feeder is suitable for cylindrical parts

II	Rotary-Centreboard Hopper Feeder 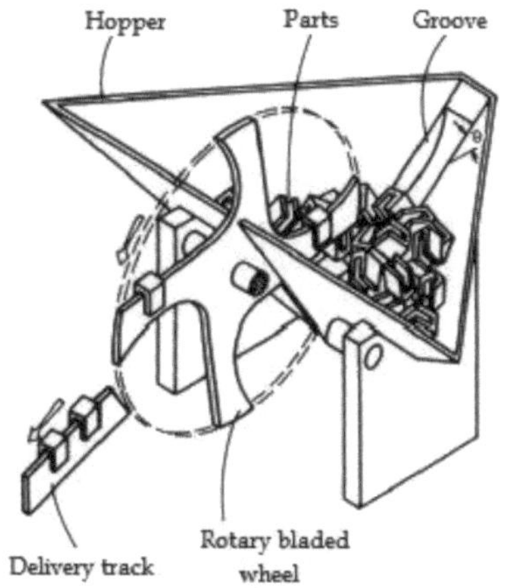	It consists of a multi-bladed wheel which can rotate inside a hopper as shown. The blades are profiled to carry parts in a desired orientation. As a blade rotates, it lifts up parts in the required orientation and subsequently during its cycle, delivers them onto a delivery track. This feeder can feed U-shaped parts.
II	Bladed-Wheel Hopper Feeder 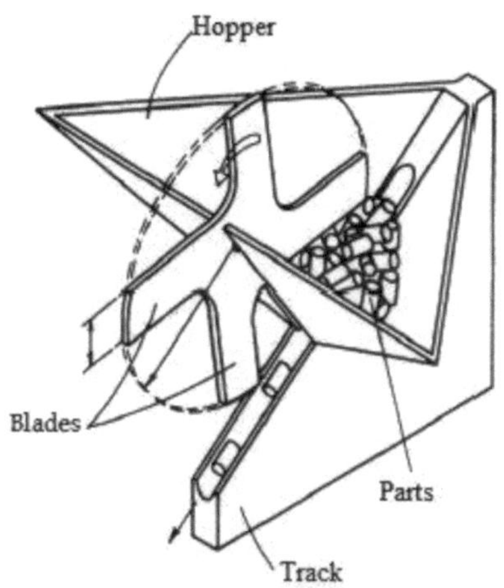	It consists of a hopper (as shown in the adjoining fig.) with a groove in its bottom to act as a track for the parts delivered by a multi-bladed wheel. The wheel, while rotating, disturbs the parts and causes the ones with right orientation to enter the track (the groove is designed to accommodate parts in one orientation only). Parts with the wrong orientation are pushed back into the mass of parts. Sufficient clearance is provided between the lowermost position of a blade of the wheel and the groove to avoid interference between the them. This feeder is best suited for feeding cylindrical parts.
II	Rotary-Disk Feeder 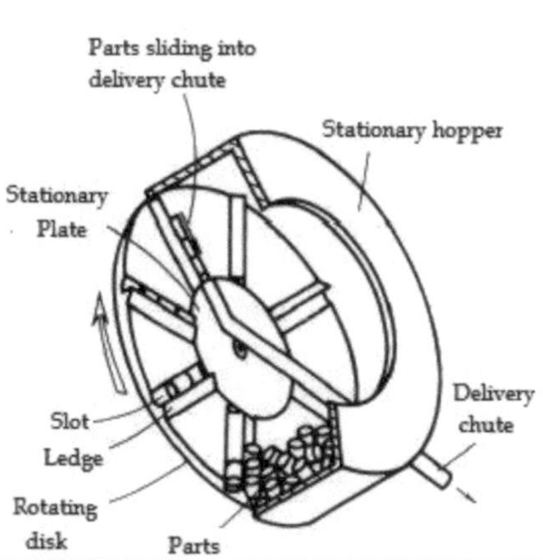	It consists of a disk with radial slots (the slot length should be sufficient to easily accommodate at least one part) abutted with ledges, with a stationary plate at the centre. The disk acts as a base to the stationary hopper, where the parts are populated. This set up is inclined at a steep angle with the horizontal. As the disk rotates, ledges disturb parts in the hopper, thereby trapping some of them in the slots. When a slot reaches its highest position, it is in line with the delivery chute and all the parts in that slot transfer into the chute. The stationary

		plate prevents the parts from falling out of slots till the time they are aligned with the chute. This feeder is best suited for cylindrical parts.
II	Centrifugal Hopper Feeder 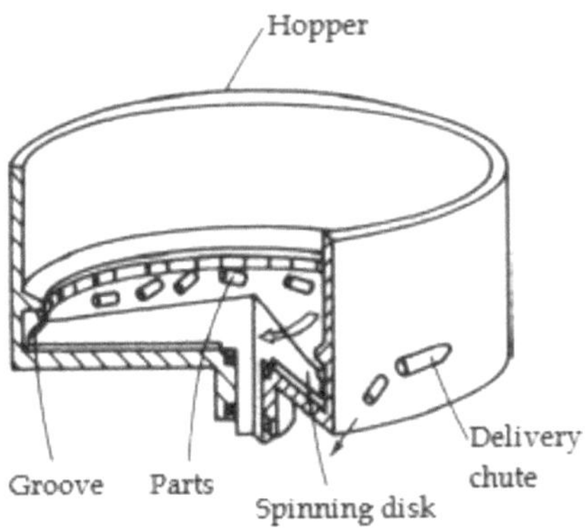	A shallow cylindrical hopper with a base (rotating at constant speed) along with a delivery chute tangential to the stationary wall of hopper constitute the feeder. The rotation (at constant speed) of base causes the parts to move around and those with the correct orientation fall into the delivery chute. Due to lack of orienting devices in the hopper, parts have to be taken off in the orientation that they naturally adopt in the hopper. This feeder is best suited for plain cylindrical parts.
II	Stationary Hook Hopper Feeder 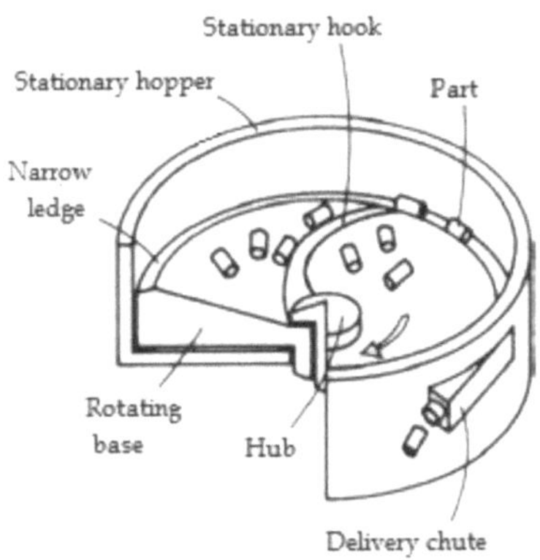	It consists of a hopper with a concave base (which can rotate), a stationary hook mounted on the base , a ledge at the periphery of the hopper, a deflector mounted on the hopper wall and a delivery chute as shown in the adjoining fig. As the concave base rotates slowly, parts move along the edge of the hook toward the ledge whereafter they are deflected by the deflector into the delivery chute. Due to its gentle feeding action, this feeder is suitable for feeding delicate parts at low speed.
II	External Gate Hopper Feeder 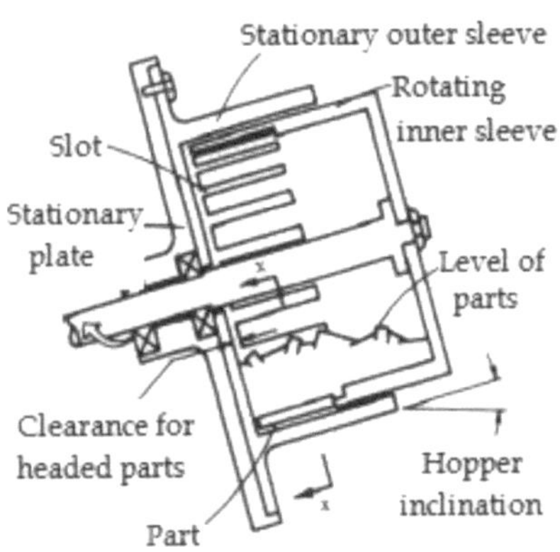	It consists of a slotted cylinder circumscribed by an unslotted sleeve consisting of an external gate. When the inner sleeve rotates, suitably oriented parts nest against the outer sleeve (which remains stationary) and undergo motion on account of the inner sleeve. As they pass over the external gate, they fall into the chute and are carried to the required location. This kind of feeder is suited for cylindrical parts.

<table>
<tr><td></td><td>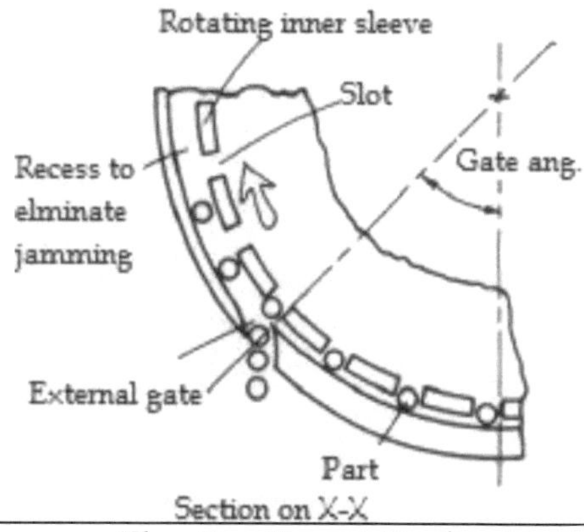
</td><td></td></tr>
<tr><td>III</td><td>Elevating Hopper Feeder
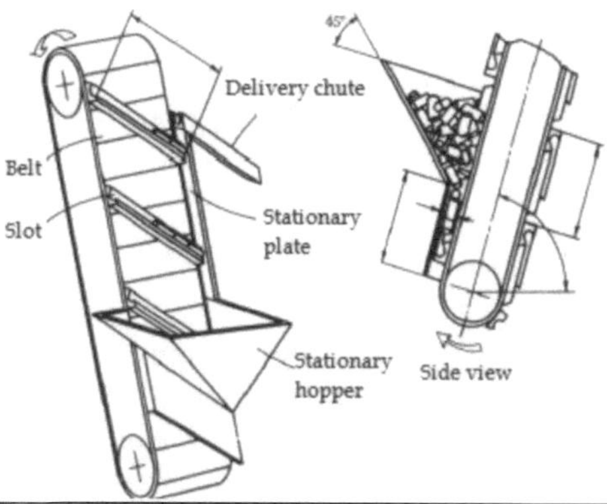
</td><td>It consists of a very large hopper with inclined sides. An endless conveyer belt fitted with selector ledges (designed to accept parts only in a specific orientation) moves as shown in the adjoining fig., thereby lifting up parts from the lowest point in the hopper to a given height where they fall off from the ledges into the delivery chute. Parts inside the hopper are pushed to the lowest point of hopper with the help of an agitating device fitted to the base. This feeder can be used for feeding any type of parts.</td></tr>
</table>

Table 1. Commonly used part feeders (adapted from Boothroyd, 2005)

For describing the optimization technique of a feeding system, a Reciprocating-Fork Hopper Feeder shall serve as an example throughout this chapter. As discussed in Table 1, it consists of a shallow cylindrical bowl that rotates about an axis inclined at a small angle (10°) to the vertical plane as it is placed on a rotating shaft of the gear box which in turn is connected to the motor shaft with the help of a belt and two pulleys (Singh et al., 2009). A two pronged fork reciprocates in the vertical plane above the rotating cylindrical bowl.

During the first stage of operation, the fork is dipped into the bowl, and due to the centrifugal force produced by rotation of the bowl, parts having the right orientation start climbing up the fork. The fork is then lifted up along with the parts nestled between its prongs. Finally, due to the effect of gravity, these parts slide down the delivery chute. For the parts being handled by this feeder to maintain their orientation throughout the operation, it is necessary to use only headed parts. The fork reciprocates by means of a pneumatic actuator.

1.3 Need for optimization

Part feeders, which singulate and orient the parts prior to packing and insertion, are critical components of automated assembly lines and one of the biggest obstacles to rapid development of assembly systems (Gudmundsson & Goldeberg, 2007). The design of part feeders is responsible for 30 % of the cost and 50% of the work cell failures (Boothroyd et. al 1982; Nevins and Whitney, 1978). Optimization of feeding systems thus becomes a critical issue in the development of assembly lines.

1.4 Analysis of feeders

The objective of this chapter is to outline methods to analyse part feeding systems on the basis of their functional specifications. The analysis of vibratory bowl feeders which require appropriate resonant frequencies to achieve optimal feeding conditions has been done by Ding & Dai, 2008. A systematic dynamic model of the bowl feeder along with the effect of various design (particularly assembly) parameters on the resonant frequencies (and thus throughput) has been developed.

But as far as the other common part feeders are concerned, the governing equations of the system are not available, and we attempt to infer the underlying structure by studying the behaviour of the system under certain conditions. Tests are designed by carefully choosing different input values, trying to design scenarios that will allow us to explore the functional relationship between the system inputs and outputs. The following sections are arranged to reflect the process of optimizing feeding systems loosely.

2. Parameter selection

2.1 Objective identification

Objective selection plays a very crucial role in overall process, by influencing the type of experiments to be run and analysis to be carried out. A poorly defined objective would lead to a poorly planned and poorly executed experiment which will not yield the information required for our purpose, and may lead to scraping of the whole effort resulting in wastage of precious time, effort, money and resources. In industrial environments such scenarios create a wrong impression about the effectiveness of the process amongst the workers, particularly if it is a first time implementation. Workers and operators would prefer to work according to traditional trial and error methods for obtaining the best performance rather than going for a systematic optimization procedure.

Depending on the system, the process of selection of objectives may vary in complexity. For a simple system, this task can be carried out by the operator based on his experience; for complex systems a panel of experts are required to identify the measures of output that require investigation. Once the objectives are defined they have to be organised in a hierarchy of their relative importance. This facilitates the selection of variables, procedures of conducting the experiment and techniques of measurement of response, so that no critically important information is lost due to lack of foresight.

The objective of an exercise in optimization involving part feeding devices can vary from maximization of feed rate, minimizing feed cycle time, minimizing cost to other relevant aspects. Depending upon the time and capital available, the amount of information required, equipment available and complexity of the system, such an exercise can have one or more than one objective. In case of feeders the output of interest if often the unrestricted feedrate, and the objective is to maximize it, or to obtain a particular pre-determined value of this output as per requirements.

In the study of reciprocating-fork hopper feeder for the example considered, the orientation of parts to be fed is not of primary importance as the fork is designed specifically to maintain the orientation of headed parts being fed by the system. Hence, the primary objective is to observe the behaviour of part flow rate by means of Regression modelling.

2.2 Process factor selection

Process factor selection depends on the stated goal of investigation. We have to identify primary factors affecting the targeted output. Sometimes these primary factors in turn depend on secondary factors.

Process variables include both *inputs* and *outputs* - i.e., *factors* and *responses*.

The factors that have a reasonable effect on the throughput of a part feeder in general are:

i. *Load Sensitivity* i.e. sensitivity in the feed rate on account of changes in the load i.e. amount of parts present in the feeder.
ii. *Part Specifications* like geometry (shape and size), weight etc
iii. *Feeder Design Specifications* like angle of inclination of hopper, depth and shapes of grooves, length of slots etc
iv. *Operating Conditions* like frequency of reciprocation, speed of rotation etc

Factors for a Reciprocating Fork Hopper Feeder are:

I. Load Sensitivity (Quantity of parts in feeder)-As has been explained earlier, the parts are circulated in the bowl of the feeder by a motor continuously during the operation of the feeder. The presence of an excessively large number of parts causes

i. overloading of the system,
ii. abrasion of feeder bowl,
iii. abrasion of parts due to friction against each other, and
iv. a large number of parts are forced onto the fork causing overloading of the fork. Overloading of the fork is a serious concern as it can disturb the designed parameter settings of the system. On the other hand if too few parts are present in the feeder bowl,
v. there will not be sufficient parts present to load the fork sufficiently,
vi. fewer parts will climb up the fork due to lack of back pressure from other parts, and
vii. scraping of the fork on bowl will cause creation of abraded tracks.

II. Part Specifications

1. **Part Size**

   The diameter of the headed part should be slightly greater than gap between the fork prongs for them to be picked up in the correct orientation. Also if the part is too long, it creates problem due to a slanted pickups at the edge of the fork, leading to fewer parts being picked up.
2. **Part Shape**

   Hexagonally shaped heads were found to have a greater probability of being picked up than circular shaped heads.
3. **Weight of Part**

   The parts should weigh between the limits prescribed during design of the fork. Too heavy parts may render the fork unable to lift the pieces or cause the fork to bend under repeated stress.

III. Feeder Design Specifications

1. **Angle of inclination of fork**

   If the angle of inclination of the fork is not appropriate the fork will pick up few pieces or simply not pick-up any pieces. The pieces picked up will not slide smoothly ahead, they may get jammed or be thrown, neither of which is a desirable condition.
2. **Angle of inclination of bowl**

The bowl has to be inclined at a small angle to its axis for facilitating a scenario where the parts come into contact with the fork such that when the fork is horizontal, the parts are at a lower level than it, so that they can climb up within the prongs of the fork.

IV. Operating Conditions

1. **Speed of rotation of bowl**

   This is a very important consideration as rotation of the bowl is what facilitates the pick-up of the pieces the fork by keeping them in constant contact with it. Too high a speed will result in the parts being thrown to the very edges of the bowl and outside the path of the fork. At high speeds even if the parts are in contact with the fork it is often unable to pick them up as the parts do not get any time to climb up the fork. At too low speeds, the rotation of the bowl ceases to have an effect on the process.

2. **Frequency of reciprocation of fork**

   The reciprocation of the fork results in picking up parts and delivering it to the next stage. Too fast a frequency of reciprocation doesn't allow the parts to climb up the prongs of the fork. At the most the fork picks up a few pieces and due to lack of time for the parts to slide down, throws them randomly. This is an extremely dangerous situation, as the flying parts can hit labour and machinery causing injuries and damage. On the other hand too slow a frequency of reciprocation results in too few pieces being picked up leading to wastage of the potential of the feeder, slowing down operations etc.

3. **Ratio of time during which fork is lifted up to pointed down**

   The time period of each cycle of the reciprocation of the fork consists of three times:

   i. Time when the fork is descending: Too fast descent will lead to the fork crashing into the rotating bowl causing damage to the fork, bowl and parts in contact. It may modify the angle of the fork. It can also cause parts to fly out of the bowl which is an extremely dangerous situation.

   ii. Time when the fork is horizontal: This time allows for the parts to climb up and nestle securely between the prongs of the fork. Too small a time will not allow sufficient parts to climb up the fork. If too large a time is allowed the number of parts that can be picked up will reach saturation and the fork will stay in position not picking up additional parts, thus holding up the operation.

   iii. Time when the fork is ascending: This time allows for the parts to slide down smoothly the delivery chute. It is very important that the parts maintain contact with the chute until they can fall down to the bin under gravity. If they break contact too soon they will fly off.

Table 2 describes some of the process factors (when the objective is maximization of feed rate) for rest of the common part feeders mentioned in Table 1. It is imperative to note that factor I (Load sensitivity) is there for all the feeders listed in the table and has thus not been mentioned due to space constraints. Preferred part shapes refer to the shapes which will give the maximum feed rate for that feeder. The factors mentioned here are collected from various results reported in literature (Boothroyd, 2005). However, lack of standard governing models mean that these cannot be considered as absolute and actual experimentation must be carried out to find out the effect of these process factors on the objective function.

Feeder	Process Factors		
	II	III	IV
Reciprocating-Tube Hopper Feeder	1. Part Shape (pref. cylindrical) 2. Part Size 3. Part Weight	1. Hopper Wall Angle (should be kept large to avoid jamming and less than a specific value to avoid a situation wherein the parts don't fall down into the tube at all) 2. Inside silhouette of delivery tube (should be designed to accept parts with right orientation, one at a time)	1. Max height of delivery tube in its cycle (just above the max level of parts is optimum) 2. Frequency of reciprocation of tube/hopper 3. Time spent in going upwards and downwards
Reciprocating-Centerboard Hopper Feeder	1. Part Shape (pref. cylindrical) 2. Part Size (esp. length) 3. Part Weight	1. Track Inclination (θ_m)(see Table 1) (increase in θ_m results in increased time for blade to complete upward motion, but lesser time for parts to slide into the chute) 2. Length of track (l) (affects max blade frequency and mean feed rate) 3. Swing radius of blade (r_b)	1. Acceleration and deceleration of blade (high deceleration value during the upward stroke results in parts leaving the track and being thrown off the feeder) 2. Frequency of reciprocation of blade 3. Overall cycle time(Time for going upward+ downward+ time for which the blade dwells at the top of its stroke)
Rotary-Centerboard Hopper Feeder	1. Part Shape (U-shaped preferably) 2. Part Size 3. Part Weight	1. Profile of blades 2. Track Inclination 3. Length of track	1. Peripheral velocity of wheel 2.Overall cycle time (Indexing + dwell time)
Bladed-Wheel Hopper Feeder	1. Part Shape (pref. cylindrical) 2. Part Size 3. Part Weight	1. Dimensions of groove (should accept parts in only a specific orientation) 2. Track Inclination 3. Length of track	1. Linear velocity of blade tip
Rotary-Disk Feeder	1. Part Shape (pref. cylindrical) 2. Part Size (esp. length) 3. Part Weight	1. Length of slots (large length leads to greater feed rate) 2. Angle of inclination of base 3. Delivery chute angle (greater angle leads to greater feed rate)	1. Rotational speed of disk 2. Overall cycle time (Indexing + dwell time)
Centrifugal Hopper Feeder	1. Part Shape (pref. plain cylindrical) 2. Part Size(esp. length) 3. Part Weight	1. Hopper diameter	1. Peripheral velocity of disk
Stationary-Hook Hopper Feeder	1. Part Shape 2. Part Size 3. Part Weight	1. Hopper diameter 2. Hook shape 3. Radius of hub	1. Rotational frequency of disk

External Gate Hopper Feeder	1. Part Shape (pref. cylindrical or rivets) 2. Part size (diameter for cylindrical; shank diameter for rivets) 3. Part weight	1. Gate Angle(θ_g) 2. Hopper inclination (λ) 3. Gap between cylinder and sleeve 4. Spacing between slots	1. Peripheral velocity of inner cylinder
Elevating Hopper Feeder	1. Part Shape 2. Part Size (esp. length) 3. Part Weight	1. Length of one slot 2. Distance between two slots	1. Conveyer belt speed

Table 2. Process factors affecting the feed rate of various kinds of part feeders

After the critical factors have been identified, their operational values have to be determined. In case of a Reciprocating Fork Hopper Feeder, the values of parameters like angle of inclination of fork, weight of parts, shape of parts, part size etc are determined and held constant. Three critical factors are taken as variable:

a. Speed of rotation of cylindrical bowl
b. Part population
c. Frequency of reciprocation of fork.

2.3 Factor level selection

Complete coverage of the entire region of operation of the system is often not possible; we have to limit our experiments to a region that is determined to be most relevant to our purpose on the basis of past experience or other preliminary analysis.

Once all factors affecting the output significantly are listed, constraints affecting them are identified. This can be done based on previous experience i.e. empirical data or based on process capability. This section will explain the process of determination of factor levels for each factor according to constraints. It is not always feasible to consider the whole set of viable values of a factor. In some cases, extreme values will give runs that are not feasible; in other cases, extreme ranges might move one out of a smooth area of the response surface into some jagged region, or close to an asymptote.

In our chosen example we have isolated the three variables of interest. For each of these variables we will identify a range of operation where conducting the investigation will be most fruitful based on previous work. This is summarised in Table 3 as well as explained in the following paragraph. To facilitate the application of transformations if required later, the high and low stages are set to -1 and 1 level respectively.

1. **Speed of Rotation of Cylindrical Bowl**
   The rotation of bowl is provided by an adjustable speed motor at the base of the feeder and its speed is measured in rotations per minute (rpm). As it has already been discussed too high or too low speeds are not conducive to obtaining a good output, the range over which speed of rotation is varied is 500 rpm to 1050 rpm.
2. **Part Population**
   The number of parts being circulated in the rotating bowl is varied from 300 to 700 parts. For less than 300 initial parts the number of parts being picked up during operation of the feeder is negligible. Whereas the number of parts transferred stagnates even if the initial part population is increased beyond 700.

3. **Number of strokes per minute**
   The number of strokes that the fork makes per minute is varied from 4 to 8. These levels are selected based on previous experience that at fewer than 4 strokes too few parts are picked up, while at more than 8 strokes the parts fly off due to too less time of contact with the fork.

Process Parameters	Code	Low Level (-1)	High Level (1)
Speed of Rotation of Cylindrical Bowl (rpm)	A	500	1050
Part Population	B	300	700
Number of Strokes per Minute	C	4	8

Table 3. Process Parameters

3. Process characterization – experimental designs

Obtaining the operating conditions at which the system output is optimal would be straightforward if theoretical expressions or model were available. In absence of such expressions, a model has to be created and the information employed to determine the performance of the device is to be obtained empirically. This can be done efficiently using Design of Experiments.

3.1 Design of experiments

The statistical Design of Experiments (*DOE*) is an efficient procedure for planning experiments so that the data obtained can be analysed to yield valid and objective conclusions (NIST, 2010). Once the objectives have been defined, variables selected and their ranges chosen, a set of conditions need to be defined that bring them together. The manipulation of these conditions in order to observe the response of the system is called an experiment.

The individual experiments to be performed can be organised in such a manner that the information obtained from each supplements the other. Thus, well planned experimental designs maximize the amount of information that can be obtained for a given amount of experimental effort. More importantly, the validity of experimental analysis is affected by the organisation and execution of the experiments. The significance, validity and confidence level of results obtained from a series of experiments can be improved drastically by incorporating elements like randomisation and replication into it. Randomisation is carried out to eliminate any existing bias or inertia in the experimental set-up. Randomisation of experiments can be done in two common ways:

a. **Completely Randomised Designs**
   As the name suggests, the sequence in which a set of experiments is performed is completely random. One of the ways in which this can be accomplished is by generating random numbers using a computer and assigning this random sequence of numbers to the run order and then performing the experiments in accordance with the new and modified arrangement.
b. **Randomised Block Design**
   In this method the experimental subjects are first divided into homogenous groups called blocks, which are then randomly assigned for experimental treatments so that all treatment levels appear in a block. (NIST,2010)

3.2 Selecting experimental design

The choice of experimental design has to be in accordance with the parameter selection i.e. process objectives, number of process variables and their levels. The possibilities for modelling data are also related to the experimental design chosen. This interconnection between all levels of the process demands that special attention is paid the step of selecting the experimental design. Depending on the requirements of the experimenter, a number of designs have been developed, like:

Completely Randomised Designs

These deigns are generally used when the experiment is small and the experimental units are roughly similar. A completely randomized design is the simplest type of randomization scheme in that treatments are assigned to units completely by chance. In addition, units should be run in random order throughout the experiment (Cochran & Cox, 1957).In practice, the randomization is typically performed by using a computer program or random number tables. Statistical analysis is carried out by one-way ANOVA and is relatively simple. It is quite flexible and missing information is not very problematic due to the large number of degrees of freedom assigned to error.

Randomised Block Designs

Randomised block designs are constructed when in addition to the factor whose effect we wish to study, there are other noise factors affecting the measured result. In order to reduce the effect of undesirable noise in the error term the method of blocking is employed, in which the experimental units are divided in relatively homogeneous sub-groups. In a block the noise factor, called blocking factor, is held constant at a particular level and the treatments are assigned randomly to the units in the block. This is done for all levels of the noise factor so that its effect maybe estimated and subsequently eliminated.

Full Factorial Designs

The full factorial design of experiments consists of the exhaustive list of treatments obtained by combination of all levels of all the factors with each other. The number of experimental runs id given by:

$$\text{Number of Runs} = n^k$$

Where,

n = number of levels of a factor

k = numbers of factors

For full factorial designs at large number of levels or for large number of factors, the number of runs required becomes prohibitive either in terms of cost or time or resources required. For this reason many other designs have been proposed for handling such cases, e.g. Plackett-Burman designs for 5 or more factors at 2 levels.

Fractional Factorial Designs

It has been observed that there tends to be a redundancy in Full Factorial Designs in terms of an excess number of interactions that are estimated (Box et al., 1978). Though full factorial designs can provide an exhaustive amount of information, in practical cases estimation of higher order interactions are rarely required. This problem can be addressed by selecting an appropriate fraction of experiments from the full factorial design. In these arrangements certain properties are employed to select a $(1/n)^p$ fraction of the complete design, and the reduced number of runs is given by:

$$\text{Number of Runs} = n^{(k-p)}$$

Where,
n = Number of levels
k = Number of factors
p = fraction to be run
The chosen fraction should be balanced, i.e. each factor occurs an equal number of times at all the levels, and the design is orthogonal. An experimental design is orthogonal if the effects of any factor balance out (sum to zero) across the effects of the other factors [4].
The interactions that have been confounded are lost, but the main effects and other interactions (not confounded) can be more precisely estimated due to reduced block size. This can be expressed in terms of the Resolution of the design:

a. **Resolution III Design**
   Easy to construct but main effects are aliased with 2-factor interactions.
b. **Resolution IV Design**
   Main effects are aliased with 3-factor interactions and 2-factor interactions are confounded with other 2-factor interactions.
c. **Resolution V Design**
   Main effects are aliased with 4-factor interactions and 2-factor interactions are aliased with 3-factor interactions.

For the example considered, a 2 level factorial design is selected to investigate the behaviour of a reciprocating fork-hopper feeder for the three factors selected. This gives rise to a 2^3 run design, i.e. the total experiment consists of 8 runs. Taking 4 replicates of each experiment we finally have 32 randomised runs that required to be conducted.

4. Analysis of data

4.1 Preliminary analysis

Once an experimental design has been selected that is commensurate with the requirements of the project, the experiments are performed under stipulated conditions and data is collected. The data obtained by performing the experiments is checked for suitability of further analysis. The best way to examine the data is by means of various plots such as Normal and Half-Normal Plots, Pareto charts, FDS graphs, etc. The right graphs and plots of a dataset can uncover anomalies or provide insights that go beyond what most quantitative techniques are capable of discovering (NIST, 2010). This analysis is further supplemented by evaluating statistics to check for outliers and errors. Some common evaluators are:

1. Response distributions

a. **Histograms**
   A histogram essentially shows the frequency with which a data point, which falls within a certain range, occurs. The ranges into which the data points are divided are called bins, in particular for histograms or classes more generally. It can also be called as the plot of frequency of response vs. response. These graphs provide information regarding presence of outliers, the range of the data, centre point of data and skewness of data. A histogram is shown in Figure 1(a)
b. **Box Plots**
   Boxplots, also called box and whisker plots, are a quick graphic approach for examining data sets. This plot provides an excellent visual summary of the important aspects of a distribution. The box stretches from the lower hinge which is drawn at the 25th

percentile to the upper hinge which is drawn at the 75th percentile and therefore contains the middle half of the scores in the distribution (Lane, 2001). The median is shown as a line across the box. Therefore 1/4 of the distribution is between this line and the top of the box and 1/4 of the distribution is between this line and the bottom of the box. Two lines, called whiskers, extend from the front and back of the box. The front whisker goes from Q1 to the smallest non-outlier in the data set, and the back whisker goes from Q3 to the largest non-outlier. A box plot is shown in Figure 1(b).

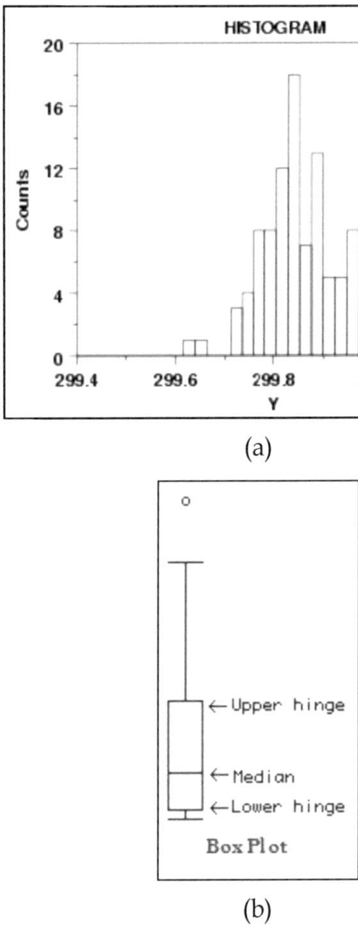

Fig. 1. (a) Histogram; (b) Box Plot

2. Typical DOE plots

a. **Pareto plots**

In terms of quality improvement, the Pareto effect states that 80% of problems usually stem from 20% of the causes. Due to this Pareto charts are extremely helpful when the goal of the investigation is to screen all possible variables affecting a system output, and to select and isolate parameters that are most significant. This is achieved by ranking the main effects and variable interactions in their descending order of contribution to the output. The chart consists of bar graph showing parameters in a prioritized order, and the bars are placed on the graph in rank order that is the bar at the left has the highest impact on output, so it can be determined which variables should be selected for further study. The purpose of the Pareto Chart is to distinguish the vital few from the trivial many, therefore, it is desirable that only a few variables are present on the left side of the Pareto Chart that account for most of the contribution to output. Then a second stage of investigations can be embarked upon, dealing with fewer parameters and thus smaller and more economical experiments. In figure 2, it can be clearly seen that feedrate has the greatest impact on the throughput of the feeder, whereas the interaction effect of speed and population has almost negligible impact.

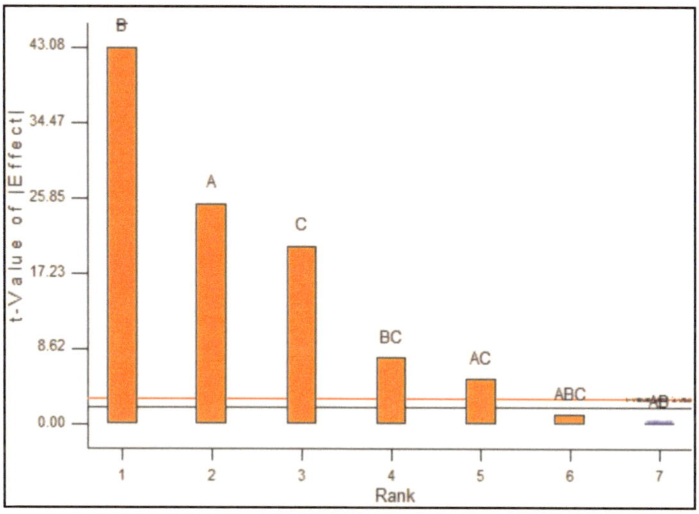

Fig. 2. Pareto plot

b. **Normal or half-normal plots of the effects**
The half-normal probability plot is a graphical tool that uses the ordered estimated effects to help assess which factors are important and which are unimportant. Quantitatively, the estimated effect of a given main effect or interaction and its rank relative to other main effects and interactions is given via least squares estimation, Having such estimates in hand, one could then construct a list of the main effects and interactions ordered by the effect magnitude. Figure 3 (a) shows a Normal Plot and Figure 3 (b) shows a Half-Normal Plot.

3. FDS Graph
When the goal is optimization, the emphasis is on producing a fitted surface as precisely as possible. How precisely the surface can be drawn is a function of the standard error (SE) of the predicted mean response—the smaller the SE the better. Figure 1 shows a contour plot of standard error for two of the factors in the. As can be seen from the graph the predictions around the perimeter of the design space exhibit higher standard errors than near the centre. To circumvent this, the design should be centred at the most likely point for the potential optimum. The fraction of design space plot is shown in Figure 4. It displays the area or volume of the design space having a mean standard error less than or equal to a specified value. The ratio of this volume to the total volume is the fraction of design space (Whitcomb, 2011).

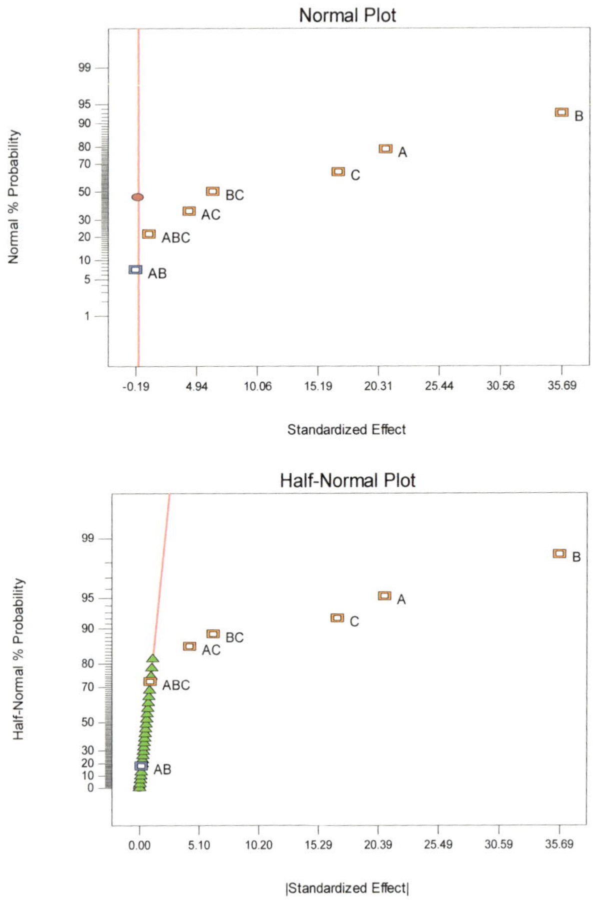

Fig. 3. (a) Normal Plot; (b) Half-Normal Plot

4.2 Theoretical model creation

The goal of this experimental investigation is to formulate an appropriate empirical model between the response, say y, and independent variables, say x1, x2, x3etc, and predict the behaviour of the system with sufficient accuracy. This can be mathematically expressed as,

$$y = f(x1, x2, x3 \ldots) + e$$

In this case the function f, which defines a relationship between the response feedrate and independent variables like part population, is not known and needs to be approximated. Usually first or second order model are sufficient to approximate the function f. The term e represents variability that is not accounted for by the function f, and is assumed to have a normal distribution with mean zero and constant variance. Often the step of model creation is preceded by coding the variables to make them dimensionless and such that they have mean zero and the same standard deviation.

The model used to fit the data should be consistent with the goal of the experiment, to the extent that even the experimental design and data collection methodology are chosen such that maximum information can be extracted from the observed data for model creation.

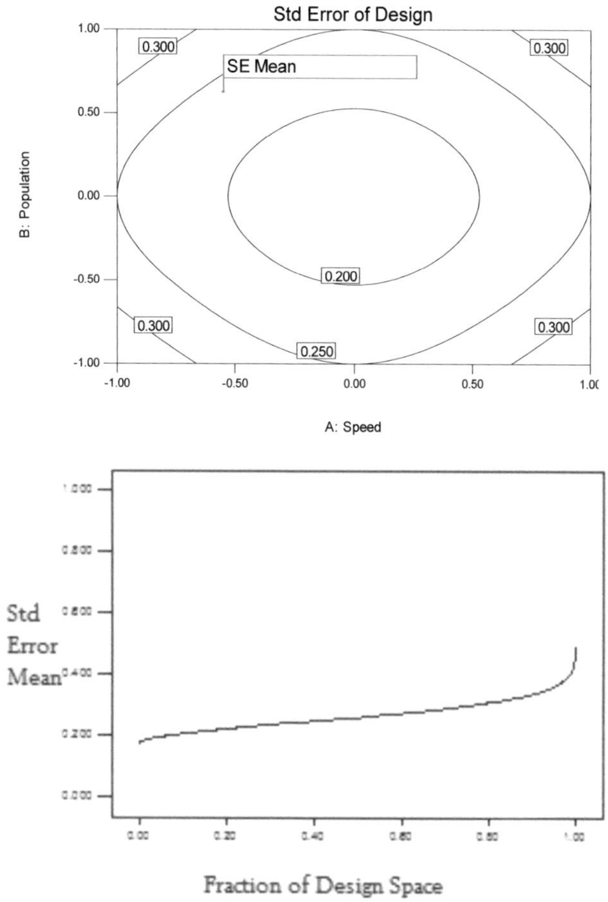

Fig. 4. (a) Contour Plot; (b) FDS Graph

The first order model is appropriate when the independent variables are varied over a relatively small region, such that the curvature of the selected region in response space is negligible. A first order model where only the main effects of the variables are deemed as significant is given by:

$$y = b0 + b1x1 + b2x2 + \dots + e$$

When the interactions of the factors amongst each other also play a significant role in the response along with the main effects, then the model is given by:

$$y = b0 + b1x1 + b2x2 + b12x1x2 + \dots + e$$

However if the curvature of the solution space is significant, then first order models are inadequate to predict the response of the system. In such cases, higher order models are used, typically second or third - order models like:

$$y = b0 + \sum bjxj + \sum bjjxj^2 + \sum\sum bijxij + e$$

$$y = b0 + \sum bjxj + \sum bjjxj^2 + \sum bjjjxj^3 + \sum\sum bijxij + e$$

Regression is a collection of statistical techniques for empirical model building (Karley et al., 2004). The independent variables are also called predictor variables or regressors, the coefficients as regression coefficients and their values are approximated using the various regression methods. A number of equations estimating the results can be formulated, larger the number of regressor variables, larger the number of possible equations. Undoubtedly evaluating all possible solutions can be computationally exacting, thus methods have been developed for evaluating only a small number of subset regression models which are built by either adding or deleting regressors one at a time. Some statistics that can help in selecting the best possible model out of the ones generated are discussed later in the section.

F-statistic is a value resulting from a standard statistical test used in regression analysis to determine if the variances between the means of two populations are significantly different. The t-statistic is the estimated coefficient divided by its own standard error. Thus, it is used to test the hypothesis that the true value of the coefficient is non-zero, in order to confirm that the inclusion of an independent in the model is significant.

The regression equation fitted to the collected data is given below and factor estimates are summarized in Table 4.

Final Equation in Terms of Coded Factors

$$
\begin{aligned}
R1 = \; & +56.78 \\
& +10.47 * A \\
& +17.84 * B \\
& +8.47 * C \\
& - 0.094 * A * B \\
& +2.16 * A * C \\
& +3.16 * B * C \\
& +0.47 * A * B * C
\end{aligned}
$$

Final Equation in Terms of Actual Factors:

$$
\begin{aligned}
R1 = \; & +56.78125 & \\
& +10.46875 & * \text{Speed} \\
& +17.84375 & * \text{Population} \\
& +8.46875 & * \text{Strokes} \\
& -0.093750 & * \text{Speed} * \text{Population} \\
& +2.15625 & * \text{Speed} * \text{Strokes} \\
& +3.15625 & * \text{Population} * \text{Strokes} \\
& +0.46875 & * \text{Speed} * \text{Population} * \text{Strokes}
\end{aligned}
$$

The coefficient of determination, R-squared, is a measure of the fraction of the total squared error that is explained by the model. By definition the value of R^2 varies between zero and one and the closer it is to one, the better. However, a large value of R^2 does not necessarily imply that the regression model is good one. Adding a variable to the model will always increase R^2, regardless of whether the additional variable is statistically significant or not. Thus it is possible for models that have large values of R^2 to yield poor predictions of new observations or estimates of the mean response. To avoid this confusion, an additional statistic called the Adjusted R-squared statistic is needed; its value decreases if unnecessary terms are added. These two statistics can, when used together, imply the existence of extraneous terms in the computed model which is indicated by a large difference, usually of more than 0.20, between the values of R^2 and Adj-R^2. The amount by which the output predicted by the model differs from the actual output is called the residual. Predicted Residual Error Sum of Squares (PRESS) is a measure of how the model fits each point in the design. It is used to calculate predicted R^2. Here, the "Pred R-Squared" of 0.9859 is in reasonable agreement with the Adj R-Squaredof 0.9897. Adeq Precision measures the signal to noise ratio. A ratio greater than 4 is desirable. These statistics are used to prevent over fitting of model. A summary of the statistics is given in Table 3.

Standard Deviation	2.34	R-Squared	0.9921
Mean	56.78	Adjusted R-Squared	0.9897
Coefficient of variation	4.13	Predicted R-Squared	0.9859
PRESS	234.22	Adequate Precision	63.594

Table 3. Regression Statistics

Inflated variances are quite detrimental to regression because some variables add very little or even no new and independent information to the model (Belsley, Kuh & Welsch, 1980). Multicollinearity causes variances to be high. A way for detecting multicollinearity is using the Variance Inflation Factor (VIF).

Coefficient	Factor Estimate	Degrees of freedom	Standard Error	Low	High	VIF
Intercept	56.78	1	0.41	55.93	57.64	1.00
A-speed	10.47	1	0.41	9.61	11.32	1.00
B-population	17.84	1	0.41	16.99	18.70	1.00
C-strokes	8.47	1	0.41	7.61	9.32	1.00
AB	-0.094	1	0.41	-0.95	0.76	1.00
AC	2.16	1	0.41	1.30	3.01	1.00
BC	3.16	1	0.41	2.30	4.01	1.00
ABC	0.47	1	0.41	-0.39	1.32	1.00

Table 4. Regression Statistics

A likelihood ratio test is a statistical test used to compare the fit of two models, the null model as compared to the selected model. The likelihood ratio numerically indicates the possibility of the selected model being the correct one to represent the data set as compared to the null model. Thus it can be used to reject the null hypothesis.
The test statistic, D is given by:

$$D = 2*\text{Ln}(\text{likelihood of null model}/\text{likelihood of selected model})$$

By default a model having more parameters included in the model will have a greater log-likelihood. Whether the fit is significantly better should be determined further before making a decision.

As has been mentioned earlier, for a set of experimentally observed data a number of regression models can be formulated. One approach through which the best model among all the possible models can be selected is through the evaluation of statistics like the Akaike Information Criterion (AIC) and Schwarz Criterion.

The Akaike information criterion is a measure of the relative goodness of fit of a statistical model (Akaike, 1974).

$$AIC = -2Ln(L) + 2k$$

Where,

k = number of parameters in the model,

L = maximized value of likelihood function of the model.

In case of a least-squares regression model,

$$AIC = n*\log(RSS/n) + 2k$$

Where, RSS is the estimated residual of the fitted model.

The first term can be called the bias term and the second the variance term. From the formula it is clear that the value of the statistic decreases with goodness of fit, and increases with increasing number of parameters in the model. This takes care of the concern of increasing the goodness of fit by increasing the number of terms included in the model by assigning them opposing roles. All possible models created using the same data set, are ranked according to their AIC values in ascending order and the best model is one with the least value of AIC.

However, the scope of AIC should not be overestimated; it does not test for model validity, it only compares the models created, even if all of them are unsatisfactory in predicting the output of the system.

Similarly Schwarz Information Criterion or Bayesian Information criterion also ranks the models created on the basis of their goodness of fit and the number of parameters included in the regression model, by assigning a penalty on increasing the number of parameters. The penalty of addition of more terms to the model is greater for Schwarz Criterion. It is given by:

$$SC = kLn(n) + 2L$$

Where, n is the sample size. The model having the smallest value of this statistic is recommended.

There may exist, correlation between different values of the same variable measured at different times; this is called autocorrelation and is measured by the autocorrelation coefficient. It is one of the requirements of a good regression model that the error deviations remain uncorrelated. The Durbin-Watson test is a statistic that indicates the likelihood that the error deviation values for the regression are correlated. This statistic also tests for the independence assumption.Autocorrelated deviations are indicators a host of possible problems in the regression model ranging from the degree of the regression model, i.e. a linear equation fitted to quadratic data to non-minimum variance of the regression coefficients and underestimation of standard error. If the true standard error is miscalculated it results in incorrect computation of t-values and confidence intervals for the analysis. The value of the Durbin Watson statistic, d always lies between 0 and 4. The ideal value of d is 2, which implies no autocorrelation. If value of d is substantially less than 2, it

indicates positive correlation and if d is greater than two, it indicates negative correlation. Values of d smaller than one indicate that particularly alarming as successive error terms are close in value to one another.
In Table5, the best regression model amongst the ones formulated for a different system is shown for which a number of the statistics discussed have been evaluated.

Variable	Coefficient	Std. Error	t-Statistic	Prob(t)
Constant	-55.72	17.83	-3.12	0.0204
Variable 1	2.52	0.30	8.31	0.0002
Variable 2	0.35	0.06	5.83	0.0011

R-squared	0.9450	Mean dependent variance	108.11
Adj R-squared	0.9267	S.D. Dependent variance	27.39
S.E. of regression	7.42	AIC	7.11
Sum of squared residuals	330.06	Schwarz criterion	7.17
Log likelihood	-28.98	F-statistic	51.56
Dubrin_watson statistic	1.53	Prob(F)	0.0001

Table 5. Regression table

An Analysis of Variance or ANOVA table provides statistics about the overall significance of the model being fitted. Table 6 displays the results of ANOVA for the system under observation. Here, Degrees of Freedom stands for the number of the independent variables of the dataset and is obtained by subtracting the number of the parameters from the number of elements in the dataset. DOF plays a very important role in the calculation and comparison of variation. The Sum of Squares and the Total Sum of Squares have different degrees of freedom and cannot be compared directly. So they are averaged such that variation can be compared for each degree of freedom.
For ANOVA, if N=total number of data points and M=number of factor levels, then:

- Df (Factor) or corresponding to between factor variance = M-1
- Df(error) or corresponding to residual = N-M
- Df total = N-1 (which is summation of above two pieces)

The F value statistic tests the overall significance of the regression model. It compares curvature variance with residual variance. It tests the null hypothesis which states that the regression coefficients are equal to zero. The value of this statistic can range from zero to an arbitrarily large number .If the variances are close to the same, the ratio will be close to one and it is less likely that curvature is significant. The Model F-value of 428.49 implies the model is significant. There is only a 0.01% chance that a F-Value this large for the model could occur due to noise. Prob(F) values give information about the probability of seeing the observed F value if the null hypothesis is true. Small probability values call for rejection of the null hypothesis that curvature is not significant. For example, if Prob(F) has a value of 0.01000 then there is 1 chance in 100 that all of the regression parameters are zero. This low a value would imply that at least some of the regression parameters are nonzero and that the regression equation does have some validity in fitting the.
The user-specified probability used to obtain the cut-off value from the F distribution is called the significance level of the test. The significance level for most statistical tests is denoted by α. The most commonly used value for the significance level is α=0.05, which means that the hypothesis of an adequate model will only be rejected in 5% of tests for

which the model really is adequate. Values of Prob(F) less than 0.0500 indicate model terms are significant. In this case A, B, C, AC, BC are significant model terms.

Source	Sum of Squares	DF	Mean Square	F-Value	p-value
Model	16465.72	7	2352.25	428.49	<0.0001
A-Speed	3507.03	1	3507.03	638.85	<0.0001
B-Population	10188.78	1	10188.78	1856.02	<0.0001
C-Strokes	2295.03	1	2295.03	418.07	<0.0001
AB	0.28	1	0.28	0.051	0.8228
AC	148.78	1	148.78	27.10	<0.0001
BC	318.78	1	318.78	58.07	<0.0001
ABC	7.03	1	7.03	1.28	0.2689
Pure Error	131.75	24	5.49		
Corr Total	16597.47	31			

Table 6. ANOVA for selected factorial model -Analysis of variance table [Partial sum of squares - Type III]

4.2.1 Testing of model assumptions using experimental data

Once a good regression model has been created for the system and its adequacy tested, it has to be ensured that the model does not violate any of the assumptions of regression. To examine how well the model selected conforms to the regression assumptions and how soundly the experimental data fits the model selected; there exist a variety of graphical and numerical indicators. However, carrying out any one of such tests is not sufficient to reach a conclusion regarding the effectiveness of the model. No statistic or test is competent in itself to diagnose all the potential problems that may be associated with a certain model. For the purpose of testing model assumptions, graphical methods are preferred as deviations and errors are easier to spot in visual representations. The task of assessing model assumptions leans heavily on the use of residuals. As already mentioned in the previous section residuals are the difference between the observation and the fitted value. Studentized residuals, i.e., residuals divided by their standard errors are rather popular for this purpose as scaled residuals are easier to handle and provide more information.

The assumptions are:

- Normality –the data distribution should lie along a symmetrical bell shaped curve,
- Homogeneity of variance or homoscedasticity - error terms should have constant variance, and
- Independence - the errors associated with one observation are not correlated with the errors of other observations.

Additionally, the influence of observations on the regression coefficients needs to be examined. In some cases, one or more individual observations exert undue influence on the coefficients, and in case, the removal of such an observation is attempted it significantly affects the estimates of coefficients.

It has already been examined how well the experimental data fits the model via some numerical statistics like R-squared and Adjusted R-squared. The plot of predicted response versus actual responses performs the same function, albeit graphically and also helps to detect the points where the model becomes inadequate to predict the response of the system. This is the simplest graph which shows that the selected model is capable of predicting the response satisfactorily within the range of data set as shown in the Figure 5 (a).

To draw evidence for violations of the mean equal to zero and the homoscedastic assumptions, the residuals are plotted in many different ways. As a general rule, if the assumptions being tested are true, the observations in a plot of residuals against any independent variable should have a constant spread.

The plot of Residuals versus Predictions tests the assumption of constant variance, it is shown in Figure 5 (b). The plot should be a random scatter. If the residuals variance is around zero, it implies that the assumption of homoscedasticity is not violated. If there is a high concentration of residuals above zero or below zero, the variance is not constant and thus a systematic error exists. Expanding variance indicates the need for a transformation.

The linearity of the regression mean can be examined visually by plots of the residuals against the predicted values. A statistical test for linearity can be constructed by adding powers of fitted values to the regression model, and then testing the hypothesis of linearity by testing the hypothesis that the added parameters have values equal to zero. This is known as the RESET test. The constancy of the variance of the dependent variable (error variance) can be examined from plots of the residuals against any of the independent variables, or against the predicted values.

Random, patternless residuals imply independent errors. Even if the residuals are even distributed around zero and the assumption of constant variance of residuals is satisfied, the regression model is still questionable when there is a pattern in the residuals.

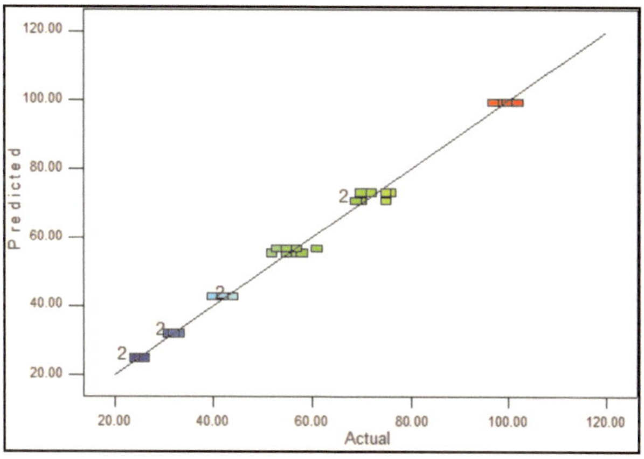

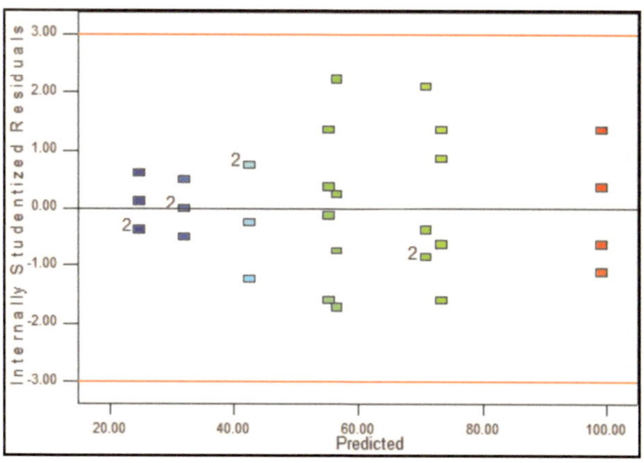

Fig. 5. (a) Actual Response vs. Predicted Response; (b) Residual vs. Predicted Response Plot

Residuals vs. Run: This is a plot of the residuals versus the experimental run order and is shown in Figure 6 (a). It checks for lurking variables that may have influenced the response during the experiment. The plot should show a random scatter. Trends indicate a time-related variable lurking in the background.

The normal probability plot indicates whether the residuals follow a normal distribution, in which case the points will follow a straight line. Definite patterns may indicate the need for application of transformations. A Normal Probability plot is given in Figure 6 (b).

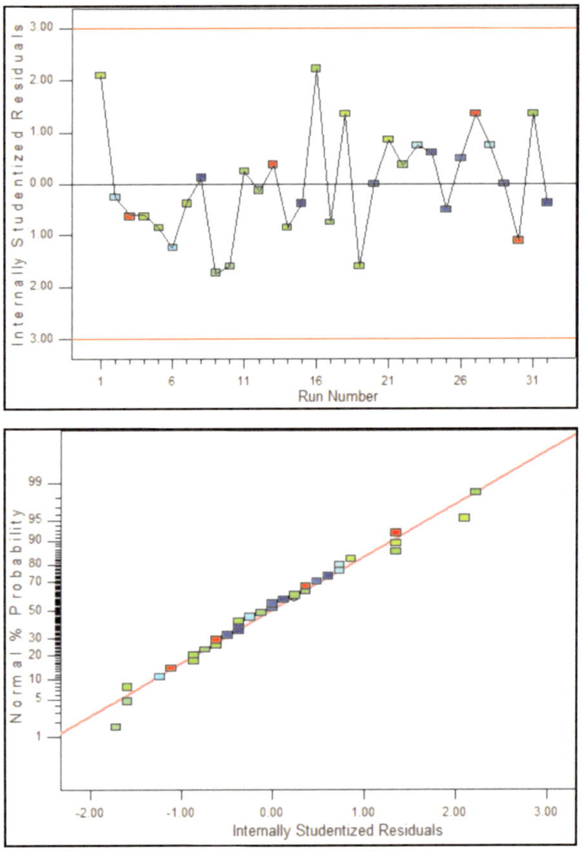

Fig. 6. (a) Residuals vs. Run Plot; (b) Normal Probability Plot

Leverage is a measure of how far an independent variable deviates from its mean. It is the potential for a design point to influence the fit of the model coefficients, based on its position in the design space. An observation with an extreme value on a predictor variable is called a point with high leverage. These high leverage points can have an unusually large effect on the estimate of regression coefficients. Leverage of a point can vary from zero to one and leverages near one should be avoided. To reduce leverage runs should be replicated as the maximum leverage an experiment can have is $1/k$, where k is the number of times the experiment is replicated. A run with leverage greater than 2 times the average is generally regarded as having high leverage. Figure 7(a) shows the leverages for the experiment.

Cook's Distance is a measure of the influence of individual observations on the regression coefficients and hence tells about how much the estimate of regression coefficients changes if that observation is not considered. Observations having high leverage values and large studentized residuals typically have large Cook's Distance. Large values can also be caused by recording errors or an incorrect model. Figure 7(b) shows the Cook's D for the investigation under discussion.

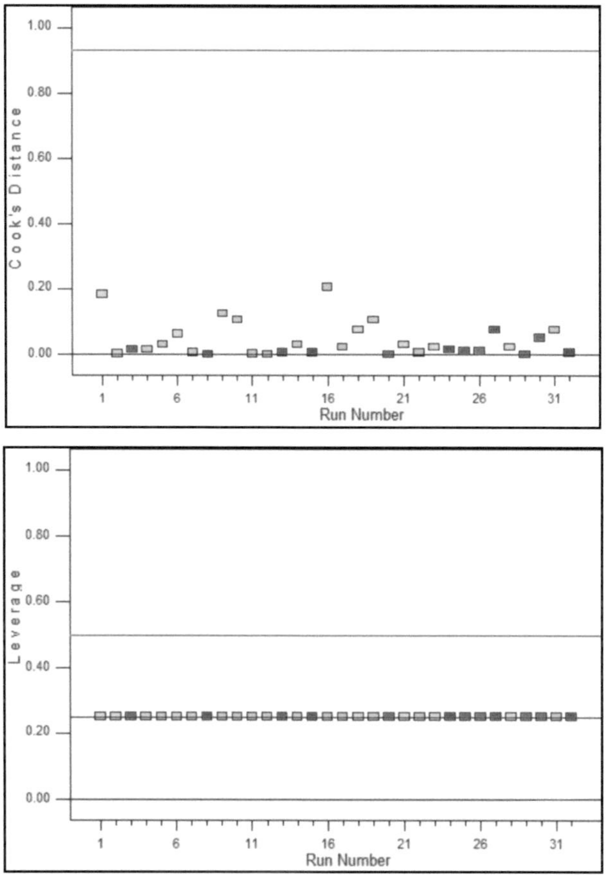

Fig. 7. (a) Leverages; (b) Cook's Distance

Lack of fit tests can be used supplement the residual plots if there remains any ambiguity about the information provided by them. The need for a model-independent estimate of the random variation means that replicate measurements made under identical experimental conditions are required to carry out a lack-of-fit test. If no replicate measurements are available, then there will not be any baseline estimate of the random process variation to compare with the results from the model.

4.2.2 Choice of transformations

Data transformations are commonly used to correct violations of model assumptions and to simplify the model. If the residuals are not randomly and normally distributed, or if the variance not constant; transformations are required to make the data suitable for statistically proper application of modelling techniques and for tests like F-test, t-Test, Chi-Squared tests etc. Significant violations of model assumptions result in increased chances of Type I and

Type II errors (Osborne, 2010). While parametric tests like ANOVA and regression modelling assuredly benefit from improved normality, their results may not be significantly affected by minor violations of these assumptions. Thus, the decision to apply transformations should be taken judiciously, as their application makes the task of interpreting results more complicated.

The need for application of transformations can be identified through residual vs. predicted value plots, Normal probability plots etc. Normal probability plots indicate whether the data is normal and in case of non-normality also give information about the nature of non-normality, thus help in selection of appropriate transforms for correcting the violation of normality assumption. To obtain the normal probability plot, the data are plotted against a theoretical normal distribution such that they form an approximate straight line. Departures from this straight line indicate departures from normality. Other commonly used tests include Shapiro-Wilk test, D'Agostino omnibus test, Kolmogorov-Smirnov test, Pearson Chi-squared test etc.

Some traditional transformations often used to correct the problem of non-normality and non-homogeneity of variance is:

1. **Logarithmic Transformation:** Logarithmic transformations can give rise to a class of transformations depending on the base used, base 10, base e or other bases. These transformations are considered appropriate for data where the standard deviation is proportional to the mean. These types of data may follow a multiplicative model instead of an additive model.
2. **Square Root Transformation:** It can be considered as a special case of power transformation, where the data is raised to one-half power. This type of transformation is applied when the data consists of small whole numbers whose mean is proportional to variance and that follow Poisson distribution. Negative numbers can be transformed by addition of a constant to the negative values.
3. **Arc Sine Square Root Transformation:** These transformations are appropriate for binomial data and involve taking arcsine of square-root of the values.
4. **Inverse Transformation:** Takes inverse of the data or raises the data to power -1. Transforms large data into small values and vice versa.

In their paper in 1964, Box & Cox suggested the use of power transformations to ensure that the usual assumptions of linear model hold. Box-Cox transformations have been shown to provide the best solutions and help researchers to easily find the appropriate transformation to normalise the data or equalise variance.

The Box-Cox power transformation is defined as:

$$x(\lambda) = (x^{\lambda} - 1) / \lambda \qquad \text{when } \lambda \neq 0$$

$$x(\lambda) = \ln(x) \qquad \text{when } \lambda = 0$$

The transformation parameter, λ, is chosen such that it maximizes the log-likelihood function. The maximum likelihood estimate of λ corresponds to the value for which the squared sum of errors from the fitted model is a minimum. This value of λ s determined by fitting a for various values of λ and choosing the value corresponding to the minimum squared sum of errors. t can also be chosen graphically from the Box-Cox normality plot. Value of λ = 1.00 indicates that no transformation needed and produces results identical to original data. A transformation parameter could also be estimated on the basis of enforcing a particular assumption of the linear model (Cochran & Cox,1957) like additivity.

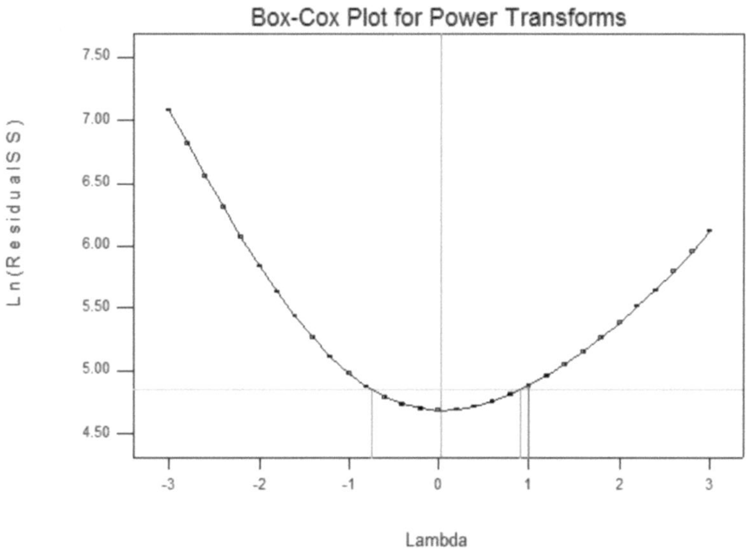

Fig. 8. Box-Cox Plot

In 1969, Cox & Draper reported that in data that could not be transformed into normal, the use of Box-Cox transformations resulted in obtaining symmetric data. Many modifications were subsequently suggested by Manly (1971), John & Draper (1980), Bickel & Doksum (1981) and Yeo & Johnson (2000).

5. Optimization

Once the best possible empirical model, that estimates the functioning of a system satisfactorily, has been formulated, the next step is to obtain the optimal region of operation for the system under consideration. From a mathematical point of view, searching for optimal region of operation is akin to finding the factor levels at which the system response(s) is maximized or minimized, depending on the nature of output being studied. Experimental optimization, as carried out under Response Surface Methods is an iterative process; that is, the model fitted to the data from one set of experiments paves the path for conducting successive experimentation which results in improved models. This process can be seen searching for the optimal region of operation. However, it should be borne in mind that the fitted responses are local approximations of the curvature of the solution space. Thus, the empirical model generated in our illustrative example holds true only for the region in which it is generated, and hence the optimization procedure should also be restricted within the space constrained. It is quite possible that outside the range investigated the behaviour of the system changes drastically and the empirical model and its predictions are no longer true. Hence, the screening experimentation should always be carried out over the maximum possible operating range.
Response Surface methodology is a collection of mathematical and statistical techniques that is useful for modelling, analysis and optimization of systems (Box and Wilson, 1951).

It is a sequential method, in which the first stage involves screening runs to identify the factors which have significant effect on the target output. Then a first order model consisting of the selected factors is formulated and examined. If the current settings do not place the system in the optimal region of operation, then the experimenter moves the system towards such a region by an iterative optimization technique called the steepest ascent technique. Once the system reaches the vicinity of the optimal region of operation, a second set of data is collected, and a model is fitted to this data. Typically near optimal region the model created is of second or higher orders due to the inherent curvature of the solution space.

5.1 Single and multi-response processes

Until now, the entire discussion has been focussed on a single response, namely throughput of a feeding system. In more complex systems and situations, engineers are often faced with problems requiring multi-response optimization. In addition to maximizing the feed rate, there can be other tasks like minimizing the cost. Single-response optimization is comparatively straightforward. On the other hand, in multi-response optimization it is not always possible to find operating conditions that simultaneously fulfil the process objectives.

Simultaneous optimization combines all the targeted response requirements into one composite requirement by assigning weights to the responses.

Dual Response Surface Method (Meyers and Carter, 1973) is based on an algorithm for obtaining the optimal solutions of to this problem by assuming a primary response and a constraint response which both of them can be fitted as a quadratic model.

Desirability function approach (Harrington, 1965) is used to simultaneously optimize multiple objectives. It is one of the more popular methods used to tackle the problem of multi-objective optimization. Each response is assigned a desirability value between 0 and 1 and its value represents the closeness of a response to its ideal value. If a response falls outside the acceptable intervals, the desirability is 0, and if a response falls within the ideal intervals or the response reaches its ideal value, the desirability is 1. If the response falls within the tolerance intervals, but not the ideal interval, or when it fails to reach its ideal value the desirability lies between 0 and 1 (Raissi & Farsani, 2009). A composite desirability function is created that combines the individual desirability values and converts a multi-response problem into a single-response one.

5.2 Confirmation of optimum

In addition to theoretical validation of the goodness of the model in prediction of response, when the analysis of the experiment is complete, one must verify that the predictions are good in practice. These experiments are called confirmation runs. This sage of conducting confirmation runs may involve a few random experiments or it may involve the use of a smaller design to carry out systematic analysis depending on the scale and complexity of the process.

Typically after the experiments have been conducted, analysis of data carried out and confirmation runs selected, there is a not insignificant probability that the value of system configurations and environmental factors may have changed. Thus it is very important for the experimenter to ensure that proper controls and checks are in place before confirmation runs are carried out.

The interpretation and conclusions from an experiment may include a "best" setting to use to meet the goals of the experiment. Even if this "best" setting were included in the design, it should be run it again as part of the confirmation runs to make sure nothing has changed and that the response values are close to their predicted values. In an industrial setting, it is very desirable to have stable process, thus multiple confirmation runs are often conducted.

The purpose of performing optimization of feed rate is to maximize the throughput for a given value the input parameters. From the regression model and the related graphs and statistics, it can be deduced the feedrate of a reciprocating fork-hopper feeder generally increases with the increase in population. Hence the population was kept at its maximum value. The optimization is carried out for both the speed and the frequency of strokes. While carrying out optimization for the speed, the frequency of strokes was kept at the maximum since throughput increases with increasing frequency of strokes and while carrying out optimization for speed the frequency of strokes was kept at maximum.

6. Conclusion

In this chapter, a general process for optimization of part feeding systems is demonstrated by taking a Reciprocating- Fork Hopper feeder as an example to clarify specific techniques. Exhaustive testing of the system is cumbersome because there exist too many treatments to be tested. Statistical methods based on experimental designs of tests, regression analysis and optimization techniques can be used to carry out this task more effectively and efficiently. To this end the implementation of a two level full factorial design and its attendant analyses are shown and explained. This work should serve as a guideline for drawing statistically valid conclusions about the system under consideration.

7. References

Boothroyd, Geoffrey(2005), *Assembly Automation and Product Design* (2nd edition),CRC Press, ISBN 1-57444-643-6, Boca Raton, FL, USA

Cochran & Cox (1957),*Experimental Designs*(2nd edition), John Wiley & Sons, ISBN 0-471-54567-8,USA

Ding X & Dai J.S.,(2008) "Characteristic Equation-Based Dynamics Analysis of Vibratory Bowl Feeders With Three Spatial Compliant Legs", *IEEE Transactions on Automation Science and Engineering*,Vol 5., No:1, pp. 164-175, ISSN 1545-5955

Gudmundsson D. & Goldberg K., (2007)"Optimizing robotic part feeder throughput with queuing theory",
Assembly Automation, Vol. 27, No: 2, pp 134-140, ISSN 0144-5154

Karley K.M., Kamneva N.Y., & Reminga J. (Oct 2004) "Response Surface Methodology", In: CASOS Technical Report, Carnegie Mellon University,Date of access: April 2011, Available from: http://pdffinder.net/(Response-Surface-Methodology)

Lane, D.M. (Aug 30 2001), HyperStat Online Textbook, Date of access: March 2011, Available from:
<http://davidmlane.com/hyperstat/ >

Li Pengfei, (Apr 11,2005)."Box-Cox Transformations: An overview", Date of access: April 2011, Available from:
<http://www.ime.usp.br/~abe/lista/pdfm9cJKUmFZp.pdf >

Mitchell Jr., John F.(Nov 11, 2010)."Variation in types of Part Feeders for Manufacturing Automation", In: *Ezine Articles*, Date of access: April 2011, Available from: < http://ezinearticles.com/?Variation-in-Types-of-Part-Feeders-for-Manufacturing-Automation&id=5367012>

Montgomery,Douglas C.(2001), *Design and Analysis of Experiments*(5th edition) , John Wiley & Sons, ISBN 0-471-31649-0, USA

NIST, (Jun 23,2010),Title: "Process Improvement", In: *Engineering Statistics handbook*; Date of access: April 2011, Available from: < http://www.itl.nist.gov/div898/handbook/pri/pri.htm>

Osborne, J.W., (2010) "Improving your data transformations: Applying the Box-Cox transformations", *Practical Assessment, Research and Development*, Vol. 15, No: 12, pp 1-9,ISSN 1531-7714, Available online: <http://pareonline.net/getvn.asp?v=15&n=12>

Raissi S. & Farsani R.E.,(2009)"Statistical Process Optimization Through Multi-Response Surface Methodology", World Academy of Science Engineering and Technology,Vol. 51,pp. 267-271, Available online: <http://www.waset.org/journals/waset/v51/v51-46.pdf >

Singh S.P., Ghosh S., Khanna P., & Tanwar A.,(2009)"Mathematical performance analysis of reciprocating fork hopper feeder", Proceedings of 2009 IEEE Student Conference on Research and Development (SCOReD), ISBN 978-1-4244-5186-9, UPM Serdang, Malaysia, Nov. 2009.

Whitcomb P (Sept 2008), "FDS-A Power Tool for Designers of Optimization Experiments", In: Stateaser *Sept 2008 issue*, Date of access: April 2011, Available from: <http://www.statease.com/news/news0809.pdf >

Part 3

Assembly Line Inspection

10

End-Of-Line Testing

Wolfgang Klippel
Dresden University of Technology
Germany

1. Introduction

Testing a manufactured unit at the end of the assembly line is a critical step in the production process. Defective products or even those not matching specification limits closely enough must be separated from the functional units shipped to the customer. End-of-line testing assesses not only the quality of the product, but also the stability and yield of the production process. Reliable detection of non-functional units is the primary objective of the test, but reducing the rejection rate and maximizing the output is the ultimate goal.
100% automatic testing replaces more and more subjective testing by human operators to shorten the production cycle and to improve the reproducibility and comparability of the results. However, objective measurements should provide a comprehensive assessment as sensitive as a human tester using his visual and aural senses. To fully compete with an experienced operator, the objective measurement instrument should also have learning capabilities to accumulate knowledge about physical causes of the fault. Furthermore it should be capable of being integrated in automated lines, robust in a harsh and noisy environment, cost effective and simple to use.

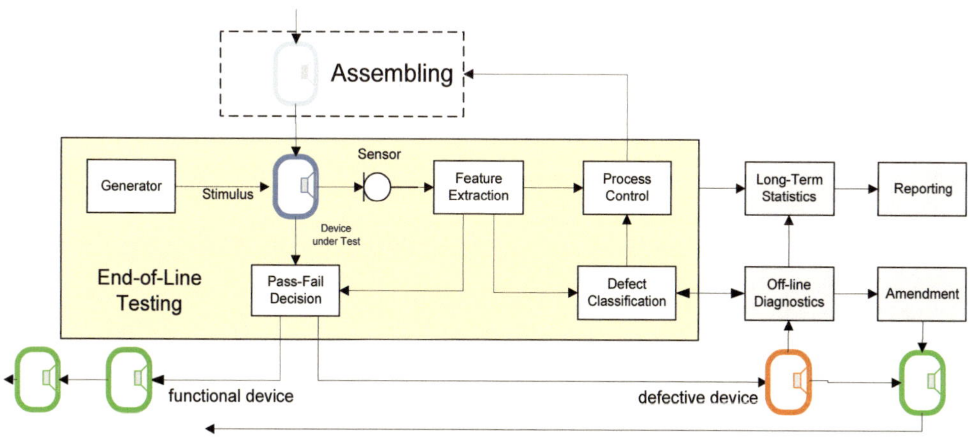

Fig. 1. Quality control at the end of the assembly line

Modern end-of-line testing (EOL) which satisfies those requirements is a complex process, as illustrated in Fig. 1. This chapter can only give a general overview on essential components, their interactions and future trends. Fig. 1 provides a roadmap of the discussion, starting with

the physical modelling of the device under test in section 2. This is the basis for ultra-fast testing, providing meaningful symptoms of the defect at high sensitivity. Important issues of the measurement will also be considered in section 3, such as the generation of a critical stimulus, the influence of the test conditions and the selection of optimal sensors. The following section 4 describes relevant features extracted by signal analysis, system identification and other kinds of transformations suppressing noise and redundant information. The Pass/Fail decision and classification of faults is the subject of section 5 and considers the problem of defining specification limits, grading the quality of the device and revealing the initial cause of the problem. The measurement results produced by end-of-line testing require a special data management and statistical analysis to support documentation, customer report and process control. The new requirements and technical possibilities in modern end-of-line testing put the future role of the human operator into question.
The discussion uses mass produced loudspeakers found in cellular phones, cars, multimedia, home entertainment and professional applications as a practical example. Some loudspeaker defects have a high impact on the perceived sound quality and will not be accepted by the customer. The loudspeaker examplealso represents other electrical, mechanical or acoustical systems manufactured at high quantities and low costs at a modern assembly line.

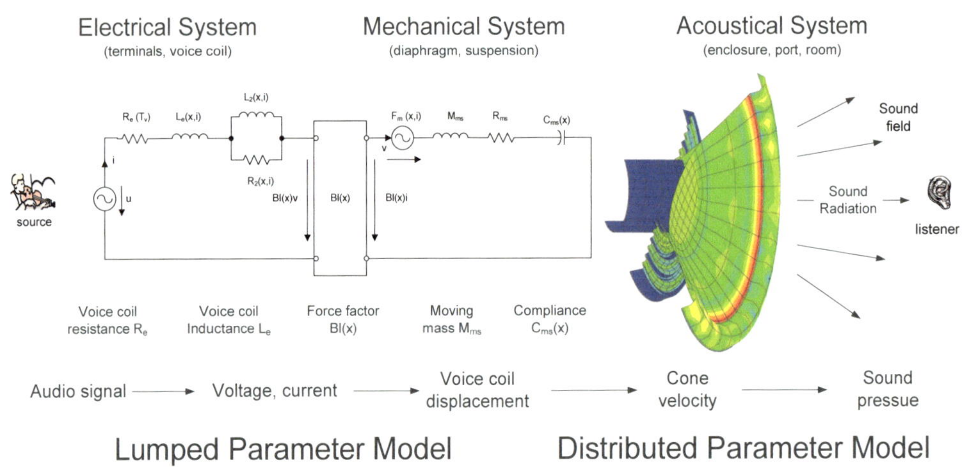

Fig. 2. Electro-dynamical loudspeakers modelled by lumped and distributed parameters

2. Physical modelling

A clear definition of the relevant properties and features of the device under test is essential for the end-of-line test to determine overall quality, perform fast measurements producing meaningful data and arrive at a correct Pass/Fail decision.

2.1 Product in the development process

The models used in the design and development process are reliable sources of this information. For example the transfer behaviour of loudspeakers can be described by two kinds of models as shown in Fig. 2.

The first model uses a small number of lumped elements representing the electrical resistance R_e, inductance L_e of the voice coil wire, force factor Bl(x) —one of the most important transducer characteristics— and other mechanical parameters such as the total moving mass M_{ms} and the compliance C_{ms} of the mechanical suspension. Those lumped parameters play an important role for the quality check of loudspeakers and can be easily identified from the electrical current i, voltage u and the mechanical displacement x. A second model is used to describe the generation of mechanical modes and acoustical waves using parameters distributed over the diaphragm and the sound field. Since the velocity and sound pressure may vary from point to point, it is impossible to measure the state of all those points on the radiator's surface and in the sound field during end-of-line testing.

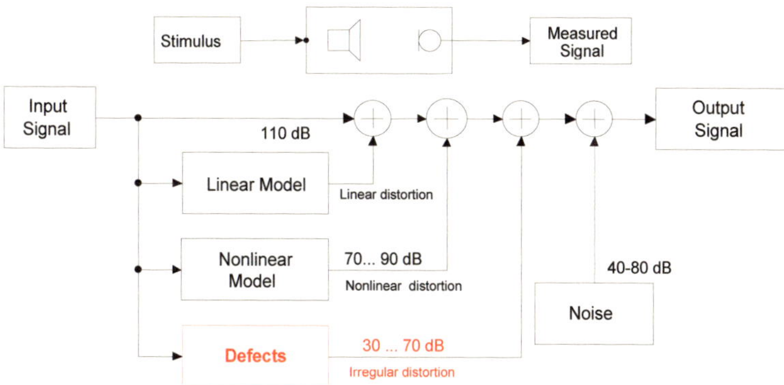

Fig. 3. Signal flow diagram showing the generation of signal distortion in a loudspeaker system

A more abstract model depicted in Fig. 3 shows the signal flow between the signal source and the sensor output, and the generation of the following distortion components in the output signal:

Linear distortions are generated by the transfer response which varies in amplitude and phase with frequency. This kind of signal distortion is independent of the amplitude of the stimulus and describes the small signal behaviour of loudspeakers. At higher amplitudes *non-linear distortions* are generated. These comprise new spectral components at multiples and combinations of the excitation frequencies. Measurement techniques assessing harmonic and intermodulation components exploit this property and play an important role in loudspeaker testing. All loudspeakers generate linear and non-linear signal distortions to a certain extent, depending on the physical limits of the electro-mechanical transducer. Those distortions are deterministic and can be predicted by numerical design tools (Klippel, 2006). The prototype at the end of the design process is a compromise between sound quality, maximal acoustical output, efficiency, size, cost and weight depending on the particular application. Those distortions are considered as regular and should be a feature of all replicated units passing the end-of-line test. Excessive signal distortions found in manufacturing are considered as *irregular distortion* and indicate a loudspeaker defect.

Ambient noise as found in a real production environment is also monitored by the test microphone and will corrupt the measurement. Those signal components differ in the sound pressure level significantly as shown in Fig. 3. Irregular distortion generated by a rubbing voice coil and other loudspeaker defects may be more than 60 dB below the total signal level and will still be detected by a human ear in the final application.

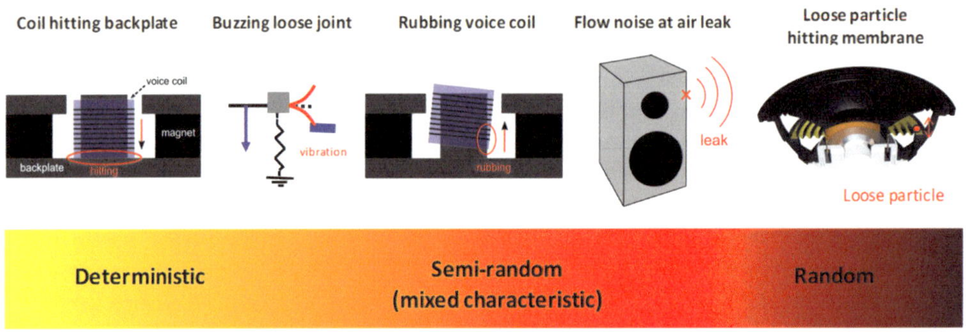

Fig. 4. Critical loudspeaker defects caused by manufacturing

2.2 Product in manufacturing

The variation in the linear and non-linear parameters provided by the physical models may be used for detecting defective units. For example, an increase of the moving mass (e.g., too much glue) or a decrease of the force factor (e.g., insufficient magnetization) reduces the sound pressure output of loudspeakers at higher frequencies. The relationship between causes and symptoms becomes more complex when initial and consecutive faults occur. For example, asymmetrical loudspeaker nonlinearities generate a dc displacement which moves the coil away from the rest position. This can cause audible distortion when the voice coil hits back-plate as shown on the left-hand side of Fig. 4. Other defects are hardly predictable, such as a poorly glued connection between the surround and the membrane behaving as an independent oscillator creating a buzzing sound. The spring-mass system performs an undesired mode of vibration at higher frequencies which is powered, triggered and synchronized by the stimulus. The faulty glue joint behaves here as a nonlinear switch activating the resonator above a critical amplitude. The beating of the braid wire on the loudspeaker diaphragm is a similar defect generating impulses at a particular position of the voice coil. The energy of those impulsive distortions is usually small and does not grow significantly with the level of the stimulus. A coil rubbing at the pole tips is a typical fault found in the production of loudspeakers generating impulsive distortion which contains deterministic and random components (Klippel, 2003). Air leaks in dust caps or in sealed enclosures emit a small air flow driven by the ac sound pressure inside the box which generates air turbulences and random noise (Klippel, 2010). Some loudspeakers defects behave randomly. For example, dust in the magnetic gap or below the dust cap in loudspeakers are accelerated by the cone displacement and hit the diaphragm at unpredictable times.

Although some irregular loudspeaker defects produce symptoms which are not predictable and cannot be modelled completely, it is still beneficial to investigate the physics of those defects and to develop sensitive measurement techniques exploiting particular features of those symptoms. Fig. 5 illustrates this approach using the example of air noise generated by a small leak at the rear of a loudspeaker enclosure. The random noise is generated by turbulences due to the high velocity of the air at the exit of the leak. However, the air volume velocity q(t) is not constant, but a function of the sound pressure $p_{box}(t)$ in the enclosure and the voice coil displacement x(t). The flow diagram on the right-hand side describes the generation by using a linear system $H_{pre}(j\omega)$ generating the sound pressure signal p_{box}, a nonlinear modulation process and a second linear system $H_{post}(j\omega)$ describing the propagation of the generated noise to any point r_a in the sound field. Information from the physical modelling provides the basis for a new demodulation technique for detecting air leaks more sensitive and reliable than the human ear.

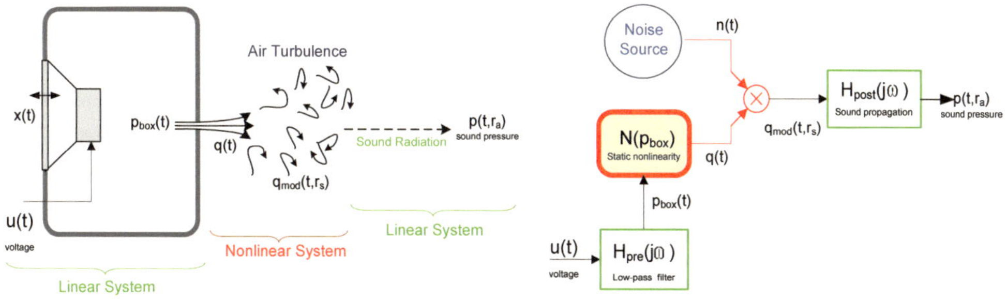

Fig. 5. Generation of turbulent air noise by a leaky loudspeaker enclosure

3. Measurement

This chapter discusses the theoretical and practical aspects of performing the basic measurements, considering the measurement condition, excitation of the device under test and using optimal sensors to monitor relevant state variables.

As expected from any other measurement process, the results of end-of-line testing should be repeatable under the same conditions, reproducible by a different operator at a different location using the same instrument, stable over time and free of bias. Repeatability and reproducibility can be tested by a gauge R&R test which reveals undefined factors that increase the variability of the measurement process. For example, loudspeaker transducers have to be clamped in the same way and at the same distance from the microphone, and should not be varied from measurement to measurement. The stability over time may be affected by climate conditions such as temperature, humidity and static sound pressure which affect the speed of sound. A critical issue in end-of-line testing is the accuracy of the measurement process producing bias-free results which agree with the "true" values and are comparable with the results of other instruments. The sound field generated by loudspeakers, for example, is influenced by the properties of the acoustical environment such as waves reflected from the table, floor, ceiling and room modes. Measurements performed in an anechoic room which are indispensable in loudspeaker design are usually not practical and too expensive in manufacturing. Measurement in a normal production environment using a simple test box is preferred but requires a special calibration routine to ensure comparability with the results of measurements in research and development (R&D). Instead of assessing absolute characteristics which might be easily a subject of a bias, it is more practical to use relative characteristics in end-of-line testing for defining the quality of the device under test. This subject will be discussed in the next chapter in greater detail.

3.1 Test environment

Fig. 6 shows a simple and cost-effective hardware setup which complies with the requirements in the manufacturing of woofers, tweeters and other transducers used in audio systems. It comprises a front-end generating the stimulus, microphones, a power amplifier, a PC for processing the data and auxiliaries integrating the instrument into the production line. The loudspeaker is clamped at a clearly defined position on a rigid test box unable to perform any parasitic vibrations. The microphone measures the sound pressure in the near field of the transducer at a fixed (local) position. The test box also provides some shielding against ambient noise generated in a real production environment. However, the enclosed air volume in the test box behaves as an additional air spring which reduces the displacement of the voice coil at lower frequencies. Although the interior of the loudspeaker box is damped by lining the inner walls with absorbing materials, the radiated sound pressure output is also affected by standing waves. Thus the acoustical measurements performed in the test box are only accurate for the particular measurement setup used at the end of the assembly line and are not directly comparable with the absolute measurements made under other measurement conditions during the design process. This discrepancy has an important consequence on defining permissible limit thresholds in the Pass/Fail detection discussed below.

The measurement setup shown in Fig. 6 uses additional sensors to monitor production noise, air temperature and humidity. This information is crucial for detecting invalid measurements corrupted by ambient noise and to ensure long-term stability under varying climate conditions.

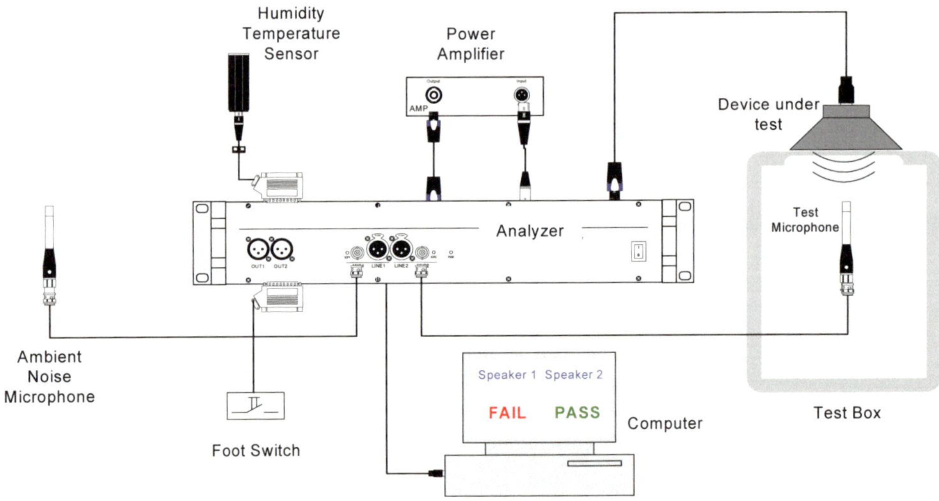

Fig. 6. Typical hardware setup for the end-of-line testing of loudspeakers

3.2 Stimulus and excitation

Most defects only produce significant symptoms if the device under test is operated in a critical state. An optimal excitation requires sufficient energy (e.g., ac main power) or a particular stimulus (e.g., analogue or digital signal). Ensuring sufficient excitation within the shortest time possible leads to ultra-fast testing which will be discussed in greater detail:

Loudspeakers are passive systems which require an electrical AC signal to produce an acoustical output. A sinusoidal stimulus excites the device at only one defined frequency and generates the fundamental component. All other frequency components found in the output spectrum can be identified as non-linear signal distortion or measurement noise. This measurement has to be applied to other frequencies within the audio band of interest (e.g., 20 Hz to 20 kHz) by stepping the frequency over discrete points with sufficient resolution (e.g., 1/10th of an octave) or by using a sinusoidal chirp sweeping continuously over all frequencies. The *stepped sine* stimulus stays at each frequency for a fixed number of periods (usually 5) to generate steady-state condition and to separate pre- and post-ringing after changing the frequency. Fig. 7 shows that the stepped sine stimulus spends most of the measurement time at low frequencies where the period length is long and both devices under the test and measurement instrument need a longer settling time. A complete measurement according to the minimal requirement can be accomplished within 1.5 s. Speeding up the measurement by reducing the number of test tones is not possible because the poor frequency resolution jeopardizes the excitation of loudspeaker defects behaving as narrow band resonators (e.g., buzzing parts not glued properly).

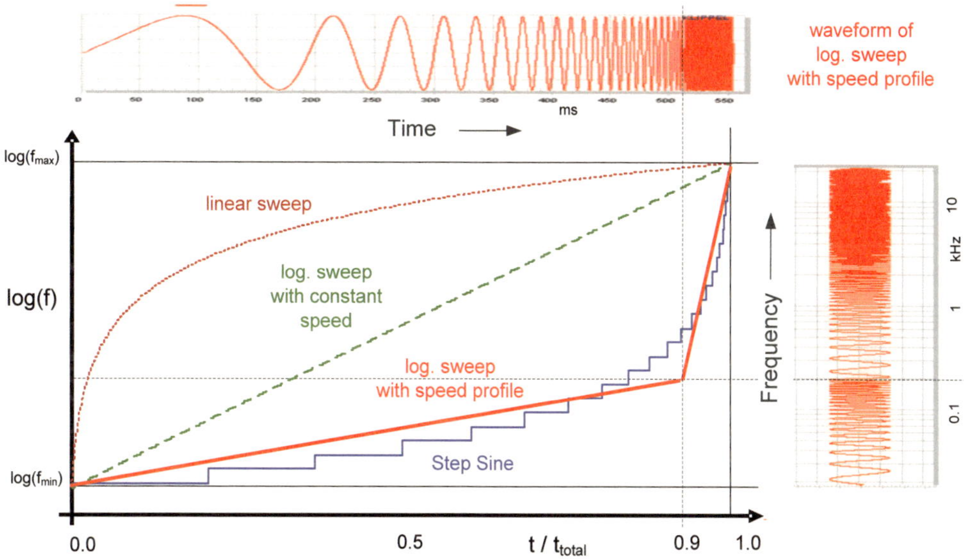

Fig. 7. Sinusoidal stimuli for measuring the transfer behaviour of loudspeakers

Continuous sweeps excite all frequencies using a defined mapping between instantaneous frequency and measurement time. The *linear sweep* passes the low frequencies quickly and spends most of the time at higher frequencies. The measurement time should be about 20 s to provide sufficient resolution at low frequencies according to minimal requirements. The *logarithmic sweep* with a corresponding frequency-time mapping reduces the measurement time at higher frequencies for the benefit of lower frequencies accomplishing the measurement within 0.4 s. However, the measurement time can be reduced even further to 0.2 s by using a *sinusoidal sweep with speed profile* as illustrated in Fig. 7. This stimulus comprises logarithmic sections with different sweep speeds and approximates the preferred frequency-time mapping of the stepped sine stimulus. It spends 90% of the measurement time below 200 Hz to activate all kinds of irregular loudspeaker defects and passes the high frequency range at a 10 times higher sweep speed.

Although the sinusoidal sweep with speed profile is a convenient stimulus for ultra-fast testing of audio equipment and other electronic devices, it cannot assess the

intermodulation distortion generated by non-linearities in the device under test. A sparse multi-tone signal comprising a multitude of distinct tones at a defined spectral distance and pseudo-random phase produces a noise-like stimulus which has properties similar to a steady-state audio signal, generating all kinds of harmonic and intermodulation distortion components in the output signal. This stimulus is ideal for assessing the large signal performance of loudspeakers and to identify motor and suspension non-linearities.
Music, speech and other natural audio signals play an important role in systematic listening tests during the design phase but play a minor role in stimulating devices during end-of-line testing because those tests are inferior with respect to sensitivity and speed.

3.3 Sensor system

Monitoring relevant state variables of the device under test is a further requirement for achieving high sensitivity in end-of-line testing. This question is closely related to the selection of optimal sensors and can be answered by using information from physical modelling as discussed in section 2. For example, the state of a loudspeaker can be observed in the electrical, mechanical or acoustical domain.
The acoustical measurement is indispensable for detecting air leakage noise and other impulsive distortion generated in the mechanical and acoustical domains which also have a high impact on the perceived sound quality. However, the sound pressure signal is less suited for assessing the properties of the electrical and mechanical system modelled by lumped parameters as shown in Fig. 2. A simple measurement of the electrical signals at the loudspeaker terminals provides results which are more reliable and easier to interpret. The direct measurement of mechanical state variables by scanning techniques is important for assessing the cone vibration during the design process, but until now has played a minor role in end-of-line testing. However, inexpensive triangulation lasers are already being used for testing spiders, passive radiators, diaphragm and other mechanical loudspeaker parts on the assembly line.
Multiple sensors combined into an array (e.g., microphone array) and parallel acquisition of the sensor output signals are required to localize the position of the defect in a sound field or a mechanical structure corresponding with a distributed parameter model. The number of sensors and locations in a sensor array depends on wavelength of the signal components and the distance to the source. The position of defects generating deterministic symptoms can also be determined by repeating the measurements while changing the position of the sensor.

4. Feature extraction

The objective of signal analysis is to extract features from the monitored signals which reveal the symptoms of the defect, to remove redundancy in the data and to suppress information not relevant for the quality assessment. This section can only give a short overview on plurality of traditional and new methods used for end-of-line testing.

4.1 Signal analysis

The first class of methods as summarized in Table 1 are applied to the time signals at the sensor outputs. There are no assumptions made as to how the device under test is excited and what properties the stimulus has. There is also no physical model of the device required. Thus the signal analysis is the most general approach which can be applied to all kinds of devices under test.

Signal Characteristics	Measurement technique and diagnostic value
Rms value	corresponds to the power of the signal (e.g., SPL) and is a simple and sensitive characteristic for detecting defects generated by deterministic processes (e.g., variation of loudspeaker sensitivity).
Peak value	is the maximal absolute value of the time signal within a time frame such as a period. This characteristic is good for assessing impulsive distortions which have a low rms value (e.g., loose particles).
Crest factor	is the ratio between peak and rms value and describes the impulsiveness of the signal independent of the magnitude of the signal (a crest factor > 10 dB indicates an irregular defect in loudspeakers).
Autocorrelation function $\psi(\tau)$	describes the correlation of a time signal $x(t)$ with itself $x(t-\tau)$ using a lag τ and gives information about repeating events (e.g., periodicity, pitch).
Coherence	describes the relationship between two signals using the cross-correlation spectrum normalized by the auto correlation of the two signals. This characteristic can be used for detecting an invalid measurement corrupted by an ambient noise source using a multi-sensor system (Bendat & Piersol, 1980).
Periodicity	can be assessed by the crest factor of the autocorrelation function $\psi(\tau)$ of the signal for $\tau \neq 0$. A high value indicates a repetitive process (e.g., sound emitted by electrical motor).
Period length T	is the time difference between adjacent maxima in the autocorrelation function corresponding with the lowest frequency component generated by a deterministic process (e.g., revolution of a car engine).
Synchronous average	is the mean value of adjacent sections of a repetitive time signal with the period length T and reveals the deterministic periodical component while attenuating all random and non-periodical components (e.g., ambient noise).
Random residual	is the difference between the original time signal and the synchronous average and reveals random and non-periodical components (e.g., loose particles, air leakage noise, external disturbances of production noise) in the time domain (Klippel, 2010).
Cepstrum	is the inverse Fourier transform of a logarithmic spectrum (Oppenheim & Schafer, 2004) which is useful for the deconvolution of source signal (e.g., engine noise) and the impulse response describing the propagation of the signal (e.g., radiation by the car body).
Signal Envelope	is the magnitude of the analytical signal calculated by using the Hilbert transform or by performing a demodulation of the signal which is useful for detecting defects generated by a semi-random process such as modulated air leakage noise (Klippel, 2010).
Discrete Fourier Transform (DFT)	provides a line spectrum of a time sequence which is assumed to be periodical. This transform is preferred for synchronous analysis of periodical signals where the length of the DFT corresponds to multiples of the period length T of the signal (e.g., hard bottoming of the voice coil at the pole plate).
Short-time Fourier Transform (STFT)	is a frequency-time analysis applying windowing to the waveform prior to a discrete Fourier transform. The transform is preferred for non-stationary signals where either a high frequency or a high temporal resolution is required (Boashash, 2003).
Discrete Wavelet Transform (DWT)	is a frequency-time analysis using wavelets with a temporal length scaled according to the analyzed frequency band. This transform is preferred for non-stationary signals where a high frequency resolution at low frequencies and high temporal resolution at high frequencies is required (Percival & Walden, 2000).

Table 1. Important characteristics provided by signal analysis

The applications of the signal analysis techniques will be illustrated on a sound pressure signal measured in the far field of a loudspeaker system having a small air leak. The diagram on the left-hand side of Fig. 8 shows the signal waveforms radiated by the defective loudspeaker as a solid line and the waveform of the same loudspeaker where the leak is sealed as dashed line.

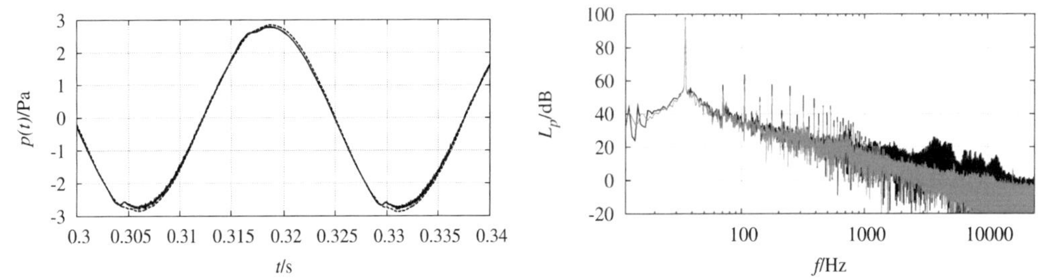

Fig. 8. Waveform (left) and spectrum (right) of the sound pressure signal measured at a leaky enclosure (solid black lines) and a sealed enclosure (dashed grey line) using a sinusoidal stimulus at 35 Hz

The loudspeaker without leak generates a smoother waveform but there is neither significant difference in the peak nor in the rms value. The autocorrelation function calculated from both signals reveals high periodicity and a period length of 28.5 ms. The discrete Fourier transform (DFT) applied to the both time signals which are exactly 128 periods long provides the line spectra shown in the right hand side of Fig. 8. There are distinct spectral components at f_0=35 Hz and at multiple frequencies f_n=nf_0 with $1<n<20$ with almost the same sound pressure level (SPL) for the functional and defective unit shown as black and grey lines, respectively. The air leak generates a higher sound pressure level for spectral components above 1 kHz. The power of the symptom is very low (-80 dB) and distributed over a wide frequency range causing a low spectral power density close to the noise floor of the measurement system. Therefore it may be more beneficial to attenuate the low frequency component below 1 kHz and to transform the high frequency component back to the time domain by using an inverse DFT. This corresponds to high-pass filtering of the original sound pressure signal revealing useful symptoms of the air leak.

4.2 System analysis

Many devices under test (such as loudspeakers) can be considered as transfer systems requiring an input signal for excitation and generating an output signal closely related to the stimulus, but containing additional signal distortion as shown in Fig. 3. Using a particular test signal with particular properties, the monitored sensor signal can be split into linear, nonlinear and irregular distortion components, providing further characteristics for end-of-line testing as summarized in Table 2.

Exploiting the properties of the stimulus, the spectral components found in the signal analysis may be interpreted as fundamental or distortion components. The sound pressure signal of the loudspeaker with and without air leak in Fig. 8 has been excited by a sinusoidal test tone at f_0=35 Hz. Thus the lowest frequency component found at this frequency is the fundamental component which dominates the total sound pressure signal. The phase of the fundamental component reveals polarity of the loudspeaker and may be used for checking

System Characteristic	Measurement technique and diagnostic value
Amplitude Response of the Fundamental Component	corresponds at small amplitudes with the magnitude of the linear transfer function and reveals failures in the small signal performance (e.g., variation of the cone mass).
Phase Response of the Fundamental Component	corresponds at small amplitudes to the phase of the linear transfer function and is important for checking time delay (e.g., microphone distance) and polarity.
Total harmonic distortion (THD)	describes the rms value of all harmonic components which are multiples of the excitation frequency. It reveals the dominant nonlinearities (e.g., nonlinear stiffness of the mechanical suspension in loudspeakers).
Total harmonic distortion and noise (THD+N)	describes the rms value of all signal components without fundamental component. It reveals the dominant nonlinearities and other disturbances which are not at multiples of the fundamental frequency (e.g., humming component at 50 Hz asynchronous to the excitation signal).
2nd-order harmonic distortion	indicates an asymmetrical shape of the nonlinearity inherent in the system (e.g., a different stiffness of the mechanical suspension for positive and negative displacement).
3rd-order harmonic distortion	indicates a symmetrical shape of the nonlinearity (e.g., symmetrical limiting of the mechanical suspension).
Peak-value of higher-order distortion	measured by using a sinusoidal stimulus of frequency f_o, high-pass filtering of the sensor signal at a cut-off frequency $f_c>10f_o$ and detecting the peak value within a period $1/f_o$ in the time domain (Irrgang, 2006). This characteristic is sensitive for all kinds of impulsive distortion generated by a random or deterministic process (e.g., voice coil rubbing).
Peak value of higher-order harmonic distortion	measured by using a sinusoidal stimulus of frequency f_o, detecting the peak value within a period $1/f_o$ in the sensor time signal comprising only nth-order harmonics with $n>10$ (Klippel, 2003). This characteristic reveals the impulsive distortion generated by a deterministic process (e.g., hitting the back plate).
Peak value of higher-order non-harmonic distortion	measured by using a sinusoidal stimulus of frequency f_o, filtering of the sensor signal to select only non-harmonic components above a cut-off frequency $f_c>10f_o$ and detecting the peak value within a period $1/f_o$ in the time domain (Klippel, 2011). This characteristic reveals the impulsive distortion generated by a random process (e.g., loose particles).
Peak value in the envelope of non-harmonic distortion	measured by using a sinusoidal stimulus of frequency f_o, filtering of the sensor signal to select only non-harmonic components above a cut-off frequency $f_c>10f_o$ and reading the peak value in the envelope of the demodulated signal averaged over multiple periods Klippel, 2011). This characteristic reveals semi-random processes having a random fine structure but a deterministic envelope (e.g., air leakage noise).
Multi-tone distortion	measured by using a sparse multi-tone stimulus and selecting the distortion components in the sensor signal found at frequencies not excited by the stimulus (Cabot, 1999). This measurement reveals harmonic and intermodulation distortion and is very sensitive for detecting a force factor nonlinearity (e.g., caused by voice coil offset).
Incoherence	describes the deviation from a linear relationship between the input and output signal (complement to coherence). This characteristic may be applied to music, speech and other arbitrary stimuli having a dense spectrum and reflects all kinds of nonlinear signal distortion and noise.

Table 2. Characteristics provided by generic methods of system analysis

the correct connection of the electrical terminals. The spectral components at multiple frequencies $f_n=nf_0$ with $n\geq2$ are the nth-order harmonic distortion components generated by the nonlinearities inherent in the device under test. The energetic summation of those components gives the total harmonic distortion THD. The symptoms of the air leak are not only higher-order harmonics (n>20) but also spectral components at other frequencies. Repeating the signal analysis for other excitation frequencies leads to the frequency response of those components.

Fig. 9 shows the frequency responses of the SPL fundamental, 2nd-5th harmonics and the total harmonic distortion in the sound pressure output using a short logarithmic sweep with speed profile as shown in Fig. 7. The roll-off of the fundamental component at the cut-off 100 Hz limits the usable audio band at lower frequencies, while the break-up modes cause the peaks and dips at higher frequencies. The motor and suspension nonlinearities cause a high value of THD at low frequencies. The higher-order distortions curve is 45-60 dB below the fundamental and shows that the device under test has no impulsive distortion generated by a rubbing coil or other irregular defects.

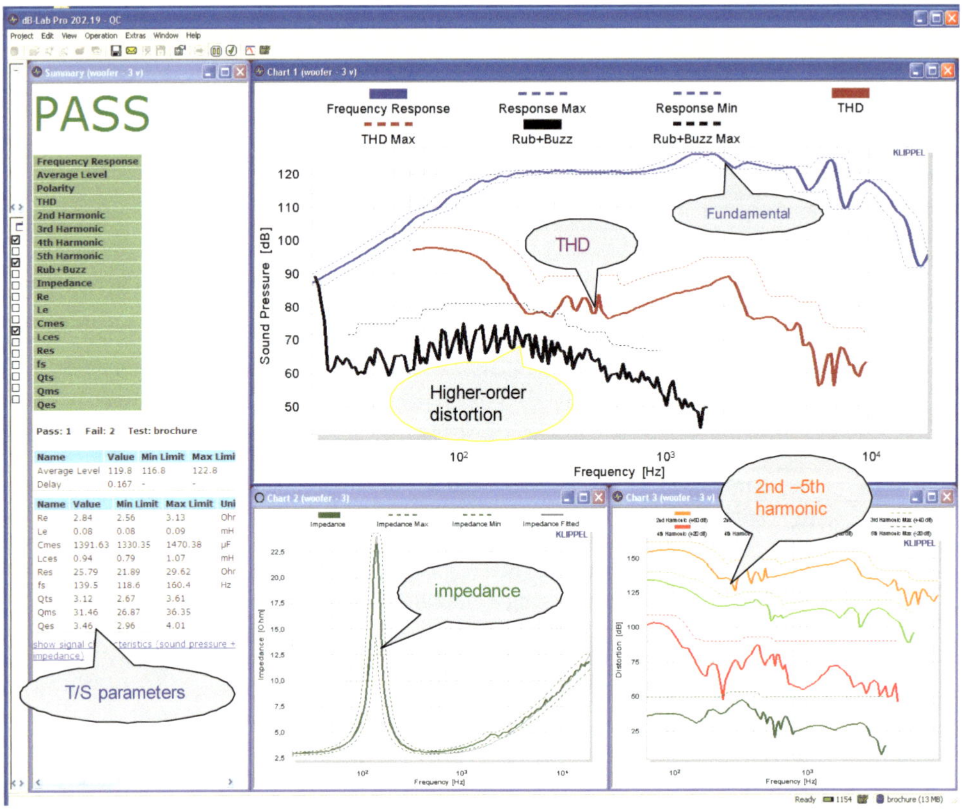

Fig. 9. Results of ultra-fast loudspeaker testing using a 200 ms stimulus

To increase the sensitivity of the end-of-line measurement and to identify the physical cause of the defect further features have to be derived from the measured signals. Synchronous averaging over adjacent periods provides the deterministic component at a higher signal-to-noise ratio while suppressing measurement noise and other random components. This can also be realized by a comb filter selecting all harmonic components in the high-pass filtered signal as shown in Fig. 10. The complementary signal comprising only non-harmonic components is a random signal.

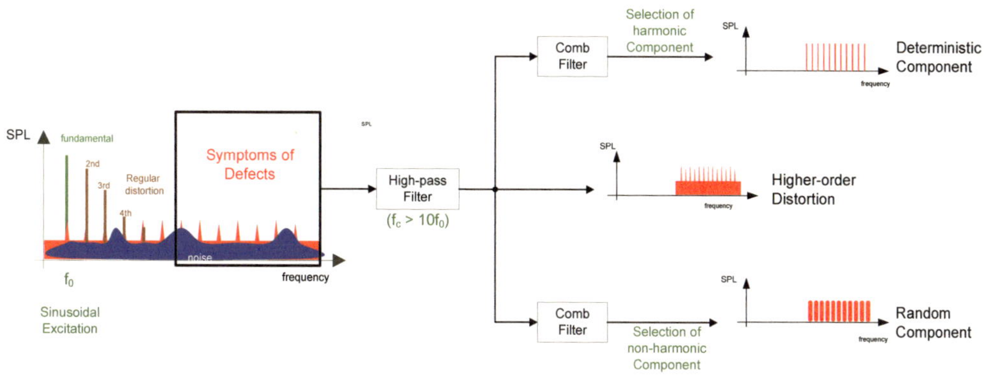

Fig. 10. Separation of the deterministic and random distortion components using a sinusoidal stimulus

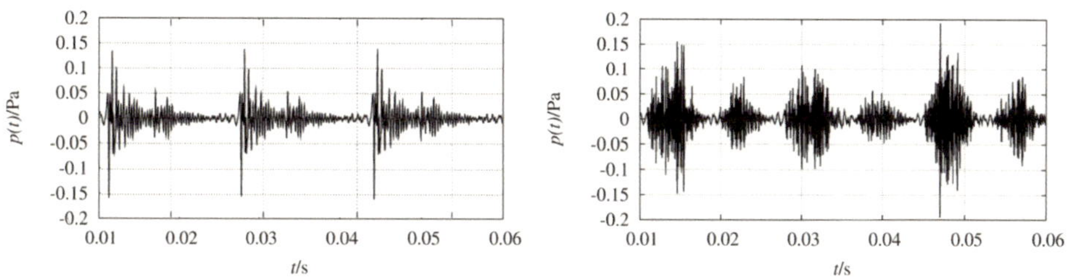

Fig. 11. Waveform of deterministic distortion component (left) caused by a buzzing loose joint and random distortion component (right) generated by an air leak

Fig. 11 illustrates the benefit of this separation on a loudspeaker having two defects. The periodic waveform in the left diagram reveals a deterministic defect caused by a buzzing loose joint while the random component on the right side is caused by an air leak in the enclosure. Turbulent noise generated at the leak has a completely random fine structure, but the envelope of the noise is deterministic due to the modulation process discussed in Fig. 5. Since the amplitude of air leakage noise is very small and in the same order as

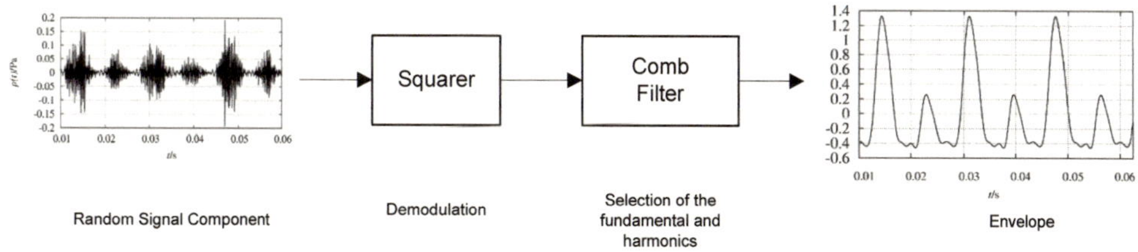

Fig. 12. Detection of turbulent noise generated by air leaks by demodulating the random signal component and synchronous averaging of the envelope

ambient noise (e.g., air conditioning) a measurement technique which accumulates the energy over time is required. Direct averaging of random distortion signals over multiple periods will reduce the noise signal and reveal no meaningful symptoms. Demodulation of the random distortion signal and synchronous averaging of the envelope as shown in Fig. 12 provides a sensitive feature for detecting semi-random noise reliably (Klippel & Werner, 2010).

4.3 Model identification

Exploiting available a-priori information on the physics of the device under test provides further features for end-of-line testing. These are more closely related to the material and geometrical properties of the device. Table 3 gives an overview on those characteristics:

Characteristic	Measurement technique and diagnostic value
Parameters	of the model are identified by reducing a cost function assessing the difference between estimated and measured signals. The estimated parameters describe the particular device under test and are independent of the stimulus used. The parameters have a physical meaning (e.g., force factor) and correspond to material parameters and geometrical variation (e.g., voice coil offset in mm) and can be used for process control.
State variables	can be calculated for the particular stimulus by using the model and identified parameters. This is beneficial for assessing state variables (e.g., voice coil displacement) which cannot be monitored at the end of line or require an expensive sensor (e.g., laser) otherwise.
Signal Distortion	can be calculated for the particular stimulus by using the identified model giving access to the linear and nonlinear signal components. This technique allows separating regular distortion found in functional units from irregular distortion generating by defects (loose particles).
Variables with reduced variance	are calculated from the original variables (e.g., resonance frequency) by compensating temperature, humidity and other external factors. This compensation of the climatic influence requires, for example, thermometer and other sensors at the assembly line and a model for predicting the behavior at defined condition. This reduces the common-cause variance in the measured variables and allows application of tighter specification limits.

Table 3. System identification techniques used for end-of-line testing

Linear system identification has a long history in loudspeaker measurement and can be realized by fitting the predicted electrical impedance based on the lumped parameter model in Fig. 2 to the measured impedance response. The best estimate of the free model parameters are the Thiele/Small parameters (T/S) are shown on the lower left hand-side of the screenshot Fig. 9. Those parameters include the voice coil resistance, resonance frequency and Q factor and are easy to interpret. However, the linear model is limited to the small signal domain and cannot explain the generation of nonlinear distortion at higher frequencies. Nonlinear system identification exploiting the nonlinear distortion measured in voltage and current at the loudspeaker terminals reveals the loudspeaker nonlinearities in the motor and suspension system. For example, an offset in the voice coil rest position can be detected in the asymmetry (e.g., shift to the right sight) of the bell shaped force factor characteristic as illustrated in Fig. 13.

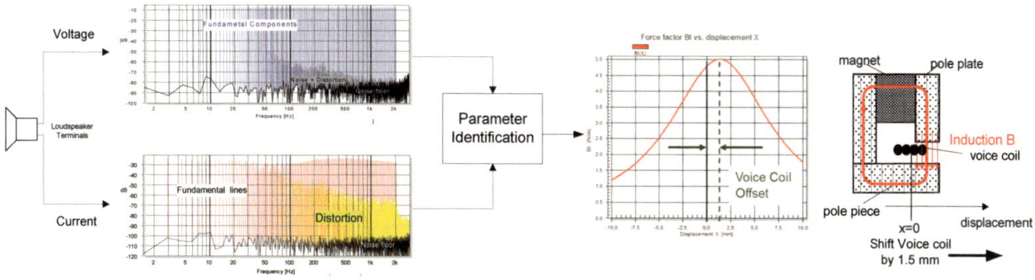

Fig. 13. Measurement of the voice coil offset in mm by identifying the force factor nonlinearity based on voltage and current monitoring at the loudspeaker terminals

A model with identified parameters also provides the state variables (e.g., displacement, voice coil temperature) for an arbitrary stimulus and dispenses with an additional sensor (e.g., laser). System identification can also be used for the measurement of irregular distortions (e.g., caused by a small loose particle) which are masked by regular distortion (e.g., caused by motor and suspension) and not detectable by a human ear or conventional signal and system analysis. Such a defective unit cannot be shipped to the customer because the defect may become worse in the final application (e.g., particle causes voice coil rubbing) eventually generating audible distortion. An adaptive model is used to learn the deterministic properties of the functional devices and synthesize the regular distortion which is subtracted from the measured signal as shown in Fig. 14. Such an active compensation (Irrgang, 2006) increases the sensitivity for irregular distortion by 10 - 30 dB and belongs to the technologies which outperform the capabilities of a human tester.

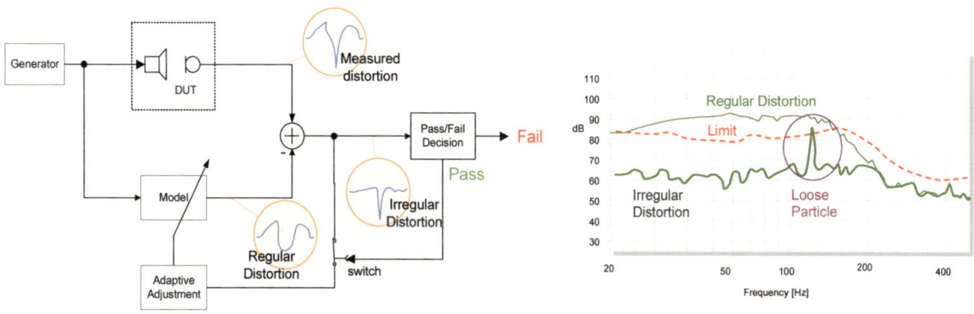

Fig. 14. Active compensation of deterministic properties found in all functional devices under test to increase the sensitivity of the measurement system for defects masked by regular distortion

4.4 Feature reduction

The previous discussion focussed on analysis techniques for increasing the sensitivity of the measurement instrument to ensure reliable detection of all potential defects in the device under test. However, features which are not relevant, redundant or having low diagnostic value should be excluded from the following classification to keep the data size small and the processing fast and robust. Table 4 provides an overview of the most important techniques for reducing the dimensions of the feature space.

Transformation	Technique and application
Relativization	refers a measured value or curve (e.g., absolute distortion) to a reference (e.g., total sound pressure) to generate a relative quantity (e.g., distortion in percent).
Curve smoothing and decimation	is the simplest way to reduce the resolution of a curve. Prior to the decimation, an energetic averaging within a defined bandwidth (e.g., third-octave) is useful for fundamental SPL response, while taking the peak value in this interval is more advantageous for higher-order distortion.
Orthogonalization	converts an original set of possibly correlated features into a set of independent features using principal component analysis (PCA). The number of principle components is less than or equal to the number of original features. The principle components are ordered to account for decreasing variability in the data. This transformation is useful for reducing the number of data points in one-dimensional curves (e.g., SPL response) and two-dimensional data sets (e.g., time-frequency responses from STFT).
Prediction of perceptive attributes	is a transformation of physical features measured by an objective measurement technique into perceptive attributes explaining the subjective sensation of a customer or human operator (e.g., audibility of signal distortion (ITU 2001)).

Table 4. Techniques to reduce the number of features submitted to the classification process

5. Classification

After providing a set of relevant features from physical measurement, the next step is the generation of Pass/Fail verdicts and the identification of the physical cause of the defect.

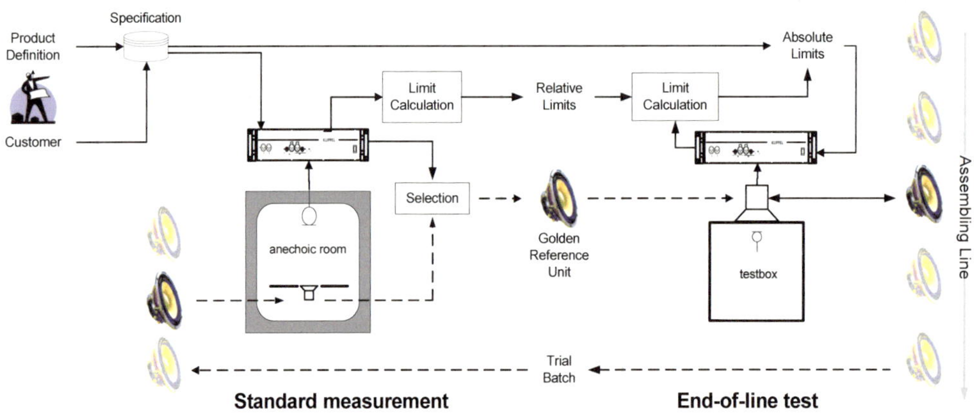

Fig. 15. Methodology for transferring specification limit defined for standard measurement condition (anechoic room) to corresponding specification limits valid for the special measurement condition (test box) used at the end of line

5.1 Pass/fail decision

Faulty units can be easily separated from functional units by comparing the measured variables with specification limits. Those limits correspond with either the target performance at the beginning of the product development, the quality found at the first prototype or the requirement defined by the customer. Specification limits referring to standard measurement condition cannot be applied to end-of-line testing if the measurement conditions are different there. For example, most loudspeaker standards define the sound pressure output at 1 m distance from the drive unit operated in a baffle under free field condition. The test box as shown in Fig. 6 is smaller, less expensive and more convenient than an anechoic room, but requires a transformation of the specification limits.

This problem can be solved by selecting a limited number of units (usually 10 - 100) and performing a measurement under standard conditions as illustrated in Fig. 15. The functional units which fulfill the specifications are subject to a statistical analysis. After calculating the mean value and variance of all features, *Golden Reference Units* are selected which represent the ensemble best. Some of the absolute specification limits corresponding to standard measurement conditions are replaced by relative limits. Now a Golden Reference Unit and the corresponding relative limits are transferred to the assembly line and measured under non-conformal conditions (e.g., test box). Finally the relative limits are transformed into absolute limits used for end-of-line testing.

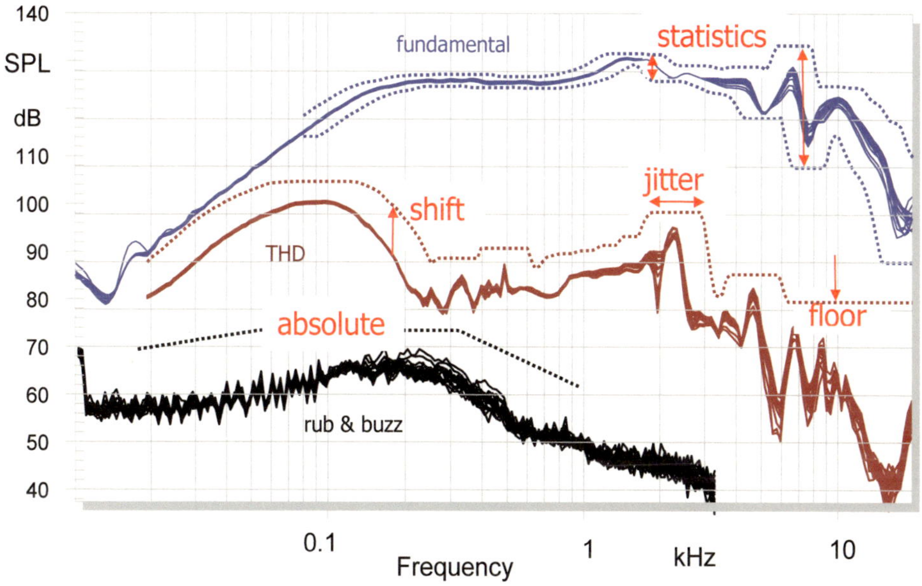

Fig. 16. Algorithms used for calculating specification limits based on statistical characteristics derived from reference units

The Golden Reference Unit is stored under manufacturing conditions and can be used to recalibrate the absolute limits at any time when temperature or humidity changes the behaviour of the device under test. This limit calculation process must also consider systematic differences between the just assembled unit where the properties are still varying (e.g., drying adhesive) and the Golden Reference Unit manufactured some time ago.

Special algorithms are useful to transfer or generate the specification limits. The shift algorithm is a simple way to generate upper and lower specification limits having a defined distance to the mean value as shown in Fig. 16. If meaningful shifting values are not known, the width of the permissible corridor can be calculated from the variance of the measured variable. For example the upper and lower limits of the fundamental SPL curve in Fig. 16 correspond to ±3 sigma and make the corridor wider at higher frequencies where the variance of the measured variable rises. The floor algorithm is a useful constraint which keeps the specification limit above a threshold. This is, for example, useful for total harmonic distortions which are acceptable if they are smaller than a defined level. A jitter algorithm increases the tolerances in the horizontal direction to cope with sharp peak and dips having a varying resonance frequency.

Statistical algorithms for limit definition have the benefit that only few setup parameters (e.g., shift value) have to be defined which are valid for similar kinds of products. The distance between the measured variable and the upper and lower specification limits may be used as a quantitative measure for grading the quality of the device under test and for assigning the device to a particular quality class.

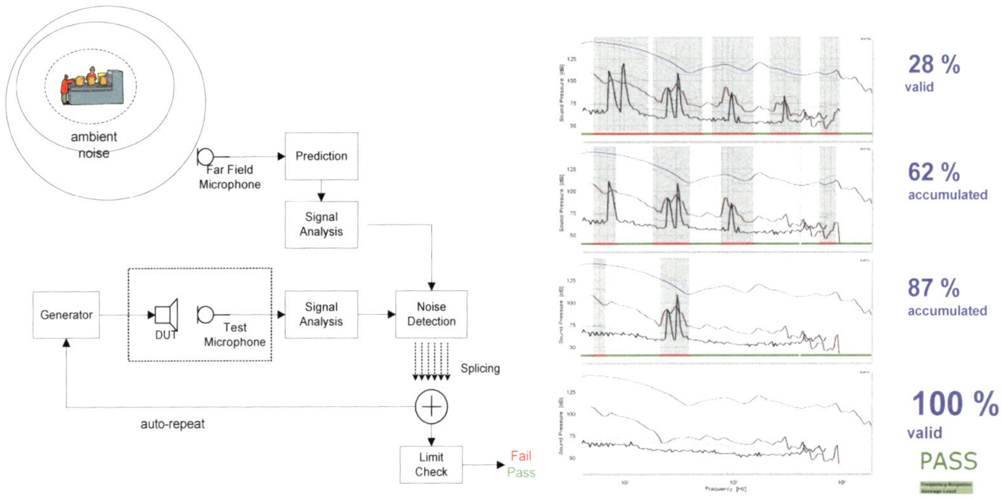

Fig. 17. Test system providing immunity against production noise by detecting invalid measurement and splicing valid parts of repeated measurements together

5.2 Detection of invalid measurements

A disturbance from an external source corrupting the monitored signal may invalidate the measurement. Acoustical measurements are especially prone to ambient noise generated in a production environment. A multiple sensor system such as a microphone array can be used for deriving the position of the source and to separate the test signal from the noise. If the test microphone is located in the near field of the loudspeaker as shown in Fig. 17 and a second microphone is located in the far field at some distance from the test microphone, an acoustical disturbance can be detected reliably. The grey sections in the SPL frequency response in the right-hand side of Fig. 17 show the corrupted parts of the measurement.

Invalid results require a repetition of the measurement, but it is beneficial to store uncorrupted parts of each measurement and to merge those parts with valid parts of the following repetitions resulting in complete valid set of data eventually. This auto-repeat and splicing technique is an important element of ultra-fast testing providing immunity against random production noise.

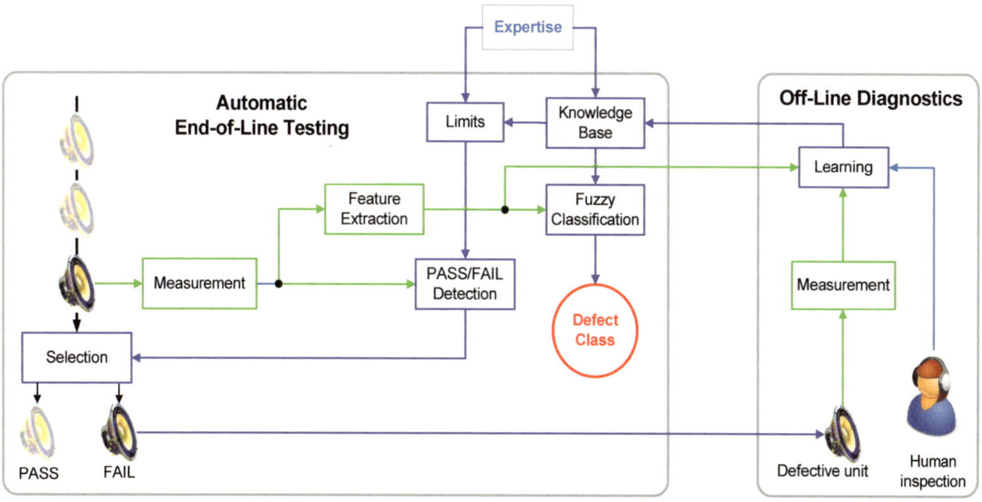

Fig. 18. Automatic classification of defects using information from an external expert and off-line diagnostic

5.3 Defect classification

If a device fails permissible specification limits, it is crucial to identify the physical cause of the defect as fast as possible to ensure timely control of the production process and to maintain a low rejection rate. To accomplish this task, an experienced operator or engineer from manufacturing or product design can perform additional measurements, visual inspection, destructive analysis or other kinds of off-line diagnostics as shown in Fig. 18. For loudspeaker evaluation, new auralization techniques are also useful, providing more sensitivity for defects than normal listening tests. Signal processing can enhance critical distortion components while protecting the operator's ears against the fundamental component's high sound pressure. However, those tools cannot replace the expertise the human investigator has accumulated over time and cannot be transferred to other co-workers.

A new objective of end-of-line testing is combining the results of objective testing with the background information about the physical nature of the defect. The first step in this process is the creation of a class, assigning it a meaningful name and storing all relevant information (symptoms, causes, remedies) in a knowledge base. Generic defect classes (e.g., "electrical shortcut") are already provided by the measurement instrument, but the particular defect classes are generated during the design process and permanently extended during manufacturing. The classificator in Fig. 18 uses this information to determine defect classes which correspond to the measured features at the highest likelihood. Fuzzy logic (Zadeh, 1987) can be used to combine the measured features with non-numeric linguistic variables such as "loudspeaker sounds distorted" and determine the membership of the defect classes. The output of the classification process is not only the name and probability value of the most likely defect class, but also a list of alternative classes at a lower rank. This information is not a strict assignment, but more a guided search process for the operator to determine the most likely defect class.

The verification of the proposed or assumed defect is part of the off-line diagnostics, usually performed on a separate measurement platform. The result of the human inspection is the most important input for the automatic learning process. The measured features of the defective device are used to update the properties of the defect prototype which represents an existing defect class. If the membership function of all existing classes is low or the QC operator detects a failure not known before, there is the opportunity to create a new defect class where the defective unit is identical to the prototype. An administrator who is usually the line leader or an experienced QC engineer supervises feedback of the operator and can unify two defect classes and may improve the verbal description of the defect. The expert system is a valuable tool for accumulating knowledge to train new, inexperienced co-workers and to simplify the communication between team members.

6. Quality monitoring and process control

Although the detection of defect units is the basic objective of end-of-line testing, the properties of the functional units passing the test also provide valuable information about the stability of the production process. Statistical analysis applied to this data reveals drifts and trends early enough to readjust the process before faulty products are produced.

6.1 Statistical analysis

The first step of quality monitoring and process control is the calculation of basic statistical characteristics of the measured variables (e.g., resistance) and classification results (e.g., counts of defects) as summarized in Table 5.

Basic characteristic	Description and application
Histogram	is used to measure the distribution of a variable by counting the number of points in equally sized, non-overlapping intervals on the data range. The normalized histogram estimates the probability density function pdf and reveals important properties of the distribution (center, spread, skew, outliers and multiple modes).
Arithmetic Mean $\overline{x}$	is the sum of data points x_i of a measured variable divided by the number N of observations.
Median $\tilde{x}$	corresponds to the value of the data point which separates the higher half of the population from the lower half. The median describes the central value of skewed populations more usefully than the mean value and suppresses the effect of outliers.
Exponentially Weighted Moving Average (EWMA)	performs a smoothing of the measured data series x_i versus unit index i (or time) by decreasing the weights of old data. The EWMA is used in control charts to detect small shifts and trends at a higher sensitivity than using the mean value calculated in subgroups.
Sample Standard Deviation σ	describes the variability of the measured single-valued variable x_i.
Range R	is calculated by subtracting the lowest value (minimum) from the highest value (maximum) found in subgroups.
Rejection Rate RR	is the ratio of non-confirmative or defective units failing the end-of-line test to the total number of units tested. The rejection rate is the complement of the process yield.
Defects Per Million Opportunities (DPMO)	is the number of defects (or non-conformities to the specifications) referred to the product of measured devices times the number of defects possible in one unit.

Table 5. Basic statistical characteristics important for end-of-line testing

The calculation of those basic characteristics does not require much processing power and can be accomplished by the computer used as part of the end-of-line tester. This provides an additional benefit as warning and alarm signals can be generated automatically and used as feedback at the end of assembly line.

6.2 Process stability

Statistical process control (SPC) separates variations arising from common causes typical to a particular production process (e.g., manual soldering of wires) from unanticipated, special causes (e.g., a new batch of parts with different properties is assembled). Common-cause variations are stable and predictable while special-cause variations have an atypical pattern and are unpredictable, even from a probabilistic point of view, and may require action to stabilize and adjust the process.

The Control Chart is a powerful tool for separating common and special causes. The chart displays the quality characteristic versus time or sample number in comparison with two horizontal lines, the upper control limit (UCL) and the lower control limit (LCL). A centre line between the UCL and LCL curves describes the long-term mean value of the process under control. There are many types of charts applicable to variables (e.g., individual data point of the measured feature, mean value, range) or attributes (e.g., counts of nonconforming units, proportions) of different subgroup sizes based on different statistical assumptions (e.g., underlying distribution) and performance (e.g., sensitivity for shifts). Table 6 describes some of the important charts (further details see Montgomery, 2005).

Control chart	Definition and application
$\overline{x}$ and σ Chart "X-Bar and S Chart"	contains a pair of charts used to monitor the (short-term) mean value and standard deviation of the variable observed over a relatively large number of units (n>10) in regularly sampled subgroups. This chart is sensitive for detecting shifts larger than 1.5σ.
$\overline{x}$ and R Chart "X-Bar and R Chart"	contains a pair of charts used to monitor the (short-term) mean value and maximal variation (range) of the variable observed over a relatively small number of units (n<10) in subgroups. This chart is sensitive for detecting shifts larger than 1.5σ.
EMWA Control Chart	uses the exponentially-weighted moving average to detect small variation (between 0.5 - 1.5σ) caused by special causes which would be otherwise assigned to the common cause variation using the $\overline{x}$ or other standard control charts. The increased sensitivity for shifts and trends is gained without increasing the chance for false alarms.
P Chart	is a control chart applied to the ratio (proportion) of nonconforming units to the total number of units using data generated by Pass/Fail decision.

Table 6. Important control charts for assessing process stability

Comparing the quality characteristic (variables and attributes) with the upper and lower control limits derived from the long term mean value and variability of the process can reveal a critical "out of control" status. Other rules (WECO, 1956; Nelson, 1984) consider additional zones at lower variance and are more sensitive to small shifts and trends. According to the WECO rules, a process is "out of control" if one of the following occurs:

- a single point is outside the ±3σ range,
- two out of three successive data points are beyond the ±2σ range,
- four out of five successive points are beyond ±1σ range or
- eight successive points are on one side of the centre line.

6.3 Process capability

If the production process is stable, it is possible to predict the output of the process by using dedicated characteristics as listed in Table. 7.

Characteristic	Definition and application
Process Capability Index C_{pk}	assesses the ability of a process considering the common causes of variation expressed by estimated mean value μ and standard deviation σ of the measured variable to produce output within given upper and lower specification limits USL and LSL, respectively. Montgomery (2005) recommends a minimal C_{pk} value of 1.33 for an existing process and 1.5 for a new process using two-sided specification.
Process Performance Index Ppk	requires a calculation similar to the process capability index Cpk, but can also be applied to an instable process not yet under control. It considers not only the common causes, but also special causes of variance in the process caused by shifts and drifts.

Table 7. Characteristics assessing the performance of the process

The yield of the production process corresponds to the process capability index Cpk, if the process is normal distributed and stable. For example a Cpk=1.33 gives a process yield of 99.99 %. To keep the rejection rate below 3 defects per million opportunities (Six sigma) the short term Cpk should be larger than 2. A first estimate can be achieved by using a few data points (> 17) but the prediction becomes more precise by using long-term estimates of μ and σ based on a larger number of observations.

6.4 Process adjustment

The control charts and the indices of process capability reveal one or more problems somewhere at the assembly line which require immediate actions to prevent an increase of the rejection rate. If the relationship between symptom and physical cause is not known, a solution for the problem can be searched by a trial and error method. The success of this approach depends on the intuition and experience of the investigator. A more systematic approach is the fault analysis techniques as listed in Table 8 collecting information on potential problems and optimal remedies.

The ability to act as quickly as possible when the process becomes instable and incapable is vital. The continuous improvement process (CIP) and the failure mode and effects analysis (FMEA) are examples of evolutionary methods for accumulating knowledge before a potential failure occurs. Because of practical experiences during the development of the product, the expertise of the R&D engineers is an important source. This information has to be documented in a format (e.g., fishbone diagram) and language suitable for application at the assembly line. If the manufacturing is outsourced to a contract manufacturer, this know-how will be transferred only if both companies are interested in a close and long-term relationship.

New measurement techniques exploiting physical modelling of the device under test and system identification make the relationship between symptoms, causes and remedies more transparent. For example, Fig. 19 shows the adjustment of the rest position of the voice coil in loudspeaker manufacturing. The rest position of the voice coil depends on mechanical suspension which is made out of impregnated fabric, rubber, foam and other material having varying properties. An offset from the optimal rest position may cause signal distortion which impairs the perceived sound quality. However, the measurement of the

Method	Description and application
Failure Mode and Effects Analysis (FMEA)	is an inductive method which captures engineering knowledge and experience with similar products, then identifies potential failure modes in future by defining causes, effects and actions and by rating severity, likelihood and detectability (Kmenta& Koshuke, 2004).
Root Cause Analysis (RCA)	attempts to find, correct or eliminate the cause of the problem as opposed to coping with obvious symptoms only (Andersen, 2006).
Fault Tree Analysis	is a deductive method which explains the undesired event by a logical combination of initial faults, external events in subsystems using Boolean logic (AND, OR gates), (IEC, 2006).
Fishbone Diagram	proposed by Ishikawa (1990) collects all causes and reasons grouped into major categories (e.g., people, material, measurements, environments).
Why-Because Analysis	is a more a posteriori analysis method developed for investigating causal relations between factors contributing to an accident.
Poerto chart	highlights the most important factors (sources) of defects according to the occurrence, cost, frequency of customer complaints or other quality criteria. This chart typically shows bars and line graphs where the bars represent the individual value of the defects in descending order and the lines show the cumulative value (Wilkinson, 2006).
Continuous Improvement Process (CIP)	is an ongoing effort to achieve incremental improvement using feedback from measurement instruments, statistical process control, ideas from customers, operators, QC and design engineers (Imai, 1997).
Design of Experiments	is a methodology (proposed by Fisher, 1971) to gather information on the effect of some controlled treatments applied to experimental units using comparison, randomization, replication and other principles.

Table 8. Methodologies used for initiating the optimal action in process control

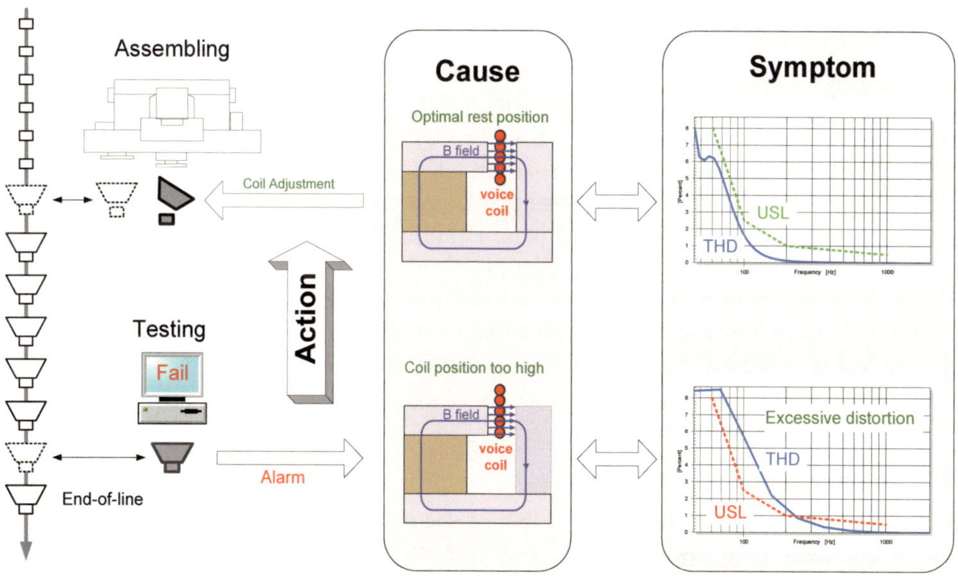

Fig. 19. Increasing the yield rate of the manufacturing process by process control (e.g., by voice coil adjustment using measured coil offset in mm using information generated during end-of-line testing)

total harmonic distortion (THD) is only a symptom and is less suitable for process control. Nonlinear identification techniques as discussed in Fig. 13 reveal the voice coil position and provide a quantitative value (in mm) for the adjustment of the voice coil position. The detection of a voice coil offset and a proper readjustment can be realized as continuous feedback control in automated assembly lines used for micro-speakers in telecommunication applications.

7. Data archiving and reporting

End-of-line testing produces an enormous amount of data. This section discusses alternative ways of storing the results in an effective format to support statistical analysis and the distribution of relevant information to different recipients.

The results of end-of-line tests reflect the quality of the total production process and are not only interesting for manufacturing, but also for the supplier of parts, design engineers in the R&D department, QC management and customers. However, each group needs a different part of the data. For example, manufacturing should immediately receive all parameters which are the basis for process control; however, those data are less important for the customer who is more interested in the overall quality of the products. The management is usually satisfied with the number of devices tested and the process yield. The designer from the R&D department is more interested in detailed measurement results of defective devices under test to understand the physical cause of the problem and discovering clues for improving design or simplifying manufacturing. The parts supplier (e.g., diaphragms in loudspeakers) only needs parameters (e.g., resonance frequency) which are important for his quality control and to address possible customer complaints. There is usually not enough time during end-of-line testing for isolating this information and generating a separate output file for each recipient in his preferred format. This problem becomes critical if the individual results of all measured devices under test are stored to ensure traceability about every step in the process chain. In this case, each device under test is identified by bar code or printing an individual label for each device under test. The computer associated with the measurement system is only used for writing a short entry for each measured device in the log file (i.e., general information such as serial number, date, time, verdicts, Pass/Fail result and selected important single-valued parameters).

In many applications, a digital format (e.g., database) is the only way to transfer a large amount of data (e.g., curves) as quickly as possible to a central computer (e.g., server) where the data is stored temporarily and the output file for each recipient is generated by a separate extraction process. The central storage of the test results allows, for example, matching loudspeaker units with similar acoustical properties which are sold as pairs to the customer.

8. Conclusion

Testing the manufactured device at the end of the assembly line differs significantly from the measurements performed during the development of the product. Most information, graphs, post-processing tools appreciated by an innovative R&D engineer are less useful in manufacturing where an identical replication of the prototype is important. End-of-line testing uses highly specialized test equipment providing limited information sufficient for quality assurance and process control. New sensor, signal analysis and system identification

techniques are used to ensure reliable detection of defects at an early stage before the product is shipped or mounted in the final application. Ultra-fast measurement techniques using an optimal stimulus are crucial for comprehensive testing within the available cycle time. End-of-line testing has to cope with measurement conditions which do not comply with R&D standards (e.g., anechoic room) and ensure comparability within the specification limits (e.g., transfer by Golden Reference Units). Invalid measurements caused by unavoidable disturbances in a production environment must be detected and repeated. This leads to new techniques (e.g., noise immunity) increasing the robustness of the test.
The measured physical variables and counts provided by end-of-line testing are the basis for process control. Trends and shifts must be detected early enough to adjust the process in time and to ensure a stable and capable production process (Six Sigma). Process control requires knowledge about the relationship between causes, symptoms and actions especially at automatic lines with continuous feedback and automatic adjustment of process parameters. One source of this knowledge is physical modeling provided by product development and applied to the particular requirements in manufacturing. A second source is the off-line diagnostic where a human operator investigates defective devices and extends the knowledge base continuously. This new task will replace manual handling and subjective evaluation which cannot meet modern requirement of 100 % testing of the products manufactured by an automated assembly line.

9. References

Andersen, B. (2006). *Root cause analysis : simplified tools and techniques,* ISBN-13 : 978-0-87389-892-4, American Society for Quality, Quality Press, Milwaukee,.

ASTM (2009). *Standard Practice for Conducting an Interlaboratory Study to Determine the Precision of a Test Method.* American Society for Testing and Materials (ASTM), E691 -09e1.

Bendat, J. & Piersol, A. (1980). *Engineering applications of correlation and spectral analysis,* ISBN 0-471-57055-9, New York, Wiley-Interscience, 1993.

Boashash,B. (2003). *Time-Frequency Signal Analysis and Processing – A Comprehensive Reference,* ISBN 0-08-044335-4, Elsevier Science, Oxford.

Cabot, R. (1999). Fundamentals of Modern Audio Measurement, *Journal of Audio Eng. Society,* Vol. 47, Number 9, Sept. 1999.

Fisher R.A. (1971). *The Design of Experiments,* ISBN 0028446909, 9th Edition, Macmillan.

IEC (2006). Fault Tree Analysis, Edition 2.0. International Electrotechnical Commission. ISBN 2-8318-8918-9. IEC 61025

Imai, M. (1997). Gemba Kaizen: A Commonsense, Low-Cost Approach to Management, 1st edition. McGraw-Hill. ISBN 0-07-031446-2.

Irrgang, S. et. all (2006). Loudspeaker Testing at the Production Line. *Preprint 6845 presented at the 120th Convention of Audio Eng. Society,* May 2006.

ITU (2001). *Recommendation BS.1387: Method for objective measurements of perceived audio quality* (PEAQ).

Ishikawa, K. (1990). Introduction to Quality Control. ISBN 4-906224-61-X, Chapman and Hall, 1991.

Klippel, W & Seidel, U. (2003). Measurement of Impulsive Distortion, Rub and Buzz and other Disturbances. *Preprint 5734 presented at 114th Convention of the Audio Eng. Society,* Amsterdam March 2003.

Klippel, W. (2006). Tutorial : Loudspeaker Nonlinearities - Causes, Parameters, Symptoms. *Journal of Audio Eng. Soc. Vol.* 54, Number 10, pp. 901-939.

Klippel, W. & Schlechter, J. (2009). Fast Measurement of Motor Suspension Nonlinearities in Loudspeaker Manufacturing. *Journal of Audio Eng. Soc.*, Vol. 58 Issue 3, pp. 115-125, March 2010.

Klippel, W. & Werner, R. (2010). Measurement of Turbulent Air Noise Distortion in Loudspeaker Systems, *presented at the 129th Convention of Audio Eng. Soc.*, San Francisco, November 2010, preprint 8174.

KLIPPEL GmbH (2011), *QC System – 100 % Testing in Manufacturing,* 21.3.2011, Available from http://www.klippel.de/our-products/qc-system.html

Kmenta, S. & Koshuke, I. (2004). Scenario-Based Failure Modes and Effects Analysis Using Expected Cost. *Journal of Mechanical Design* 126 (6): 1027, November 2004.

Montgomery, D. (2005). *Introduction to Statistical Quality Control.* Hoboken, New Jersey: John Wiley & Sons, Inc. ISBN 9780471656319.

Nelson, L. (1984). Technical Aids, *Journal of Quality Technology* 16, no. 4 (October 1984), 238-239.

Oppenheim A. & Schafer R. (2004). From Frequency to Quefrency: A History of the Cepstrum, *IEEE Signal Processing Magazine*, Vol. 21, Issue 5, Sept. 2004, S. 95-106

Percival, D & Walden, A. (2000). *Wavelet Methods for Time Series Analysis,* Cambridge University Press, 2000, ISBN 0-5216-8508-7.

WECO (1956). *Statistical Quality Control handbook.* (1 ed.), Western Electric Company, Indianapolis, Indiana: Western Electric Co., p. v, OCLC 33858387

Wilkinson, L. (2006). Revising the Pareto Chart. *The American Statistician* 60: 332–334.

Zadeh, L. (1987), *Fuzzy Sets and Applications, Selected Papers.* Wiley & Sons, New York, 1987.

11

Multi-Classifier Approaches for Post-Placement Surface-Mount Devices Quality Inspection

Stefanos Goumas[1] and Michalis Zervakis[2]
[1]*Technological Educational Institute of Kavala*
[2] *Technical University of Crete*
Greece

1. Introduction

In the continuing effort to shrink the electronics components and assemblies, the need for streamlined production processes and quality assurance is emerging stronger than ever. Surface Mounted Devices (SMD) is one of the breakthrough techniques that drove printed-circuit board production to a new level, increasing substantially the component density and reducing the size of produced circuits. Quality inspection of SMD is recognized as a critical and complex task in the production process (Bartlet et al., 1988). Speed is also important as to reduce the overall production costs (Lecklider, 2004). Specific SMD defects have been reported in the literature (Loh & Lu, 1999) including component misplacement and absence, component with wrong polarity, solder joint defects and component shifting. Much of the current research efforts are concentrated on detecting solder joint defects. The types of solder joint defects include surplus solder, insufficient solder and lacking solder. Component shifting has also been reported as a special solder joint defect.

Among the methods employed by industrial automatic solder-joint quality inspection systems are laser infrared signatures, digital radiography, laser Doppler vibrometry or laser acoustic microscopy. However, the high cost, low throughput and sampling loss of the above approaches call for research on non-destructive machine vision systems. Several optical inspection systems for solder joints have been reported (Bartlet et al., 1988; Capson & Eng, 1988; T-H. Kim et al., 1999; Ryu & Cho, 1997), using different illumination techniques and defect classification schemes. There have also been efforts to evaluate data fusion to combine data from various sensors for quality inspection of soldering processes (Lacey et al., 1993).

In an earlier work (Zervakis et al., 2004), our effort focused on overcoming the degrading effects of illumination and/or inaccurate measurements by exploiting stochastic modeling of lead displacement and its effects. As part of the aforementioned work, we have provided a novel framework to inspect the placement quality of SMD immediately after they have been placed in wet solder paste on a Printed Circuit Board (PCB). This type of inspection has the advantage that the critical data is available immediately after placement, so no extra time and components are spent on an already faulty PCB. Three measures of quality placement from individual lead images are of general interest, namely overlap, insulation distance and slump gap. Under general geometric conditions and using simple geometric relations, it can be shown that these measures are only affected by the displacement (i.e., shift and rotation) of the component, relative to its pad region (Goumas et al., 2002; Zervakis et al., 2004). Furthermore, positioning measures can be inferred from quantitative analysis of the inter-lead images. Instead of concentrating in one and every (poorly imaged) lead, we may fuse complementary information from all leads into a Bayesian estimation framework. In this work we attempt to further improve the positioning measurements of individual leads, by means of information fusion. To our knowledge this is the first time higher level (classifier) fusion is applied to the problem of automated solder joint inspection. More specifically, the quantification of positioning measures is viewed as a classification problem, where the lead displacement is inferred from characteristic features associated with image analysis for optical inspection. In order to overcome inaccuracies due to the poor optical quality of the component images, we use a variety of multiple classifiers fusion strategies based on statistical and soft computing methods to improve the performance of the classification task on individual leads.

Hyper classifiers or classifier ensembles have been intensively studied with the aim of overcoming the limitations of primary classifiers (Kittler et al., 1998; Xu et al., 1992). The most often used classifiers fusion approaches include the majority voting (Xu et al., 1992); the weighted combination (weighted averaging) (Kuncheva, 2004); the probabilistic schemes (Kittler et al., 1997; Kittler et al., 1998) ; various rank-ordered rules, such as the Borda count (Ho et al., 1994; E. Kim et al., 2002); the sum rule (averaging), product-rule, max-rule, min-rule, median rule (Kittler et al., 1998); the Bayesian approach (naïve Bayes combination) (Altincay, 2005; Kuncheva, 2004; Xu et al.,1992); the Dempster-Shafer (D-S) theory of evidence (Denoeux, 1995; Xu et al., 1992); the behavior-knowledge space method (BKS) (Huang & Suen, 1995; Shipp & Kuncheva, 2002); the fuzzy integral (Chi et al., 1996; Kuncheva, 2004; Mirhosseini et al., 1998); fuzzy templates (Kuncheva et al., 1998); decision templates (Kuncheva, 2001, 2004); combination through order statistics (Kang et al., 1997a, 1997b); combination by a neural network (Ceccareli & Petrosino, 1997). In a recent review paper (Oza & Tumer, 2008) a summary of the leading ensemble methods and a discussion of their application to four broad classes of real-world classification problems is provided. In addition, two novel information fusion approaches are presented recently in (Giacinto et al., 2008; Parikh & Polikar, 2007).

From the point of view of training and the form of *input pattern representation*, there are basically two classifier combination scenarios. In the first scenario, all the classifiers use the same representation of the input pattern (*identical pattern representation*). In the second scenario, each classifier uses its own representation of the input pattern (*distinct pattern representation*) (Kittler et al., 1997; Kittler et al., 1998; Mashao & Skosan, 2006; Rodriguez-Linares et al., 2003). In this case, the measurements extracted from the pattern are unique to each classifier, i.e., each individual classifier uses a different set of features. This poses practical encoding and statistical difficulties much in the sense of presenting a number of experts (classifiers) with different representations of the same phenomenon and deriving an unbiased overall decision. In fact one can argue that the primary classifiers that use distinct feature representations are inclined to be more biased than others operating on a common set of features. However it has the added benefit that the classification results tend to be uncorrelated and complementary, thus contributing to improved ensemble accuracy.

The objective of this chapter is to highlight the aspects of classifier fusion in solder-joint inspection and identify under which conditions it improves positioning measurements. Both fusion schemes, using identical and distinct pattern representations are considered. In the former case, the features for classification are obtained directly from the lead images. The latter scheme uses features that encode "reduced content" of the original images, i.e. the edge structure and the projection profile of leads. Notice that all features are obtained from the same primary source of information, i.e. the lead images. In this form, only a subset of features can be viewed as complementary. In fact, the edge and projection characteristics can be considered as only slightly correlated and, thus complementary, but the set of optical features are highly correlated to the rest. We elaborate on two schemes for distinct pattern representations. In the former scheme we use only reduced dimensionality features (i.e., topological and projection features), whereas the latter enriches the topological and projection features with optical ones, in order to improve the classification rates and robustness across all lead-displacement classes. By combining the power of the individual classifiers through multi-modular architectures we attempt to improve the classification results and enhance the robustness of the overall classification system. Furthermore, through the use of distinct feature representations we test the potential of using and combining reduced-content data towards increasing the speed of inspection.

With respect to the identical pattern representation scheme, we apply four representative schemes for soft fusion of multiple classifiers, ranging from simple majority voting to Bayesian, possibilistic and fuzzy schemes. The hyper classifiers used with the distinct pattern features are non-parametric methods. We use the product, sum, min and max combiners which are among the simplest classifier fusion rules, yet provide adequate results. We should notice here that following the process of classification of individual leads, we can further proceed with the Bayesian estimation approach developed in (Lacey et al., 1993; Zervakis et al., 2004) to accurately estimate component displacements based on the measurements from many individual leads.

The rest of this chapter is organized as follows. Section 2 involves a brief description of the experimental set up. In Section 3 we present all feature extraction schemes from individual lead images, along with the corresponding primary classifiers utilized. The formulation of various classifier combination strategies is described in Section 4. The results of the classifier fusion methods applied on SMD post placement quality inspection are presented in Section 5. Section 6 concludes this comparative study on several classifier combination methodologies with relevant observations and future research directions.

2. Experimental set up

A high dynamic range CMOS camera equipped with a simple LED illumination device and a general purpose processor performs the acquisition of the component image. For the purpose of presenting our results, QFP (Quad Flat Pack) SMD components with 120 leads (30 leads per side) are employed. Following image acquisition, the four regions of interest (ROIs), one at each side of the component, are isolated and all 120 small lead-images are extracted. The density of the CMOS sensor is 1024x1024 pixels, deriving an image resolution of 20x20μm per pixel. To capture the entire area of interest around each lead the size of the lead images is set to 36x56 pixels. Each lead image captures information about four material areas of interest, namely lead, pad, paste and background. Notice that the extraction of lead images can be easily customized to any conventional SMD component. Our approach aims to estimate each lead displacement over its ideal position centered at the pad/paste region. For the purposes of estimating other quality measures, only the displacement along the side of the component is essential (Lacey et al., 1993). Thus, our problem is restated as estimating lead displacement on the direction perpendicular to the lead axis. Essentially, we consider quantized displacement estimations organized at multiples of a pixel displacement. The displacement classes considered are {-6, -3, 0, +3, +6} and {-6, -4, -2, 0, +2, +4, +6}, in pixel displacements over the lead's central position.

Visual inspection techniques proceed by extracting features from segmented lead images. An example of input lead image and its segmented version is illustrated in Fig. 1.

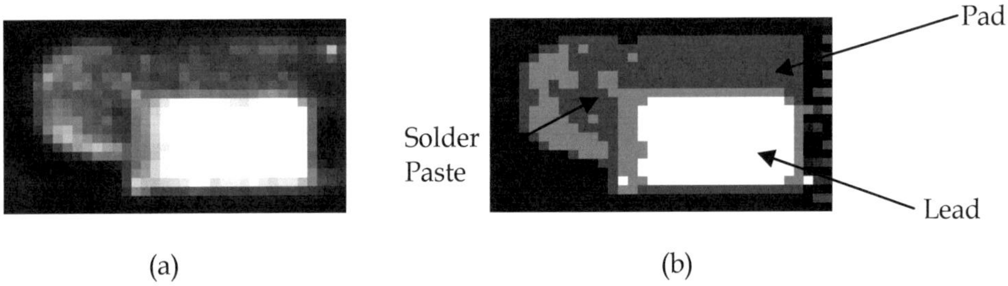

Fig. 1. Segmentation of lead image based on the 4-level Otsu algorithm: (a) original lead image, (b) segmented lead image.

In this case a four-level Otsu algorithm (Otsu, 1979) is applied on each ROI as to segment the lead images that are included in the examined ROI image. The 4-level Otsu-segmented image has been further processed by region growing/merging, labeling and line fitting approaches at the lead sides to arrive at the segmented result e.g. as in Figure 1b. The outcome of the segmentation algorithm is a four-level image that corresponds to the regions of lead, pad, solder paste and background. A labelling algorithm that relies on certain criteria, such as intensity, shape, location, and size, is subsequently applied in order to define (label) the four areas of interest.

3. Feature extraction and primary classifiers

Three sets of features are extracted from each segmented lead image, which encode different characteristics of this image. The first set encodes optical characteristics, by means of simple area measures that sustain the most desirable image attributes. The second set reflects only the edge information, whereas the third set pertains to features derived from the one-dimensional projection profile of the lead image in one direction. Even though all three sets are eventually derived from the same primary source (i.e. the original lead image), each set reflects different attributes of this source. The optical set may be seen as reflecting information from its full representation, whereas the latter two capture attributes of a reduced-attribute representation of the source. In fact, the second set is based on a reduced dynamic range representation of the binary edges, whereas the third set is based on a reduced dimensionality representation of the optical image through its intensity projection on only a single spatial dimension. In this form, the first set may be viewed as a "complete" characterization of the source image, whereas the other two can be interpreted as "incomplete", uncorrelated and complementary attributes of the same source. The last two approaches have been extensively evaluated in (Goumas et al., 2004).

3.1 Optical features

The feature extraction process is geared towards a single lead region. Our method requires only the roughly segmented lead, as in Fig. 2, which is simply enclosed by its bounding rectangle. For each lead we define two sub-regions presented in Fig. 2, based on the bounding rectangle. One region (L1) concerns the area where the lead is located and the other (L2) spans the pad area in front of the lead outwards the component. The area backwards the lead towards the body of the component is disregarded, since it contains misleading (non-useful) information. The features of each sub-region are appropriately normalized to the length of the corresponding region, in order to make them independent of the axial (u-direction) shift of the lead within the area of its pad.

From lead sub-region-1 (L1) we extract 7 features, which are the area of pad, the area of solder paste, the center-of-gravity distance on v axis between all (non-background) areas and the lead, the center-of-gravity distance on v axis between solder paste and lead, the center-of-gravity distance on v axis between solder paste and pad, the pad mean width on v axis, the pad total length on u axis. Furthermore, from lead sub-region-2 (L2) we extract the following 5 features: the area of pad, the area of solder paste, the center-of-gravity distance on v axis between all non-background regions and the lead, the center-of-gravity distance on v axis between solder paste and pad, and the elongation of pad.

For more details regarding the computation of the aforementioned features, the interested reader is referred to (Jain et al., 1995). The above 12 features constitute a feature vector for pattern classification of each lead. Any classifier can be utilized to perform this task. In our work we use a Bayes classifier, a multilayer Perceptron (MLP) neural network classifier and a learning vector quantization (LVQ) neural network classifier as primary classifiers for optical features.

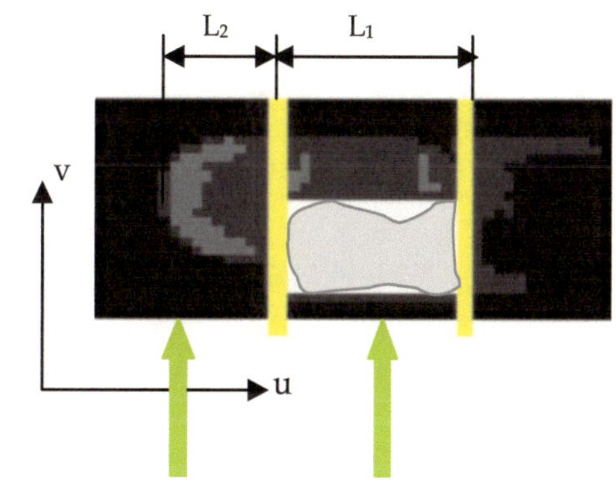

Fig. 2. The lead sub-regions used in the feature extraction process.

3.2 Reduced dynamic-range features

In our first approach related to data-space reduction, we utilize the edge structure extracted from the input lead image for classification purposes. We employ a Laplacian edge detector followed by simple thresholding. In most cases, the derived edge structure is partially deformed or destroyed. Thus, the major task is to relate edge patterns so that we can recall a class assignment for each test pattern that may be presented for classification. We exploit the concept of associative memories (AMs) as stored patterns representing the desirable classes and the Hamming distance for quantifying the distance between the input pattern and each one of the stored memories (Fausset, 1994). For classification of input patterns we employ the Hamming network, which is used to determine the proximity of an input vector to several exemplar vectors or prototype patterns. An input pattern that partially resembles the stimulus of an association invokes the associated response pattern by means of the shortest Hamming distance. Thus, an associative memory can retrieve a stored pattern given a reasonable subset of information embedded in that pattern. Moreover, an associative memory is error correcting in the sense that it can override inconsistent information in the cues presented to it. The input pattern to the network is a binary edge pattern obtained from the grayscale input lead image through segmentation and edge detection. The stored AM patterns reflect the edge structure of the "typical" edge image representing each class of lead displacements. Thus, the reduced-dimensionality binary edge image is fed into the Hamming network to determine pattern similarities and implement the desirable classifier. The classifier is trained for 5 and 7 classes, corresponding to integer lead displacements from -6 to +6 pixels per 3 and 2 pixel displacements, respectively. Its operation aims at selecting one of the stored patterns (or classes) that is at a minimum Hamming distance (HD) from the binary input vector. The Hamming network consists of two layers. The first layer calculates the M distances between the input vector $\mathbf{p}^{probe}$ and the stored $\mathbf{p}^1, \mathbf{p}^2, ..., \mathbf{p}^M$ fundamental memories in a feed-forward pass. The strongest response of neurons in this layer is indicative of the minimum HD between the input and the fundamental memories. In our implementation the input in Hamming neural network is a binary image 36×56=2016 pixels. Thus, the input vector of Hamming neural network has dimension 2016, i.e., the first layer of Hamming neural network is constituent of 2016 neurons. The second layer of the Hamming network is a winner-take-all network (MAXNET), implemented as a recurrent network. The MAXNET's ε parameter was set to ε=0.0385. The MAXNET suppresses all of its input values except the one at the maximum node of the first layer.

Given a set of binary prototype (exemplar) vectors p_i^k, $i = 1,...,N$ and $k = 1,...,M$, the operation of the Hamming network is summarized as follows.

1. For storing the M prototype vectors, initialize the weights:

$$w_{ij} = \frac{p_i^j}{2}, \ (i = 1,...,N; j = 1,...,M)$$

and the bias terms:

$$b_j = \frac{N}{2}, (j = 1,...,M)$$

2. For each unknown N-dimensional input vector $\mathbf{x}$ compute the input to each unit Y_j of second layer:

$$Y_j = b_j + \sum_{i=1}^{N} w_{ij} x_i(t), (j = 1,...,M)$$

3. Initialize activations for MAXNET:

$$y_j(0) = Y_j, (j = 1,\ldots,M)$$

4. MAXNET iterates to find the best-match exemplar pattern based upon the equation:

$$y_j(t) = f\left(y_j(t-1) - \varepsilon \sum_{k \neq j} y_j(t-1) \right)$$

where f is the activation function : $f(x) = \begin{cases} x, x > 0 \\ 0, x \leq 0 \end{cases}$ and ε is a small parameter $0 < \varepsilon < \frac{1}{M}$.

In our application we set $\varepsilon = \frac{1}{2}\left(\frac{1}{M}\right) = 0.0385$.

Since we exploit the concept of associative memory, the input pattern must have a structure similar to its closest one of fundamental memories. In order to enforce such pattern similarity, we fix the location of the lead in both the test image and the fundamental memories, so that large pattern differences in the comparison of two images can only be attributed to different shifts of the outside boundaries over the fixed lead location.

An important issue of associative memories is the definition of its *fundamental memories*. Each fundamental memory comprises the specific characteristics for discriminating its class. Moreover, the fundamental memories used in lead displacement must assess the standard characteristics of the problem, such as same image size, uniform lead position, etc. To satisfy these requirements, we first select the memory for one displacement (0 pixels) and then construct the memories associated the other classes by shifting the outside edge structure with respect to the fixed structure of the lead. The basic fundamental memory at shift 0 is selected from a number of test images reflecting exactly this specific case through statistical analysis of the mean pattern in this class. The resulting memories are depicted in Fig. 3.

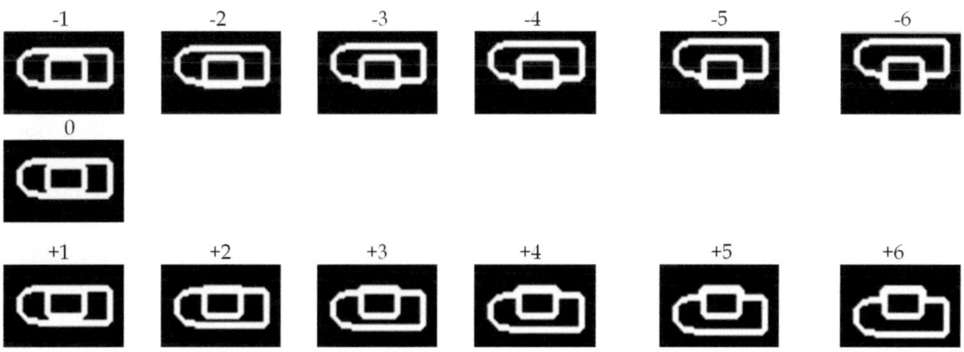

Fig. 3. Fundamental memories after dilation.

Based now on the design of the fundamental memories, we train the network so that it recovers the closest stored pattern in response to each test-input. An operation example of the associative memory in the case of a test image with +3 pixels lead-shift is illustrated in Fig. 4.

(a) (b)

Fig. 4. Associative memory operation: (a) test input image, (b) output response.

3.3 Reduced input-dimension features

In this approach we exploit the structure of the lead image profile (projection) along one, the most descriptive direction vertical to the lead axis for extracting meaningful features related to displacement measurements. The important component of this classification scheme is its feature extraction unit. We propose a complete feature extraction and classification approach that consists of three distinct modules. The first module receives the lead projection function at its input and utilizes a nonlinear filter based on a high-order neural network (HONN) for feature extraction. The second module implements feature reduction and de-correlation of the feature space by using the Karhunen-Loeve transform (KLT). The third module comprised by the Bayes classifier serves as a classifier that assigns each feature vector to one of the predetermined classes for lead displacements.

HONNs are fully interconnected single-layer networks, containing high order connections of sigmoid functions in their neurons. If we define as x, y its input and output respectively, with $x \in R^n$ and $y \in R^m$ the input-output representation of a HONN is given by:

$$y = \mathbf{W}^t \mathbf{S}(x) \quad (1)$$

where $\mathbf{W}$ is a $q \times m$ matrix of adjustable synaptic weights and $\mathbf{S}(x)$ is a q-dimensional vector of sigmoids. For sufficient high order terms, there exist weight values $\hat{\mathbf{W}}$ such that the HONN structure $\hat{\mathbf{W}}^t \mathbf{S}(x)$, can approximate an unknown function $f(x)$ to any degree of accuracy, in a compact domain (Rovithakis et al., 2001).

The KLT is used to de-correlate and reduce the dimensionality of feature vectors, disjoint class spaces in the new (reduced) feature space and aid the classifiers in performing accurate discrimination. The KLT projects the feature vector to the *K* most important directions. In essence, the KL transform projects feature vectors on the directions that best preserve class properties. Two different forms of the KLT are studied. In the first form only one KLT matrix (1 KLT) is created for the entire data set, whereas in the second form a KLT matrix is created for each class (each displacement). Thus, the first approach computes the most significant directions of the entire problem space and preserves directions where the data set expresses the largest diversion. In the second approach each individual class is represented by its most significant directions. Thus, it encompasses class specific characteristics and uses them to better isolate and discriminate classes by avoiding class mixing in irrelevant directions. The theoretical background of the multiple KLT approach is given in (Cappeli et al., 2001), whereas its application as a general analytic tool is established in (Goumas et al., 2002).

3.3.1 HONN based feature extraction

The HONN based feature extraction module receives as input a normalized projection function of the tested lead image (as in Fig. 5) and updates its weights by stable Lyapunov learning laws as to approximate this input function. Prior to entering, the input function is linearly transformed in the range [0, 1], as to avoid the appearance of destabilizing mechanisms caused by purely numeric issues, (i.e., large variations in the image projection data). Moreover, for uniformity reasons the rising point of this function is shifted to the origin. The position of the lead on the pad region determines the location of the main lobe in the projection function. Hence, the location of the main lobe and the overall structure of the projection function become the main characteristics that can be exploited for classifying the lead shift.

In the following we study the construction of the feature extraction system through an approximate modeling of the projection function and we rigorously analyze its performance. Let $x \in R_+$ be the data point on the projection axis, $y \in R_+$ be the projection value of lead image (R_+ denotes the set of positive real numbers), and f represent the actual but unknown projection function. Obviously the projection profile is modeled as a function $y = f(x)$. Moreover, let $\hat{y} = \mathbf{W}^t\mathbf{S}(x)$ be a HONN approximation of the actual projection function $f(x)$. Due to the one-dimensional structure of the problem, the HONN is designed for scalar input/output pairs linked at a higher dimension with a weight vector $\mathbf{W}$. Define the projection approximation error as

$$e = f(x) - \mathbf{W}^t\mathbf{S}(x) = y - \hat{y} \tag{2}$$

Observe that e is directly measured even though $f(.)$ is unknown. The nonlinear adaptive filter

$$\dot{z} = -\alpha z + y - \mathbf{W}^t\mathbf{S}(x), \ \alpha > 0, z \in R \tag{3}$$

equipped with the update law

$$\dot{\mathbf{W}} = -\gamma\mathbf{W} + z\mathbf{S}(x), \ \gamma > 0 \tag{4}$$

guarantees the uniform ultimate boundedness of its output $z \in R$ with respect to the arbitrarily small set

$$Z = \left\{ z \in R : |z| \leq \frac{\varepsilon}{2\alpha} + \frac{1}{2}\sqrt{\left(\frac{\varepsilon}{\alpha}\right)^2 + \frac{2\gamma|\hat{W}|^2}{\alpha}} \right\} \tag{5}$$

as well as the boundedness of the optimal HONN weights $\hat{\mathbf{W}}$ $\forall x \geq 0$ (Rovithakis et al., 2001). In the aforementioned relations α, γ are design constants and $\varepsilon \geq 0$ is an unknown but small bound on the HONN reconstruction error.

After convergence of the HONN, the feature vector $\mathbf{F}$ is formulated by the vector of trained weights $\mathbf{W}$ augmented with the approximation error e. In our approach, we form the feature vector $\mathbf{F}$ to be

$\mathbf{F} = \begin{bmatrix} \mathbf{W} \\ e \end{bmatrix} = \begin{bmatrix} w_1 \\ w_2 \\ \vdots \\ w_N \\ e \end{bmatrix}$, where $\mathbf{W} = \begin{bmatrix} w_1 \\ w_2 \\ \vdots \\ w_N \end{bmatrix}$ is the HONN weights vector (of specific dimension N=12) and e is the approximation error. This selection allows $\mathbf{F}$ to encode all HONN variables that characterize the projection function. An obvious feature is the approximation error e. Furthermore, since the HONN possesses a linear-in-the-weights property, the existence of a unique optimal vector $\hat{\mathbf{W}}$ different for each different projection function is guaranteed. Thus, the weights vector $\hat{\mathbf{W}}$ also serves the purpose of a relevant feature.

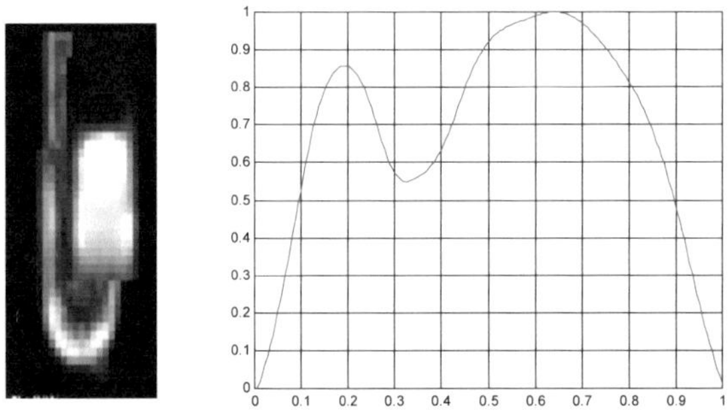

(a)

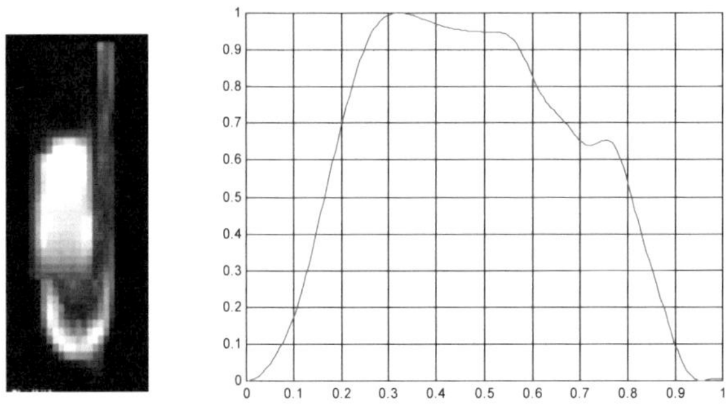

(b)

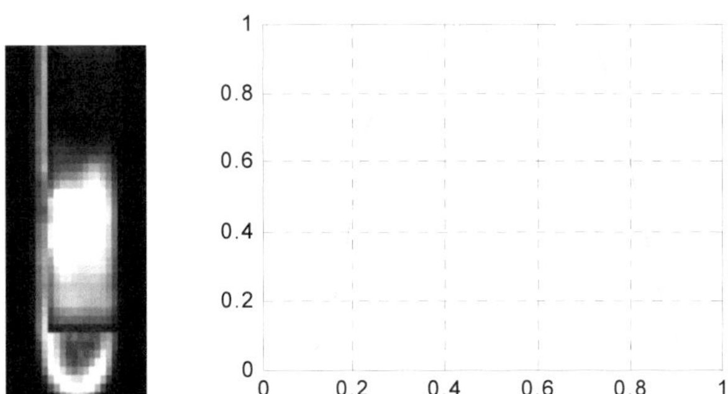

(c)

Fig. 5. Original lead images and projection functions for (a) +3 pixels, (b) -3 pixels, (c) zero pixels lead shift.

To approximate the unknown projection function the following HONN structure is used:

$$y = \mathbf{W}^T \mathbf{S}(x) = \sum_{i=1}^{3} w_i s_1^i(x) + w_4 s_2^4(x) + \sum_{i=5}^{8} w_i s_3^{(i-4)}(x) + \sum_{i=9}^{12} w_i s_4^{(i-8)}(x)$$

with

$$s_1(x) = \frac{0.9571}{1+e^{-35.703(x-0.076)}} + 0.2245, \quad s_2(x) = \frac{0.3838}{1+e^{-0.3598(x-1.488)}} - 0.2607$$

$$s_3(x) = \frac{0.9625}{1+e^{-22.4438(x-0.7927)}} + 0.5625, \quad s_4(x) = \frac{1.2906}{1+e^{-51.468(x-0.3287)}} - 0.3572 \ .$$

The HONN weights are updated according to:

$$\dot{w}_i = -0.000534 w_i + z s_1^i(x), i = 1,2,3, \quad \dot{w}_4 = -0.000756 w_4 + z s_2^4(x),$$

$$\dot{w}_i = -0.000825 w_i + z s_3^{(i-4)}, i = 5,6,7,8, \quad \dot{w}_i = -0.000407 w_i + z s_4^{(i-8)}, i = 9,10,11,12 \ .$$

The parameter α that appears in (5) is fixed to $\alpha = 8.0913$ through the use of a genetic algorithm. For the training of HONN, which derives 13 features from projection profiles, the network architecture involves 13 neurons. The use of Karhunen-Loeve transform (KLT) reduces the dimensionality down to 11.

4. Multiple classifier combination methods

4.1 Formulation of the combined classifier problem

In this chapter we assume that a small set of trained classifiers is available operating on the same dataset and we are interested in combining their outputs aiming at the highest possible accuracy. Let $C = \{C_1, C_2, \ldots, C_K\}$ be a set of classifiers and $\Omega = \{\omega_1, \omega_2, \ldots, \omega_M\}$ be a set of class labels. Each classifier gets as input a feature vector $\mathbf{x} \in R^n$. The classifier output is an *M*-dimensional vector $\mathbf{y}_i = C_i(\mathbf{x}) = [c_{i,1}(\mathbf{x}), \ldots, c_{i,M}(\mathbf{x})]^T$, where $c_{i,j}(\mathbf{x})$ is the degree of "support" given by classifier $C_i, i = 1, \ldots, K$ to the hypothesis that $\mathbf{x}$ comes from class $\omega_j, j = 1, \ldots, M$. Without loss of generality we can restrict $c_{i,j}(\mathbf{x})$ within the interval $[0,1]$ and call them "soft labels", with 0 meaning "no support" and 1 implying "full support", $i = 1, \ldots K, j = 1, \ldots M$ (Shipp & Kuncheva, 2002). Most often $c_{i,j}(\mathbf{x})$ is an estimate of the posterior probability $P(\omega_j \mid \mathbf{x})$. The process of combining classifiers attempts to combine the K classifier outputs $C_1(\mathbf{x}), \ldots, C_K(\mathbf{x})$ as to obtain a soft label for $\mathbf{x}$, denoted $C(\mathbf{x}) = [\mu_1(\mathbf{x}), \ldots, \mu_M(\mathbf{x})]^T$, where $\mu_j(\mathbf{x})$ denotes the overall degree of support for ω_j

given by the ensemble classifier. If a crisp class label of $\mathbf{x}$ is needed, we can use the maximum membership rule, which assigns $\mathbf{x}$ to class ω_s if,

$$c_{i,s}(\mathbf{x}) \geq c_{i,j}(\mathbf{x}) \ \forall\, j = 1,\ldots,M,\ i = 1,\ldots,K, \text{ for individual crisp labels and} \tag{6}$$

$$\mu_s(\mathbf{x}) \geq \mu_l(\mathbf{x}),\ \forall\, l = 1,\ldots,M,\ i = 1,\ldots,K, \text{ for the final crisp label} \tag{7}$$

The minimum-error classifier is recovered from (7) when $\mu_j(\mathbf{x}) = P(\omega_j \mid \mathbf{x})$. In the following we briefly address more advanced combination methods used in this chapter.

4.2 Non-trainable and probabilistic combination schemes for identical pattern representations

This type of combiners constitutes a group of simple, yet often surprisingly effective, methods for fusing the primary classifiers' soft labels. Their key advantage, apart from speed, is the fact that, having no tunable parameters, they do not impose a second training phase on the model. Additionally they belong to the group of class-conscious hyper classifiers since they utilize only one column of the ensemble's Decision Profile (Kuncheva, 2004). More specifically non-trainable combiners use only soft labels $c_{1,j}(\mathbf{x}), c_{2,j}(\mathbf{x}),\ldots,c_{k,j}(\mathbf{x})$ corresponding to class ω_j to estimate the fused maximum support value $\mu_j(\mathbf{x})$ for this class. A combination function F maps the primary to the fused labels (Kuncheva, 2004)

$$\mu_j(\mathbf{x}) = F\left[c_{1,j}(\mathbf{x}), c_{2,j}(\mathbf{x}),\ldots,c_{k,j}(\mathbf{x})\right] \tag{8}$$

Some popular choices for the functional F are the sample mean (average), min, max, median and the product rules. Furthermore, majority voting is a popular and easy to implement method (Xu et al., 1992). The primary classifiers "vote" with their class labels and the class label with most votes is assigned to $\mathbf{x}$.

Whereas the voting method only considers the result of each classifier, the approach of Bayesian formalism (E. Kim et al., 2002; Kuncheva, 2004) considers the error of each classifier. The "naïve Bayes" scheme assumes that the classifiers are mutually independent given a class label (conditional independence).

Consider the crisp class labels obtained from the K classifiers and let $L_1,\ldots,L_K$ be the class labels assigned to $\mathbf{x}$ by classifiers $C_1(\mathbf{x}),\ldots,C_K(\mathbf{x})$, respectively and denote by $P(L_j)$ the probability that classifier C_j labels $\mathbf{x}$ in class $L_j \in \Omega$. The conditional independence allows for the following representation

$$P(\mathbf{L} \mid \omega_k) = P(L_1,\ldots,L_K \mid \omega_k) = \prod_{i=1}^{K} P(L_i \mid \omega_k) \tag{9}$$

Then, the posterior probability needed to label $\mathbf{x}$ is given by

$$P(\omega_k \mid \mathbf{L}) = \frac{P(\omega_k)P(\mathbf{L} \mid \omega_k)}{P(\mathbf{L})} = \frac{P(\omega_k)\prod_{i=1}^{K} P(L_i \mid \omega_k)}{P(\mathbf{L})}, \quad k = 1,\ldots,M \tag{10}$$

Since the denominator does not depend on ω_k and can be ignored, the support for class ω_k by the set of classifiers can be computed as

$$\mu_k(\mathbf{x}) \propto P(\omega_k)\prod_{i=1}^{K} P(L_i \mid \omega_k) \tag{11}$$

Naïve Bayes fusion is applied as follows. Assuming *M* classes labeled 1 through M, the error for the *i*th classifier, $i = 1,\ldots,K$, can be represented by a two-dimensional *confusion matrix* as follows:

$$CM^i = \begin{pmatrix} \alpha^i_{1,1} & \cdots & \alpha^i_{1,M} \\ \vdots & \ddots & \vdots \\ \alpha^i_{M,1} & \cdots & \alpha^i_{M,M} \end{pmatrix} \tag{12}$$

For each classifier C_i, a $M \times M$ confusion matrix CM^i is calculated by applying C_i to the training data set. The *(k,L)* th entry of this matrix, $\alpha^i_{k,L}$ is the number of elements of the data set whose true class label was ω_k and were assigned to class ω_L by C_i. By N_L we denote the total number of elements of **S** that truly belong to class ω_L. Taking $\alpha^i_{k,L_i} / N_k$ as an estimate of the probability $P(L_i \mid \omega_k)$, and N_k / N as an estimate of the prior probability of class ω_k, Eq. (11) is equivalently written as:

$$\mu_k(\mathbf{x}) \propto \frac{1}{N_k^{K-1}} \prod_{i=1}^{K} \alpha^i_{k,L_i} \tag{13}$$

4.3 Multi-classifier combination based on fuzzy Integral

The theory of fuzzy measures and fuzzy integrals was first introduced by Sugeno (Mirhosseini et al., 1998) and has been successfully used in decision fusion (Chi et al., 1996). Contrary to fuzzy sets, in fuzzy measures a value is assigned to crisp subset of the universal set signifying the degree of evidence or belief that a particular element belongs in the subset. In this form, fuzzy measures are used to solve *ambiguity* associated with making a choice between two or more alternative decisions. In classifier fusion, the fuzzy measure relates to a measure of competence of each classifier. The ensemble support $\mu_j(\mathbf{x})$ for class ω_j, $j = 1,\ldots,M$ is obtained from the support values of individual classifiers $c_{i,j}(\mathbf{x}), i = 1,\ldots,K$, but also taking into account the competences of experts expressed through a *fuzzy measure* denoted by $g_i(\mathbf{x}), i = 1,...,K$. This form of fusion is implemented by means of a *fuzzy integral*. The Choquet fuzzy integral is often utilized, which is based on Sugeno's λ-fuzzy measure (Chi et al., 1996; Mirhosseini et al., 1998).

4.4 Multi-classifier combination based on Dempster–Shafer theory of evidence

The Dempster-Shafer theory of evidence (Xu et al., 1992), also known as the theory of belief functions, is a generalization of the Bayesian theory for subjective probability. This theory is more flexible than Bayesian when our knowledge is incomplete and we have to deal with uncertainty and ignorance. Belief functions allow us to assign degrees of belief (or evidence) for one event (i.e. support for a class from one classifier) based on evidence for a related event (i.e. support of this class from another classifier). In the context of measurement-level classifier combination, a method for evidence combination is presented in (Rogova, 1994) and is also adopted here.

For the input vector $\mathbf{x}$ the output of the ith classifier is $\mathbf{y}_i = C_i(\mathbf{x}), i = 1,...,K$. Let $\{\mathbf{t}_j\}$ be the training set for class ω_j and $\mathbf{r}_{i,j}$ be the mean output vector of the ith classifier on this training set. The support function for class ω_j by the classifier C_i can be obtained by using the Euclidean distance between $\mathbf{r}_{i,j}$ and $\mathbf{y}_i$:

$$d_{i,j} = \varphi(\mathbf{r}_{i,j}, \mathbf{y}_i) = \frac{\left(1 + \left\|\mathbf{r}_{i,j} - \mathbf{y}_i\right\|^2\right)^{-1}}{\sum_{k=1}^{M}\left(1 + \left\|\mathbf{r}_{i,k} - \mathbf{y}_i\right\|^2\right)^{-1}} \tag{14}$$

and the overall evidence for class j from this classifier is computed as:

$$e_j(\mathbf{y}_i) = \frac{d_{i,j} \prod_{k \neq j}(1 - d_{i,j})}{1 - d_{i,j}\left[1 - \prod_{k \neq j}(1 - d_{i,j})\right]} \tag{15}$$

Finally, evidences for all classifiers may be combined (according to a simplified Dempster-Shafer rule) to obtain a measure of confidence for each class ω_j for the feature vector $\mathbf{x}$ as $e_j(\mathbf{x}) \propto \prod_{i=1}^{K} e_j(\mathbf{y}_i)$, so that we may assign class ω_k to the feature vector $\mathbf{x}$ if $e_k(\mathbf{x}) = \max_{j=1}^{M}\{e_j(\mathbf{x})\}$.

4.5 Combination rules for distinct pattern representations

The case of distinct pattern representation poses an additional burden to the design of the combination rules, since the representations of information (features) are quite inhomogeneous. Nevertheless, based on a Bayesian framework for relating the available information similar simple rules can be derived for classifier combination under distinct pattern representations. Assume that K classifiers are available, each representing the given pattern by a distinct feature vector. We consider K conditionally independent feature subsets (or distinct pattern representations). Each subset defines a part of the feature vector, $\mathbf{x}^{(i)}$, so that $\mathbf{x} = \left[\mathbf{x}^{(1)}, \mathbf{x}^{(2)}, \ldots, \mathbf{x}^{(K)}\right]^T$, $\mathbf{x} \in R^n$. Notice that there is a one-to-one correspondence between each feature vector $\mathbf{x}^{(i)}$ and its underlying classifier $C_i, i = 1, \ldots, K$. Using a Bayesian framework, Kuncheva (Kuncheva, 2004) derives the posterior class probability using the entire information from all representations as:

$$P\left(\omega_j \mid \mathbf{x}\right) \propto P^{(1-K)}\left(\omega_j\right) \prod_{i=1}^{K} P\left(\omega_j \mid \mathbf{x}^{(i)}\right) \tag{16}$$

so that we can assign

$$\mu_j(\mathbf{x}) = \hat{P}^{(1-K)}\left(\omega_j\right) \prod_{i=1}^{K} c_{i,j}\left(\mathbf{x}^{(i)}\right) \tag{17}$$

Using product expansion, we can further expand this assignment as in (Kittler et al., 1998) to derive the sum combination rule:

$$\mu_j(\mathbf{x}) = \hat{P}\left(\omega_j\right)(1 - K) + \sum_{i=1}^{K} c_{i,j}\left(\mathbf{x}^{(i)}\right) \tag{18}$$

Furthermore, for equal prior probabilities, the assignment of Eq. (17) can be reduced to:

$$\mu_j(\mathbf{x}) = \prod_{i=1}^{K} c_{i,j}\left(\mathbf{x}^{(i)}\right) \tag{19}$$

and the sum combination rule (18) can be viewed as the average a posteriori probability for each class over all the classifier outputs (Kittler et al., 1998) :

$$\mu_j(\mathbf{x}) = \frac{1}{K} \sum_{i=1}^{K} c_{i,j}\left(\mathbf{x}^{(i)}\right) \tag{20}$$

The aforementioned combination rules (17) and (18) or their simplified versions (19) and (20) constitute the fundamental schemes for combining classifiers, each representing the given pattern by a distinct feature vector. Some additional non-trainable fusion strategies can be developed from these rules by considering the inequalities:

$$\prod_{i=1}^{K} c_{i,j}\left(\mathbf{x}^{(i)}\right) \le \min_{i=1}^{K}\left\{c_{i,j}\left(\mathbf{x}^{(i)}\right)\right\} \le \frac{1}{K} \sum_{i=1}^{K} c_{i,j}\left(\mathbf{x}^{(i)}\right) \le \max_{i=1}^{K}\left\{c_{i,j}\left(\mathbf{x}^{(i)}\right)\right\} \tag{21}$$

The relationship (21) suggests that the product and sum combination rules can be approximated by their upper or lower bounds deriving the max and min combination rules, respectively.

5. Experimental results

5.1 Generation of test data

In this research we present two types of results. The first one deals with simulated data operating in an external leave-one-out validation scheme. The results presented show the average accuracies attained for each lead displacement through this recursive cross validation scheme. The second type of results refers to testing on the real data. The training of classifiers is performed on the entire set of simulated data, whereas testing is performed on the completely independent set of real images. For the generation of simulated data we use the *Monte Carlo simulation* process (Bremaud, 1999; Fishman, 1996; W. Martinez & A. Martinez, 2002; Robert & Casella, 1999), in order to generate lead samples with appropriate size and intensity distributions for training the classifiers. Our purpose is to simulate after placement the four Regions of Interest (ROIs), namely left, right, bottom and top ROI of an entire QFP component consisting of four sides with 30 leads on each side.

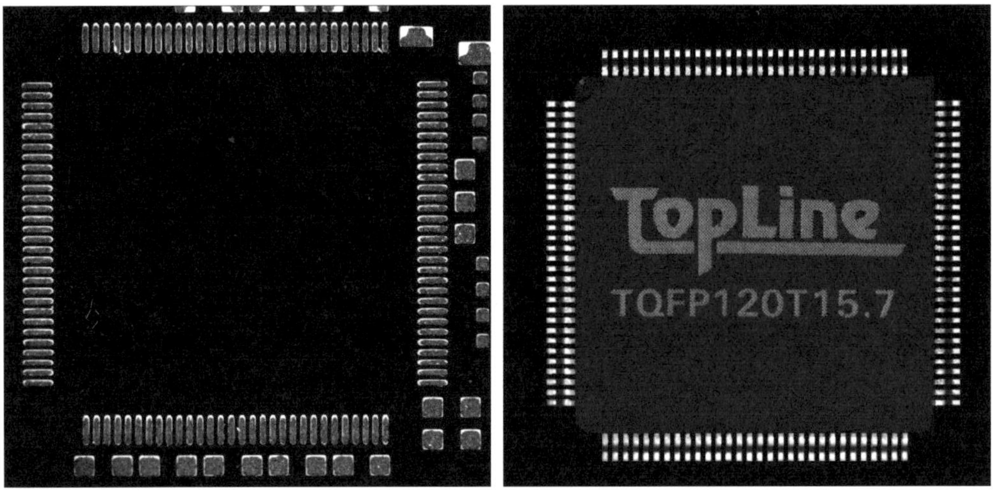

(a) (b)

Fig. 6. Pad and QFP component images; (a) pad and smeared solder paste, (b) the component QFP 120.

The data available consist of one set of 4 actual images of pad and smeared solder paste and another set of 5 actual images with only the component QFP 120 in front of a dark background (as in Fig. 6a and b, respectively). Thus, the process of simulation of new images with controlled component translations is based on the constituent images of individual pad regions with solder paste and individual lead regions. By superimposing individual lead images on pad regions (as in Figures 7b and 7a, respectively) we can control both the displacement and varying illumination conditions of the actual placement environment.

It is apparent that the leads inside a ROI may have varying dimensions and intensity levels. Similarly, the pads may have varying intensity levels due to the different distribution of the smeared solder paste. In order to simulate these varying factors of a ROI, we first estimate the distributions of these characteristics. For every lead-image we estimate length, height and mean intensity and for every pad-image we compute mean intensity. The distributions of these characteristics are used in the Monte-Carlo process that each time extracts a lead and pad image controlled by these distributions. Then, the displacement is implemented by shifting the lead on the pad region accordingly.

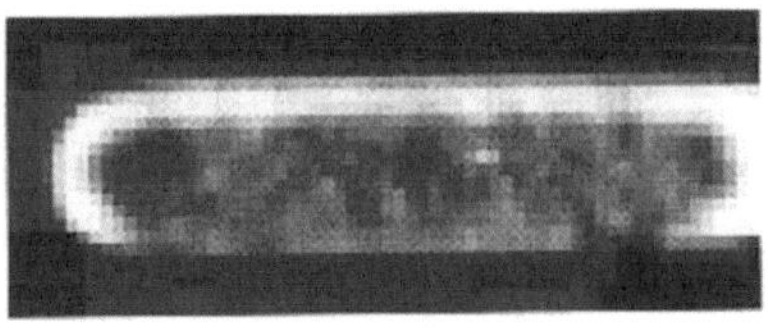

(a)

(b)

Fig. 7. Individual pad and lead images: (a) individual pad, (b) individual lead.

The Monte Carlo process simulates variable size and illumination conditions for individual leads and implements entire component displacements on the pad regions, which are then employed in training. For class labeling we define 13 classes of component displacements i.e., {-6, -5,..., 5, 6} pixels and each displacement involves three neighbouring values of displacement in simulations; e.g., class –4 involves displacements {-4.2, -4, -3.8}. Both directional displacements have been considered, namely horizontal and vertical.

From the total set of 13 classes we form two groups for training and testing the classifiers, composed of 5 and 7 classes, respectively. These two cases study the ability of the classifiers to discriminate classes in the feature space separated by 3 and 2 pixels apart, respectively. We do not consider training on all 13 classes, since all these classes are hardly separable in the feature spaces defined. A larger training sample size might improve class bounds and allow 13-class mapping. By increasing the target displacement step from 1 pixel (13 classes) to 2 pixels (7 classes) to 3 pixels (5 classes), we aim at balancing the classifiers' mapping capability to the overall displacement estimation accuracy. This means that although it is unfeasible to estimate small pad shifts using the given dataset, it is useful to derive larger shift estimates with high confidence. All the above apply for the given resolution level of the current dataset.

For testing with real images, a set of 20 real component images are kindly provided from the actual placement environment of Philips, The Netherlands. Ten actual boards with different shifts are provided, with two images from each case. Each individual case is controlled by the placement machine and conveys the limited accuracy of placement. The sides of each component are located and the individual lead areas are extracted. These images are used for testing of our developed algorithms; the training stage of classifiers is performed with the simulated data.

5.2 Classification results obtained from primary classifiers

For the simulated data, the training set consists of 120 lead images randomly generated from each class. To overcome the problem of statistical significance of the results caused by the rather small data set (for a quite high dimensional feature space) we apply a jack-knifing validation process. Regarding the design of the primary classifiers used in identical feature representation, the LVQ neural network architecture was defined by the feature vector size, training set size and output class mapping. In particular, for use with the 12 geometric (optical) features the LVQ input layer consisted of 12 neurons. In accordance to LVQ theory the hidden competitive layer contained a number of neurons equal to the number of training

set cases. In the output layer for 5 classes (2 pixel shift precision) 5 output neurons were used. Accordingly, discrimination of 7 classes required 7 output neurons. The model was trained for 1000 epochs with a learning parameter a=0.09. The MLP neural network was designed with 50 hidden layer neurons and 5 or 7 output neurons depending on the required output classes. The input layer was as above defined by the dimensionality of the feature vector. As stated before, 12 features per lead formulate the feature vector that forms the input to each classifier.
The design of primary classifiers for distinct pattern representation uses the topological features as input vectors to a Hamming neural network. As already mentioned, the first level of Hamming neural network is constituted of 2016 neurons. The second layer of the Hamming network is a winner-take-all network (MAXNET), implemented as a recurrent network. The MAXNET's ε parameter was set to ε=0.0385. For the case of projection features, which uses 13 features from projection profiles, the training of HONN is performed with a network architecture involving 13 neurons. The use of Karhunen-Loeve transform (KLT) reduces the dimensionality down to 11. These final 11 projection features are used by a non-parametric Bayesian classifier to test the efficiency of this representation.
The classification rates of primary classifiers on 5 classes of simulated lead-images are shown in Table 1a. Table 1b presents the classification rates of individual classifiers on 7 classes. The classification accuracies from the jack-knifing process are presented along with the 95% confidence interval. From the classification results of primary classifiers, we can initially conclude that the Bayes classifier provides better results than the MLP and LVQ classifiers on the 5-classes case. However, competing performances of Bayes and MLP classifiers are observed on the 7-classes case. Furthermore, as can be observed in Tables 1a and 1b, the discrimination between different classes becomes easier as we move to larger displacement intervals; the distinction of 3-pixel difference in Table 1a is more efficient than that of 2-pixel difference in Table 1b. Overall, we observe a large variance of each classifier's performance along the classes of interest.

Features type	Classifier type	- 6 pixels shift	-3 pixels shift	0 pixels shift	+ 3pixels shift	+6 pixels shift
optical	Bayes	97.75 (±0.73)	94.35 (±0.88)	94.35 (±1.35)	93.14 (±0.86)	98.25 (±0.75)
optical	MLP	95.32 (±0.97)	93.87 (±1.05)	92.28 (±1.40)	95.83 (±0.98)	97.63 (±0.99)
optical	LVQ	93.27 (±0.60)	90.67 (±0.76)	78.24 (±0.74)	95.42 (±0.80)	94.78 (±0.32)
topological	Hamming	86.67 (±1.20)	79.17 (±1.03)	95.00 (±1.42)	92.50 (±1.12)	93.33 (±0.84)
projection	Bayes	92.50 (±0.90)	93.33 (±0.96)	86.70 (±0.54)	93.33 (±0.88)	97.63 (±0.79)

Table 1. a. Classification rates of primary classifiers on 5 classes using Monte Carlo simulated images.

In the sequel we test the primary classifiers on the set of 20 real component images from the actual placement environment. The testing set consists of 120 lead-images obtained from the components of the corresponding class. The classification rates of primary classifiers on 5 classes for the real lead-images are shown in Table 2a, whereas Table 2b presents the classification rates of individual classifiers on 7 classes. As we observe by comparing the results for real and simulated data, there is a small decrease (ranging from 0.30 to 1.30 in different classes) in classification rates for the real data, which are used as an independent test set. Nevertheless, the results on real data are only slightly inferior to those from cross validation, indicating the robustness of developed techniques in realistic operation.

Features type	Classifier type	- 6 pixels shift	- 4 pixels shift	- 2 pixels shift	0 pixels shift	+2 pixels shift	+4 pixels shift	+6 pixels shift
optical	Bayes	93.00 (±0.45)	81.82 (±0.67)	75.72 (±0.75)	85.87 (±0.92)	77.80 (±0.97)	85.87 (±0.70)	91.86 (±0.57)
optical	MLP	91.18 (±0.83)	80.46 (±1.16)	80.36 (±2.94)	79.74 (±1.30)	82.41 (±1.46)	86.83 (±0.90)	88.27 (±0.89)
optical	LVQ	83.47 (±0.56)	76.42 (±1.43)	78.90 (±0.93)	57.06 (±0.13)	80.40 (±0.74)	74.35 (±0.37)	81.16 (±0.30)
topological	Hamming	85.00 (±0.94)	84.17 (±0.74)	77.50 (±0.69)	92.50 (±2.79)	82.50 (±1.85)	82.50 (±1.55)	88.33 (±1.07)
projection	Bayes	76.70 (±0.98)	80.83 (±1.03)	86.07 (±2.01)	68.64 (±0.66)	83.30 (±0.41)	90.00 (±1.91)	93.10 (±1.29)

Table 1. b. Classification rates of primary classifiers on 7 classes using Monte Carlo simulated images.

Features type	Classifier type	- 6 pixels shift	-3 pixels shift	0 pixels shift	+ 3pixels shift	+6 pixels shift
optical	Bayes	96.43	93.61	93.17	92.08	97.33
optical	MLP	94.56	93.19	91.84	94.42	96.37
optical	LVQ	92.73	91.24	79.66	94.77	94.19
topological	Hamming	85.83	78.56	94.13	92.06	92.79
projection	Bayes	91.46	93.76	87.24	92.95	97.18

Table 2. a. Classification rates of primary classifiers on 5 classes using real images.

Features type	Classifier type	- 6 pixels shift	- 4 pixels shift	- 2 pixels shift	0 pixels shift	+2 pixels shift	+4 pixels shift	+6 pixels shift
optical	Bayes	91.87	80.26	74.66	85.26	77.32	85.42	91.03
optical	MLP	90.33	79.74	79.23	78.44	81.92	86.05	87.49
optical	LVQ	83.16	77.28	79.57	56.59	79.63	73.94	80.71
topological	Hamming	84.25	83.79	76.63	91.16	81.59	81.37	87.75
projection	Bayes	75.68	79.96	84.75	67.60	82.45	89.26	92.53

Table 2. b. Classification rates of primary classifiers on 7 classes using real images.

The classifiers designed on reduced dimensionality features achieve lower classification rates than those designed on optical features. There are some exceptions to this general trend, especially associated with the class of zero pixel-shift. Nevertheless, the variance of these classifiers within and among classes is quite large, showing unstable performance. Regarding classification on 7 classes, the Hamming network based on topological (edge) features achieved high rates for the 0 pixel class, but this is rather incidental due to the large variation of results within this class.

On classes corresponding to negative lead shifts we observe slightly worse classification rates than in the positives shifts. This effect is more evident in the cases of reduced dimensionality features. Following extensive experimentation with the data set we conclude that this is attributed to lighting effects during acquisition of the test images, which result in better contrast in one direction.

5.3 Results of combined classifiers using identical pattern representations

The methods of section 4 are used here to combine the three primary classifiers (Bayes, MLP, LVQ) using different methodologies, but operating on the same feature sets (optical). For such a three-classifier combination, majority voting (MV) assigns classification to one class if two or three classifiers produce this same class. Otherwise, the input pattern is rejected. To apply the naïve Bayes (NB) combination, the conditional probabilities $P(L_i \mid \omega_k)$, $i = 1,2,3$, $k = 1,\ldots,5$ or $k = 1,\ldots,7$ are obtained from the resulting confusion matrices of individual classifiers on the training set. In a same manner, to fuse the results using the Choquet fuzzy integral (CFI), the initial fuzzy densities $g_i, i = 1,2,3$, are computed from the resulting confusion matrices of individual classifiers on the training set. The Dempster-Shafer (D-S) fusion is performed in the context of measurement-level classifier combination based on the proposed method in section 4.4.

The classification results obtained from the above four combiners are presented in tables 3a and 3b on 5 and 7 classes, respectively. As we observe from these tables, all combinations of classifiers achieve better performance than any individual classifier used for fusion. Examining deeper their performance, we conclude that the naïve Bayes and the Dempster-Shafer combiners achieve better overall performance than the other schemes, with the naïve Bayes reaching the best performance of all combining classifiers employed. The largest improvement achieved by the combined classifiers over the best individual classifier performance is also depicted in Tables 3a and 3b for the naive Bayes scheme. In fact, the maximum improvement is achieved by this fusion approach for the class of -3 pixels shift. The advantage of naïve Bayes combiner over the others fusion schemes, along with the advantage of primary Bayes classifier over the others individual classifiers, cannot be generalized. The ranking of classification schemes observed in this application is partially attributed to the stochastic properties of the data set, supporting the assumption that the distribution of our experimental data follows the normal (Gaussian) distribution.

Combiner type	- 6 pixels shift	-3 pixels shift	0 pixels shift	+ 3pixels shift	+6 pixels shift
MV	98.25 (±0.38)	95.13 (±0.98)	94.78 (±1.08)	96.41 (±0.33)	98.43 (±0.82)
NB	99.20 (±0.52) (>1.45)	98.33 (±0.81) (>3.98)	97.21 (±0.64) (>2.86)	98.84 (±0.70) (>3.01)	99.87 (±0.41) (>1.62)
CFI	98.34 (±0.34)	95.89 (±0.49)	95.44 (±1.02)	96.90 (±0.23)	98.79 (±0.67)
D-S	98.67 (±0.22)	97.71 (±0.19)	96.63 (±0.68)	97.56 (±0.95)	99.36 (±0.31)

Table 3. a. Classification rates of combined classifiers on 5 classes using identical (optical) features based on simulated images

Combiner type	- 6 pixels shift	- 4 pixels shift	- 2 pixels shift	0 pixels shift	+2 pixels shift	+4 pixels shift	+6 pixels shift
MV	94.12 (±0.75)	82.53 (±0.95)	81.05 (±0.83)	86.38 (±0.71)	83.44 (±0.53)	87.85 (±0.28)	92.48 (±0.90)
NB	96.82 (±1.13) (>3.82)	85.49 (±0.94) (>3.67)	83.87 (±0.80) (>3.51)	88.79 (±0.46) (>2.92)	85.21 (±0.55) (>2.80)	90.35 (±0.38) (>3.52)	95.53 (±0.61) (>3.67)
CFI	95.27 (±0.40)	83.76 (±0.69)	82.64 (±0.65)	87.30 (±0.98)	84.37 (±1.33)	88.54 (±1.21)	93.86 (±0.77)
D-S	95.79 (±0.89)	84.88 (±0.32)	83.26 (±0.76)	88.14 (±0.83)	84.69 (±0.20)	89.87 (±0.19)	94.64 (±0.34)

Table 3. b. Classification rates of combined classifiers on 7 classes using identical (optical) features based on simulated images

In the sequel we derive the classification results using the four combination schemes based on the classifiers Bayes, MLP and LVQ employing real images, given in Tables 2a and 2b. These results are presented in Tables 4a and 4b on 5 and 7 classes, respectively. By comparing these results with Tables 3a and 3b, we can detect a small decrease (ranging from 0.30 to 1.30 in different classes) in classification rates from the case of testing simulated data, which can be attributed to small differences in the formation of the training and the testing data. Nevertheless, by comparing them with the results of individual classifiers on real image data (Tables 2a and 2b), we observe a consistent increase of the success rate achieved by any fusion methodology.

Combiner type	- 6 pixels shift	-3 pixels shift	0 pixels shift	+ 3pixels shift	+6 pixels shift
MV	97.16	94.42	93.56	95.73	98.04
NB	97.70 (>1.27)	96.97 (>3.36)	95.56 (>2.39)	97.61 (>2.84)	98.61 (>1.28)
CFI	96.95	94.88	95.00	96.49	98.40
D-S	97.26	95.76	94.63	96.15	98.27

features based on real images.

Table 4. a. Classification rates of combined classifiers on 5 classes using identical (optical)

Combiner type	- 6 pixels shift	- 4 pixels shift	- 2 pixels shift	0 pixels shift	+2 pixels shift	+4 pixels shift	+6 pixels shift
MV	92.65	80.69	79.94	85.90	82.73	86.89	91.61
NB	95.33 (>3.46)	83.50 (>3.24)	82.76 (>3.19)	87.84 (>2.58)	84.29 (>2.37)	89.33 (>3.28)	94.44 (>3.41)
CFI	93.69	81.80	81.43	86.45	83.67	87.53	92.68
D-S	94.23	82.98	80.94	87.30	83.77	88.48	93.61

features based on real images.

Table 4. b. Classification rates of combined classifiers on 7 classes using identical (optical)

The high accuracies of fusion schemes can be partially attributed to the diversity of the three primary classifiers. It is the authors' opinion that an additional improvement can be achieved in the 7-class case by enriching the primary classifier's pool. This would require a very careful choice of additional classifiers that would contribute to the ensemble's diversity, if possible. The 5-class case is less likely to benefit since the obtained accuracies are already nearly maximized. Such a refinement might also render 1-pixel resolution shift estimation (13 classes) manageable. In any case, one has to keep in mind that model complexity should not outweigh possible minimal gains and that results have to be extended to other datasets.

The increased computational complexity of fusion in a real time inspection system was also a factor considered. The overhead in a multiple classification process of this type is additive. This problem is addressed in three ways towards minimizing this overhead. Firstly the number of classes is kept to a minimum required for quality inspection by quantizing the output displacements. Secondly the features used were chosen so that no intensive image processing or costly transformations are involved in their computation. Thirdly a minimal primary classifier pool is used whilst maintaining a decent misclassification rate. From a different point of view, classifiers in this application area can benefit from certain symmetries and prior knowledge inherent to the problem. Limiting the displacements to one axis (for the corresponding component side) reduces the degrees of freedom in problem specification and classifier design. Additionally, the shape and size of the areas is roughly known or can be easily inferred for any new dataset and thus geometry metrics can be used reliably.

5.4 Results of combined classifiers using distinct pattern representations

In this section, the combination rules of Section 4.5 are used to combine the two primary classifiers (Hamming classifier and Bayesian classifier), using two sets of distinct pattern representations (topological and projection) for individual lead classification. Four different combination rules are tested under the assumption of equal priors and their results are compared. Each combiner uses the outcomes of primary classifiers as estimates of a posterior class probability, in a soft-level combination manner.

The classification results obtained from the above four combiners are presented in tables 5a and 5b for 5 and 7 classes, respectively. As we observe from these tables, the sum combination rule achieves better performance than any individual classifier alone with the exception of the class of -3 pixels shift on the 5 class formulation and the class of 0 pixels shift on the 7 class formulation. The max combination rule follows closely in performance, whereas the worst results are achieved when using the product and min combination rules. These results are in close agreement with the findings of (Kittler et al., 1998), based on a theoretical error sensitivity analysis, where the sum combination rule is found to be much more resilient to estimation errors of the posterior probabilities $P\left(\omega_j \,|\, \mathbf{x}^{(i)}\right)$ than the product combination rule. In particular, the product combiner is oversensitive to classification estimates close to zero. Presence of such estimates from one classifier has the effect of veto on that particular class, regardless the outcome of other classifiers.

We should further emphasize that fusion may not improve the classification results for each and every lead displacement compared to the individual classifiers, but it rather improves the overall classification ability for all lead-shifts examined. Even though fusion increases the classification accuracy for lead shifts where individual classifiers generally lag in performance, there are a few cases where one or the other individual classifier (based on topological or projection features) by chance achieves extremely high accuracy. The results of primary classifiers show a large variance of performance across the lead displacements, as in Tables 1a and 1b or 2a and 2b for simulated and real data, respectively. From these results, we cannot claim that one individual classifier, either Hamming based on topological of Bayes based on projection features, surpasses the other in performance. Each one attains maximum performance by chance at some specific lead displacement. We cannot generalize such results of individual classifiers due to the limited number of available data. Notice that this large variation is reduced by the fusion approaches. Thus, fusion using distinct, reduced-content representations not only boost the overall classification performance, but also makes the overall classification performance more consistent across all lead-displacements examined.

Combination Rule	- 6 pixels shift	-3 pixels shift	0 pixels shift	+ 3pixels shift	+6 pixels shift
Product	83.08 (±0.46)	78.25 (±0.62)	84.23 (±0.70)	88.76 (±0.32)	89.68 (±0.09)
Sum	94.64 (±0.37)	90.68 (±0.78)	95.67 (±0.39)	97.42 (±0.57)	98.73 (±0.87)
Max	90.78 (±0.06)	87.14 (±0.79)	91.62 (±0.94)	93.46 (±0.34)	96.09 (±0.56)
Min	84.33 (±0.45)	79.04 (±0.58)	85.59 (±0.75)	89.94 (±0.83)	90.81 (±0.47)

Table 5. a. Classification rates of combined classifiers on 5 classes using distinct features (topological & projection) based on simulated images

Combination Rule	- 6 pixels shift	- 4 pixels shift	- 2 pixels shift	0 pixels shift	+2 pixels shift	+4 pixels shift	+6 pixels shift
Product	71.24 (±0.49)	75.81 (±0.32)	74.41 (±0.65)	70.93 (±0.97)	76.13 (±0.23)	78.95 (±0.31)	83.76 (±0.94)
Sum	85.08 (±0.46)	86.63 (±0.58)	86.12 (±0.77)	85.91 (±0.24)	87.38 (±1.22)	90.53 (±0.45)	94.97 (±0.75)
Max	81.59 (±0.64)	83.43 (±0.21)	82.36 (±64.)	81.22 (±0.57)	83.87 (±0.59)	87.11 (±0.88)	91.55 (±0.98)
Min	72.09 (±0.44)	76.88 (±0.23)	75.23 (±0.38)	71.64 (±0.55)	77.39 (±0.16)	79.83 (±0.29)	85.03 (±0.72)

Table 5. b. Classification rates of combined classifiers on 7 classes using distinct features (topological & projection) based on simulated images

Considering the classification of real data, the results of these four combination rules are presented in Tables 6a and 6b for the 5 and 7 class formulations, respectively. We recall that the individual classifiers used at first level are the Hamming neural network operating on topological features and the Bayes classifier operating on projection features extracted from the set of real images considered. As can be observed in Tables 6a and 6b, the sum combiner again achieves overall better results, but there is a small decrease (ranging from 0.30 to 1.30 in different classes) in classification rates in comparison with Tables 5a and 5b for the simulated data.

In general, the classification scores achieved using reduced dimensionality features are inferior to those obtained using the optical features. Furthermore, the combination of topological and projection features in a distinct representation fusion scheme also lags in performance to the combination of classifiers trained with optical features alone in Section 5.3. This is expected since all feature sets are obtained from the same primary source (original lead images), so that the information captured by topological and projection features does not add much to the information already conveyed by optical features.

Furthermore, the primary data in reduced content representation (1-bit edge images and 1-D projections) are inter-related, rendering the corresponding features (topological and projection) not quite independent. At this stage we do not perform any feature selection process, since we are focusing on the nature of primary data (edges and projections) and the information conveyed in these forms, rather than the nature of features. Nevertheless, it is worth mentioning that reduced dimensionality features using just a portion of information available can still attain acceptable results, especially through the employment of fusion. Reduced dimensionality features have the benefit of summarizing the required information for adequate shift detection in a compact format that can significantly reduce processing time. It is the authors' opinion that such features should be used when balance is required between speed and effectiveness. In addition, the particular reduced dimensionality features possess conceptual attributes that can instigate further speed-up and improvement in component inspection systems. More specifically, the topological features (edges) may be used for appropriate modeling of the component placement process and can be directly obtained from a number of commercial cameras, eliminating the need of preprocessing. The projection features on the other hand may eventually enable the use of faster and cheaper line sensors instead of area cameras for component inspection.

Combination Rule	- 6 pixels shift	-3 pixels shift	0 pixels shift	+ 3pixels shift	+6 pixels shift
Product	81.89	77.34	85.00	87.93	89.44
Sum	93.33	91.38	94.61	92.77	94.55
Max	89.28	87.94	90.47	94.38	95.35
Min	83.09	78.38	85.79	88.94	90.55

Table 6. a. Classification rates of combined classifiers on 5 classes using distinct features (topological & projection) from real images

Combination Rule	- 6 pixels shift	- 4 pixels shift	- 2 pixels shift	0 pixels shift	+2 pixels shift	+4 pixels shift	+6 pixels shift
Product	69.97	75.14	73.06	69.54	75.54	77.45	82.94
Sum	84.68	85.98	85.12	84.75	86.08	89.63	93.95
Max	80.41	82.77	81.31	86.10	81.67	86.01	91.24
Min	70.85	76.00	73.98	70.04	77.03	77.86	84.93

Table 6. b. Classification rates of combined classifiers on 7 classes using distinct features (topological & projection) from real images

Elaborating on the use of distinct feature representations and its potential in increasing accuracy and robustness for all classes of lead displacements, we further consider a combination of optical, topological and projection features. We define the resulting distinct features set (i.e., *optical* and *topological* and *projection features)* as ***distinct features-2***. This set of features captures information form many different aspects of the problem and contains features that are more likely to be independent than the set used before employing only topological and projection features. The quite diverse nature of information handled by each approach justifies the assumption of class conditional independence (at least approximately) for the distinct representations used by the individual classifiers. Motivated by good results of the sum combination rule we also use it as a fusion rule on the Bayes classifier with optical features, Hamming classifier with topological features and Bayes classifier with projection features. The classification rates based upon Monte Carlo simulated and real images are presented in Tables 7a and 7b on 5 and 7 classes, respectively. We observe that the sum combination rule achieves better performance than any individual classifier alone based on distinct features-2, with the exception of the class of 0 pixels shift on the 7 class formulation. It improves the results of the first level Bayes classifier and derives quite uniform results across all classes. Comparing this distinct feature combination with the one in Section 5.3 using identical pattern representation, we can claim that the former achieves comparable and at cases (7-class formulation) even better performance than the latter. This result further supports the potential of the distinct representation scheme, requiring however further investigation on the appropriate selection of distinct features, which is out of the scope of this work.

Sum Combination Rule	- 6 pixels shift	-3 pixels shift	0 pixels shift	+ 3pixels shift	+6 pixels shift
Simulated Images	98.37	97.14	96.27	97.45	99.73
Real Images	97.49	95.53	95.04	95.52	98.92

Table 7. a. Classification rates of Sum Combination Rule on 5 classes using distinct features-2 (optical & topological & projection) on simulated and real images

Sum Combination Rule	-6pixels shift	- 4 pixels shift	- 2 pixels shift	0 pixels shift	+2 pixels shift	+4 pixels shift	+6 pixels shift
Simulated Images	93.65	88.25	86.92	92.70	87.53	91.46	95.19
Real Images	92.45	86.66	85.44	92.28	86.18	90.27	94.12

Table 7. b. Classification rates of Sum Combination Rule on 7 classes using distinct features-2 (optical & topological & projection) on simulated and real images

5.5 System performance

With respect to time requirements, the tested approaches achieve the following performance using a fast Intel Core 2 Duo workstation. The optical feature approach takes about 0.34 sec for processing an entire QFP chip of 120 leads. The reduced dynamic-range approach requires 0.15 sec, less than half of the computation time of the conventional approach. Finally, the reduced input-dimension processing requires about 0.22 sec for the entire QFP-120 component. In comparison with existing industrial systems for PCB inspection, the proposed fusion system can achieve better throughputs, even though it considers each lead separately. The results of our survey of industrial systems are outlined on Table 8. The performance data for commercial products have been obtained through the vendors' online available product datasheets. In our case, the speed of the fusion algorithms was mapped to throughput (cm^2/sec) by simulating performance on a 120-lead QFP component of roughly 3.3×3.3cm surface at a sampling resolution of 20 μm/pixel. The speed of each algorithm was estimated with respect to the chip's total lead area. The reported times refer to processing alone, without including the board placement/ adjustment times required by the mechanical operation of the production line.

System	speed (cm^2/sec)	resolution (μm/pixel)
optical feature	32.4	20
reduced dyn range	72.6	20
reduced imput dim	49.1	20
Agilent Medalist SJ50 3	38.7	16
Orbotech Symbion P36	22-60	20
Viscom S3088	20-40	15

Table 8. Inspection speed comparison

High abstraction features are generally less descriptive than pixel-based features for classification purposes. With respect however to computational complexity, the distinct features in cooperation with a fusion scheme can yield appreciable reduction at the cost without compromising the effectiveness of inspection

6. Conclusion and future work

In this research, we tested several combination methods for soft fusion of the outputs of multiple classifiers. The aim is to improve the performance of primary classifiers used for individual lead-image classification in post-placement quality inspection of components. Two different schemes of classifier fusion are considered. The first one refers to identical feature representations, where the primary classifiers operate on the same feature set. The second scheme uses distinct pattern representation, where each of the primary classifiers operates on a different set of features. Comparing the classification results of the proposed combined classifiers, we can derive that all combiners have better performance than any individual classifier alone. In addition, it is verified that both the naïve Bayes and the Dempster-Shafer combiners on identical feature representations achieve better overall performance, with the naïve Bayes reaching the best performance improvement over the primary classifiers. The combiners based on distinct feature representations present lower overall performance and higher variability of their results. This is expected due to the

reduced content of information exploited. Despite that, their performance is still better than that of most primary classifiers, showing a good potential for accelerating the inspection process when speed needs to be balanced against effectiveness.
According to market studies (Frost & Sullivan, 2005), the PCB inspection field is in need of reliable systems in order to sustain growth as component densities get higher. Use of exhaustive solder paste inspection helps reduce the contribution from the print process to solder joint defects, in-turn saving money by reducing the cost of scrap with minimal cost to rework (i.e. wash boards) and with no penalty in solder joint reliability (Lecklider, 2004). Some companies claim this number to be as high as 80% of their overall defect Pareto chart (Mendez, 2000). Furthermore, the total misclassification cost in an automated optical inspection system is the product of the production volume, cost-per-defective PCB and accuracy. Taking into account the ranges of the first two variables it is evident that even a minor, yet consistent, improvement in classification accuracy is translated to amplified profits.
Overall, classifier fusion can contribute to the visual solder-joint inspection domain by improving accuracy and speed. One of the conditions under which fusion is favorable is the high diversity in features and primary classifier outputs. Evaluation of a number of diversity metrics indicated that using distinct representations (different feature sets) of leads, in most cases leads to a reduction in the correlation between the outputs of individual classifiers. This is attributed to the reduced correlation in the input vectors of distinct information content. Since this is a desirable feature in fusion, a further research is required to establish the effects of combining truly different input representations besides exploiting different attributes of the same primary source of information (as with the use of the same optical images to obtain the different features sets). Fusion at different levels (measurements, features, and outputs) can then be evaluated overall.

7. Acknowledgement

The authors would also like to thank Philips, The Netherlands, for the provision of test images.

8. References

Altincay, H. (2005). On naïve Bayesian fusion of dependent classifiers. *Pattern Recognition Letters*, 26 (2005) 2463 – 2473.

Bartlet, S., Besl, P., Cole, C., Jain, R., Mukherjee, D., & Skifstand, K. (1988). Automatic Solder Joint Inspection. *IEEE Trans. Pattern Analysis and Machine Intelligence*, Vol. 10, No. 1, pp. 31-41.

Bremaud, P. (1999). *Markov Chains, Gibbs Fields, Monte Carlo Simulation, and Queues*, Springer Science + Business Media, New York, USA.

Cappelli, R., Maio, D. & Maltoni, D. (2001). Multispace KL for Pattern Representation and Classification. *IEEE Trans. on Pattern Analysis and Machine Intelligence*, Vol. 23, No. 9, pp. 977-996.

Capson, D. & Eng, S. (1988). A tiered-colour illumination approach for machine inspection of solder joints, *IEEE Trans. Pattern Analysis and Machine Intelligence,* Vol. 10, No. 3, pp. 387-393.

Ceccareli, M. & Petrosino, A. (1997). Multi-feature adaptive classifiers for SAR image segmentation. *Neurocomputing*, 14 (1997) 345-363.

Chi, Z., Yan, H. & Pham, T. (1996). *Fuzzy algorithms: With Applications to Image Processing and Pattern Recognition, Advances in Fuzzy Systems – Applications and Theory, Vol. 10,* World Scientific Publishing Co. Pte. Ltd., USA.

Denoeux, T. (1995). A k-nearest neighbor classification rule based on Dempster- Shafer theory. *IEEE Transactions on Systems, Man and Cybernetics - Part B*, Vol. 25, No. 5, pp. 804-813.

Fausset, L. (1994). *Fundamentals of Neural Networks Architectures, Algorithms, and Applications*, Englewood Cliffs, NJ: Prentice-Hall, USA.

Fishman, G. (1996). *Monte Carlo: Concepts, Algorithms and Applications*, Springer-Verlag, New York, USA.

Frost & Sullivan. (2005). Growth Opportunities for the World SMT Inspection Equipment Markets. Market survey report, May 2005.

Giacinto, G., Perdisci R., Del Rio, M. & Roli, F. (2008). Intrusion detection in computer networks by a modular ensemble of one-class classifiers. *Information Fusion*, 9 (2008) 69 - 82.

Goumas, S., Zervakis, M. & Stavrakakis, G. (2002). Classification of Washing Machines Vibration Signals Using Discrete Wavelet Analysis for Feature Extraction. *IEEE Trans. on Instrumentation and Measurement*, Vol. 51, No. 3, (June 2002), pp. 497-508.

Goumas, S., Rovithakis, G. & Zervakis, M. (2002). A Bayesian Approach to Post Placement Quality Inspection of Components, *Proceedings of IEEE Int. Conf. on Image Processing*, 2002.

Goumas, S., Zervakis, M. & Rovithakis, G. (2004). Data-Space Reduction in Component Quality Inspection, *Proceedings of ESANN'04*, pp. 275-280.

Ho, T., Hull, J. & Srihari, S. (1994). Decision combination in multiple classifier systems. *IEEE Transactions on Pattern Analysis and Machine Intelligence*, Vol. 16, No. 1, pp. 66-75.

Huang, Y. & Suen, C. (1995). A Method of Combining Multiple Experts for the Recognition of Unconstrained Handwritten Numerals. *IEEE Transactions on Pattern Analysis and Machine Intelligence*, Vol. 17, No. 1, (January 1995), pp. 90-94.

Jain, R., Kasturi, R. & Schunck, B. (1995). *Machine Vision,* McGraw-Hill International Editions, Singapore.

Kang, H., Kim, K. & Kim, J. (1997a). Optimal approximation of discrete probability distribution with *k*-order dependancy and its application to combining multiple classifiers. *Pattern Recognition Letters*, 18 (1997) 515-523.

Kang, H., Kim, K. & Kim, J. (1997b). A Framework for Probabilistic Combination of Multiple Classifiers at an Abstract Level. *Engng Aplic.Artif. Intell.*, Vol. 10, No. 4, pp. 379-385.

Kim, T-H., Cho, T-H., Moon, Y. & Park, S. (1999). Visual Inspection System for the Classification of Solder Joints. *Pattern Recognition* 32 (1999) 565-575.

Kim, E., Kim, W., & Lee, Y. (2002). Combination of multiple classifiers for the customer's purchase behavior prediction. *Decision Support Systems*, 34 (2002) 167-175.

Kittler, J., Hojjatoleslami, A. & Windeatt. (1997). Strategies for combining classifiers employing shared and distinct pattern representations. *Pattern Recognition Letters*, 16 (1997) 1373-1377.

Kittler, J. Hatef, M., Duin, R. & Matas, J. (1998). On Combining Classifiers. *IEEE Transactions on Pattern Analysis and Machine Intelligence*, Vol. 20, No. 3, (March 1998), pp. 226-239.

Kuncheva, L., Bezdek, J., & Sutton M. (1998). On combining multiple classifiers by fuzzy templates, *Proceedings of NAFIPS'98*, Pensacola, FL, 1998, pp. 193-197.

Kuncheva, L. (2001). Using measures of similarity and inclusion for multiple classifier fusion by decision templates. *Fuzzy sets and systems*, 122 (2001) 401- 407.

Kuncheva, L. (2004). *Combining Pattern Classifiers: Methods and Algorithms*, Wiley-Interscience publication, New Jersey, USA.

Lacey, G., Waldron, R., Dinten, J-M. & Lilley, F. (1993). Flexible Multi-Sensor Inspection System for Solder-joint Analysis. *Proceedings of SPIE Machine Vision Applications, Architectures, and Systems Integration {II}*, 1993, No. 2064.

Lecklider, T. (2004). PCB Inspection Outlook for 2005. *Evaluation Engineering*, (Dec 2004).

Loh, H-H. & Lu, M-S. (1999). Printed Circuit Board Inspection using Image Analysis. *IEEE Trans. Industry Applications*, Vol. 35, No.2, pp. 426-432.

Martinez, W. & Martinez, A. (2002). *Computational Statistics Handbook with MATLAB*, Chapman & Hall / CRC, Florida, USA.

Mashao, D. & Skosan, M. (2006). Combining Classifier Decisions for Robust Speaker Identification. *Pattern Recognition*, 39 (2006) 147-155.

Mendez, D. (2000). An Integrated Test and Inspection Strategy, *Proceedings of* APEX, 2000.

Mirhosseini, A., Yan, H., Lam, K-M. & Pham, T. (1998). Human Face Image Recognition: An Evidence Aggregation Approach. *Computer Vision and Image Understanding*, Vol. 71, No. 2, (August 1998), pp. 213-230.

Ng, G. & Singh, H. (1998). Data equalization with evidence combination for pattern recognition. *Pattern Recognition Letters*, 19 (1998) 227 - 235.

Otsu, N. (1979). A Threshold Selection Method from Grey-Level Histograms. *IEEE Trans. On Systems, Man and Cybernetics*, Vol. 9, No. 1, pp. 62-66.

Oza, N. & Tumer, K. (2008). Classifier ensembles: Select real-world applications. *Information Fusion*, 9 (2008) 4-20.

Parikh, D., & Polikar, R. (2007). An Ensemble-Based Incremental Learning Approach to Data Fusion. *IEEE Trans. On Systems, Man, and Cybernetics - Part B: Cybernetics*. Vol. 37, No. 2, (April 2007), pp. 437-450.

Robert, C. & Casella, G. (1999) *Monte Carlo Statistical Methods*, Springer-Verlag, New York, USA.

Rodriguez-Linares, L., Garcia-Mateo, C. & Alba-Castro, J-L. (2003). On Combining Classifiers for Speaker Authentication. *Pattern Recognition*, 36 (2003) 347-359.

Rogova, G. (1994). Combining the Results of Several Neural Networks Classifiers. *Neural Networks*, Vol. 7, No. 5, pp. 777-781.

Rovithakis, G., Maniadakis, M., Zervakis, M., Filippidis, G., Zacharakis, G., Katsamouris, A. & Papazoglou, T. (2001). Artificial Neural Networks for Discriminating Pathologic From Normal Peripheral Vascular Tissue. *IEEE Trans. on Biomedical Engineering*, Vol. 48, No. 10, (Oct. 2001), pp. 1088-1097.

Ryu, Y. & Cho, H. (1997). A Neural Network Approach to Extended Gaussian Image Based Solder Joint Inspection. *Pergamon: Mechatronics*, Vol. 7, No.2, pp. 159-184.

Shipp, C. & Kuncheva, L. (2002). Relationships between combination methods and measures of diversity in combining classifiers. *Information Fusion*, 3 (2002) 135-148.

Xu, L., Krzyzak, A. & Suen, C. (1992). Methods of Combining Multiple Classifiers and Their Applications to Handwriting Recognition. *IEEE Transactions on Systems, Man and Cybernetics*, Vol. 22, No. 3, (May/June 1992), pp. 418-435.

Zervakis, M., Goumas, S., & Rovithakis, G. (2004). "A Bayesian Framework for Multi-lead SMD Post-Placement Quality Inspection", *IEEE Transactions on Systems, Man and Cybernetics - Part B: Cybernetics*, Vol. 34, No. 1, (February 2004), pp. 440-453.

12

Machine Vision for Inspection: A Case Study

Brandon Miles[1] and Brian Surgenor[2]
[1]*University of Western Ontario*
[2]*Queen's University*
Canada

1. Introduction

Automated inspection systems have the potential to significantly improve quality and increase production rates in the manufacturing industry. Machine vision (MV) is an example of one inspection technology that has been successfully applied to production lines. A wide variety of industrial inspection applications of MV systems can be found in the literature. For example, Lee et al. (2007) applied a MV system to bird handling for the food industry. Reynolds et al. (2004) looked at solder paste inspection for the electronics industry. Gayubo et al. (2006) developed a system to locate tearing defects in sheet metal.

There has also been numerous laboratory based work on MV systems with practical applications. Jackman et al. (2009) used a vision system to predict the quality of beef. Kumar (2003) worked on the detection of defects in twill weave fabric samples. Garcia et al. (2006) checked for missing and misaligned electronics components. Hunter et al. (1995) confirmed circularity in brake shoes. Kwak et al. (2000) identified surface defects in leather.

Although the range of applications is broad, they all tended to adopt the same image processing system with four main stages. The first is image acquisition. This is followed by preprocessing of the image, including applying various filters and selecting regions of interest. The third stage is feature extraction where individual features are extracted from the image. Finally a classifier is used to determine whether a given part is acceptable or not.

The automotive industry presents a particularly challenging environment for MV based inspection. With changing lighting conditions in a dirty environment there is a need for robust and accurate classifiers to perform accurate inspection. Feature selection routines can be used to improve the results of an ANFIS based classifier for automotive applications (Miles and Surgenor, 2009). Although there is potential for good results with this approach, it can take hours of processing time to compute an accurate solution (Killing et al., 2009).

This chapter presents the results of a project where six classification techniques were examined to see if development time could be reduced without sacrificing performance. As a case study, the problem of fastener insertion to an automotive part known as a cross car beam was investigated. Images taken from a production assembly line were used as the source of the data. The types of classifiers under investigation were: 1) a Neural Network based processor, 2) Principle Component Analysis to reclassify the input feature set and 3) a direct Eigenimage approach to avoid the need to extract features from each image. These methods were compared in terms of classification accuracy. An additional data set was also used to test the performance of these classifiers in detecting orientation defects in addition to presence and absence of clips. The results of these investigations are presented with a comparison of the performance on different datasets.

2. Techniques

Data for classifiers can be generated from input images in a variety of ways. The first is a feature based classifier. In this approach, features such as lines, holes and circles are extracted from an image. Numerical values are then obtained from these features, such as the x, y coordinates of a circle. These values can then be used as inputs to either a traditional Neural Network or a Neuro-Fuzzy System such as ANFIS. Principle Component Analysis (PCA) can be used to reclassify features to generate a data set with fewer inputs.

A second method to generate input values involves the application of Eigenimages. Eigenimages are generated from a set of training images. New images can be expressed as combinations of these Eigenimages. Coefficients are assigned to all the Eigenimages used to express a given set of input images. These coefficients can then be used to train either a Neural Network or an ANFIS system.

The six specific types of classifiers under investigation for this study are summarized in Table 1. They are grouped into Feature Based classifiers and Eigenimage Based classifiers along with ANFIS versus Neural Networks. This allows for a comparison between different classification techniques namely Neural Networks and ANFIS. However more importantly, it compares the Eigenimage based approach, which works directly on the pixels to produce classifier inputs, with the feature based approach, where features are first identified and then a classifier is trained.

	ANFIS	**Neural Network**
Feature Based Methods	Feature based with ANFIS	Feature based with a Neural Network
	Feature based with PCA and ANFIS	Feature based with PCA and a Neural Network
Eigenimage Based Methods	Eigenimage based with ANFIS	Eigenimage based with a Neural Network

Table 1. Summary of the six classifiers under investigation.

2.1 ANFIS

ANFIS possesses a full Fuzzy Inferencing Structure. A fuzzy structure is established as a preliminary model of the system. It can then be updated with additional inputs. Thus, it is trainable like a Neural Network. In this application, the ANFIS system was implemented using the MATLAB Fuzzy Toolbox. The reader is referred to (Roger Jang, 1993) for further background on this technique. The ANFIS technique was used as the benchmark in previous work reported in Killing et al. (2009).

2.2 Neural networks

The Neural Network used in this study was a Multi Layer Perceptron (MLP) network with one hidden layer. The hidden layer had a sigmoidal activation function and the output layer had a linear activation function. This approach was taken because the output value needed to be a scalar instead of the more common binary value. Twenty hidden nodes were used to minimize the possibility of over training the system while giving it sufficient size to be useful.

2.3 Principal component analysis

PCA can be used to reclassify the data in terms of maximized variance between input data and output data. By reorganizing the inputs into components ranked by variance the input data set can be reduced in size by keeping only the components with a high degree of variance. In this case, as chosen by experience, only the components comprising the top 95% of the variance were kept. This was typically five to seven inputs.

To calculate the principle components Singular Value Decomposition (SVD) was used. The sample covariance matrix **S** (calculated from the supplied data) was used to find the principle components:

$$S = \overline{X}\overline{X}^T \quad (1)$$

where $\overline{X}$ is X calculated around the sample means or $\overline{X} = (X - c)$ and c is the mean of X. Once S is calculated the Eigenvalues and eigenvectors of S can be calculated and the principle components can subsequently be found based.

2.4 Eigenimages

The Eigenimage classifier offers a more direct approach without the need to extract features from images. Grey scale images can be represented as vectors of pixels. In this way the dataset X can be generated by vectors of images where $X = [x_1, x_2, x_3 \dots x_n]$ for n sample images. SVD can then be performed identifying the principle components $[e_1, e_2, e_3, \dots, e_n]$. These Eigenimages or principle components are a series of images that when combined are able to represent the entire dataset of images. It is often desirable to only select the principle components that have the largest variances $E = [e_1, e_2, \dots e_k]$. In this case, it is possible to project a given image onto the Eigenspace constructed from these principle components. This projection generates an Eigenpoint from the image **x** with a set of project coefficients $p = [p_1, p_2, \dots p_n]$. This projection is calculated by:

$$p = E(x - c) \quad (2)$$

Once this projection is known the set of points p can be used to classify the images. Note that p presents a smaller set of input data than the entire image, which presents benefits computationally. In this way the PCA technique acts like a feature detector, producing numerical values from an image. The theory of EigenImages is discussed further in Sun et al. (2007) and Ohba and Ilkeuchi (1997). Both a Neural Network and ANFIS can be trained based on these principle components.

2.5 Features

Lines, Holes and Circles can be found in an image using a Hough transform. Additionally large colour blobs can be located based on locations of certain colours of pixels. Also given a specific region of interest in the images the average red, green, blue and grey scale intensity values can be found. A radial hole method as detailed in Miles and Surgenor (2009) has also been used. These features are found relative to the centre of the beam where possible.

Two new features have been introduced in order to help improve the accuracy of the results. These are a Generalized Hough Rectangle feature and PCA colour feature.

2.5.1 Hough rectangle feature

Lines, Holes and Circles can be found in an image using a Hough transform. Additionally One of the extensions of the generalized Hough transform (GHT) is using it to find rectangles. The symmetry that rectangles have can be employed to help locate them. First the gradient direction and magnitude of all the pixels are calculated. Say the rectangle has side lengths A and B, where A>B. The sides A are oriented in the direction of the major axis of the rectangle. The sides B are oriented in the direction of the minor axis of the rectangle. Assume that the angle of A is between 0 and 90°, then for a given edge pixel, if it's gradient is between 0 and 90° cast votes for a rectangle with a centre on a line at ±B/2 pixels away from the edge pixel in the direction of the edge pixel's gradient. Alternatively if the directional gradient of the edge pixel is between 90° and 180° then cast votes for a rectangle centred on a line at ±A/2 from the edge pixel.

Then in another plane accumulate votes assuming that the angle of the major axis is between 90° and 180°. If the directional gradient of an edge pixel is between 90° and 180° cast votes for rectangles centred on a line A/2 pixels away from the edge pixel in the direction of the edge pixel's gradient. Alternatively if the directional gradient of the edge pixel is between 0 and 90° then cast votes for a rectangle centred on a line at ±B/2 from the edge pixel.

This approach will produce two prominent peaks in one of the two accumulators. The largest one is for the major axis and the smaller one is for the minor axis. Where these two peaks intersect should be the centre of the rectangle and the direction should be known because of the slope of the line. Figure 2 illustrates this technique. See Chapter 14 in Davies (2005) for further details.

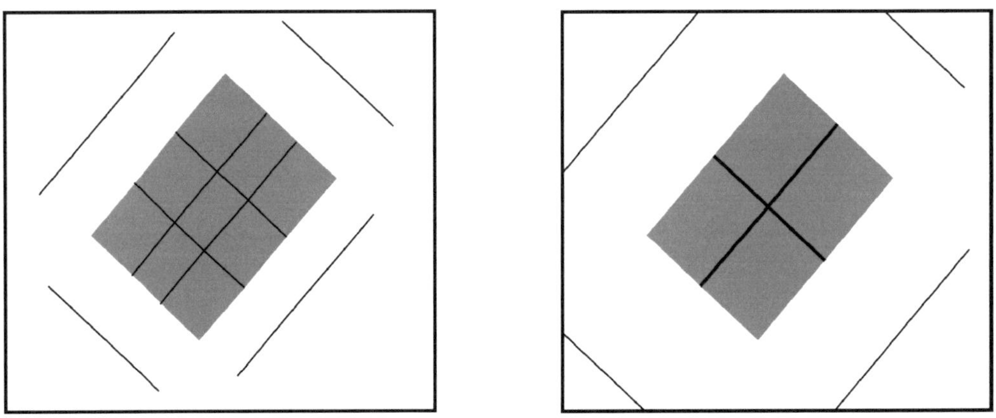

Fig. 1. Illustration of the Hough rectangle method.

2.5.2 PCA colour feature

As an additional source of information, it is possible to apply PCA to the three red, green and blue elements of an image. The result is three new colour components that **can be** reclassified in terms of maximum order of intensity (Lee et al., 2007). The image is represented as a vector of pixels each with red green and blue values. By applying PCA to this vector new colour components can be generated. Applying SVD to the sample correlation matrix will generate eigenvectors and Eigenvalues. These eigenvectors can then be used to translate the image into its new colour components. The Eigenvalues of these new colour components can be used as numerical inputs.

3. Case study

The part in question is a cross car beam, which is the metal support located behind the dashboard of an automobile. See Figure 2. The beam is a stamped metal part and a stamped metal radio bracket is welded to this beam. As part of the assembly process four small rectangular clips must be inserted into the radio bracket. These four clips serve as locations for securing the radio unit to the cross car beam. Figure 3 illustrates the manufacturing cell used to inspect each part.

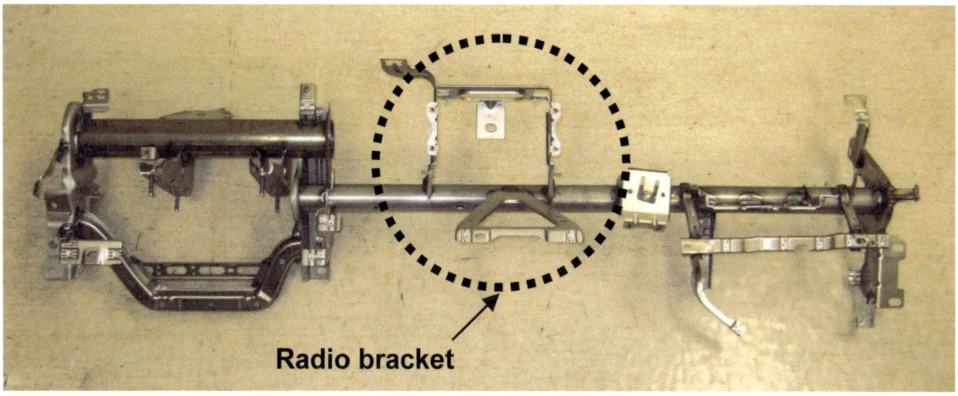

Fig. 2. Cross car beam containing the radio bracket.

This is a safety critical part. It is essential to ensure that these clips are properly installed, because of the consequences that could result from of an unsecure radio unit during a collision. In order to ensure the presence of these clips a machine vision system was developed to automatically inspect each part before it leaves the manufacturing cell that is dedicated to attached the clips. As illustrated in Figure 3, the PLC that controls the manufacturing cell communicates with a standalone PC that controls the vision system.

3.1 Setup

A two camera system was installed on the production assembly line in order to inspect the part. This is shown in the Figure 4(a). The part is pictured in Figure 4(b). The two digital firewire cameras capture 1024 x 768 images. For a lighting solution an LED ring light with a diffuser was chosen. The ring lights are visible on the front of the cameras in Figure 4(a). The decision to use ring lights was made based on the need for a lighting solution that could illuminate the part, but did not have a large footprint due to space restrictions in the cell. It also needed to be easily mountable. Because of this it was not feasible to install an off angle lighting source or to use backlighting. The lights were manually aligned to illuminate the bracket and be centred on the beam.

Fig. 3. PLC controlled manufacturing cell with PC controlled vision sytsem.

Fig. 4. Orientation of (a) cameras looking at bracket and (b) radio bracket in holder.

3.2 Software

QVision a custom software system based on MATLAB® has been used to classify the data. This software is capable of loading library images, selecting and extracting features or regions of interest for Eigenimages, training classifiers through to final inspection of new images. It was integrated to the PLC running the manufacturing cell for online inspection of parts. Figure 5 shows an image of the classification results of the program.

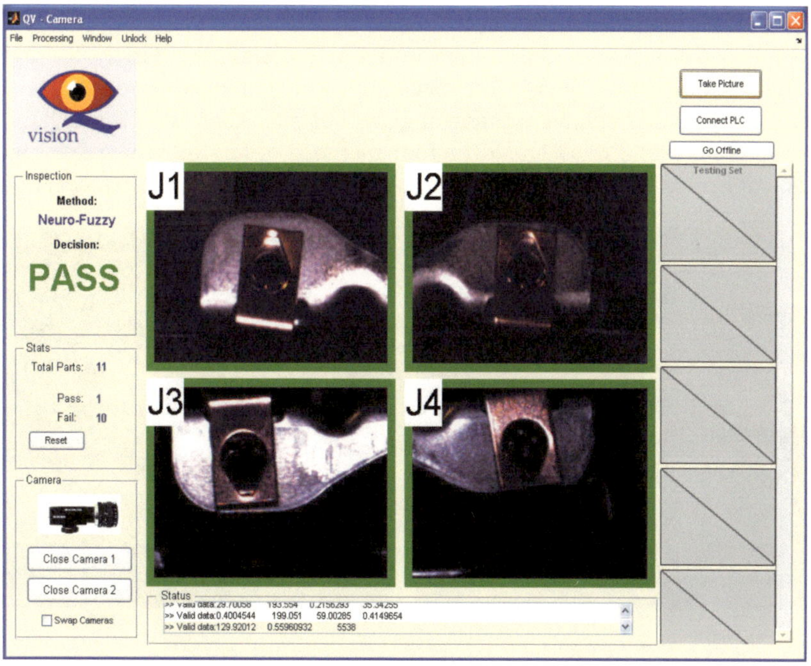

Fig. 5. QVision software GUI used to detect the presence of clips.

3.3 Datasets

Two different data sets were used for comparing the classification techniques. They contain pass images and fail images taken from parts on the assembly line. A pass images consists of the clips present and a fail image consists of the clips missing. Figures 6 and 7 illustrate sample cases of these images. Since there are two cameras looking at the top and bottom clips, there are also top and bottom images.

The first data set which will be referred to as the "original set" consisted of 150 pass images and 114 fail images for the top bracket, and 147 pass images and 125 fail images for the bottom bracket.

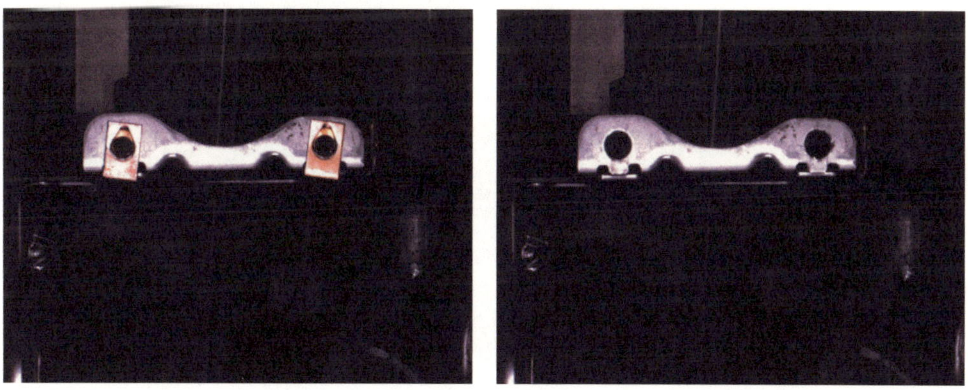

(a) (b)

Fig. 6. Sample "clean" image from manufacturing cell for top bracket: (a) pass and (b) fail.

The second data set (which will be referred to as the orientation set) has been examined to determine the performance of the system under two conditions. Firstly the lighting has changed in this orientation image set. Secondly 3 new failure modes have also been introduced. The clips do not have significant glare, but the background is much more visible in these images. Figures 8 to 11 show the four methods of failure, which are: missing clip, backwards and upside down clip, backwards clips and upside down clip respectively.

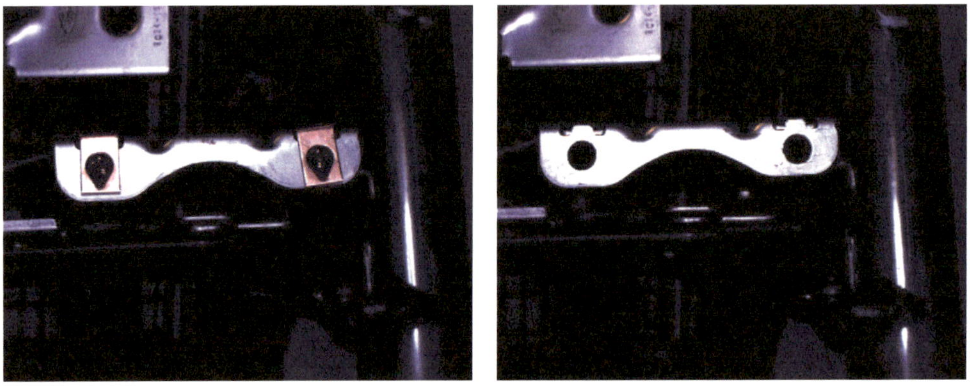

(a) (b)

Fig. 7. Sample "clean" image from cell for bottom bracket: (a) pass and (b) fail.

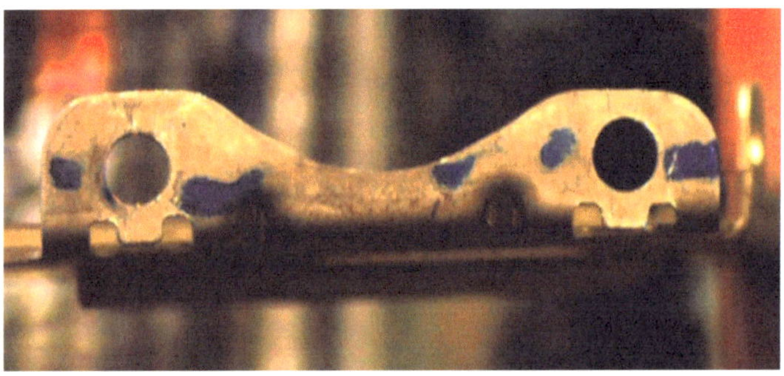

Fig. 8. Sample fail (missing clip) for the orientation image set.

Fig. 9. Sample fail (backwards and upside down clip) for the orientation image set.

Fig. 10. Sample fail (backwards) for the orientation image set.

Fig. 11. Sample fail (upside down) for the orientation image set.

3.4 Performance criteria

The industrial partner has specified a performance goal in terms of false positives (FPs) and false negatives (FNs). An FP is a defective part classified as good and hence shipped to the customer. This is a safety hazard and hence can't be tolerated. There must be no FPs. An FN is a good part classified as bad. The industrial partner has set this at a maximum rate of 2%. It is a measure of the scrap rate. It should be noted that Receiver Operating Characteristic graphs (Fawcett, 2005) can also be generated as a measure of performance.

For the purposes of this study, the root-mean-squared (RMS) error E_{rms} for a set of images is used as a performance measure. The root-mean-square of the output error defined as:

$$E_{rms} = \sqrt{\frac{\sum_{i=1}^{n}(Z_i^d - Z_i)^2}{n}} \quad (3)$$

where, Z_i^d is the desired (correct) classification for the ith image, Z_i is the output of the classifier algorithm (Z_i^{nf} or Z_i^{th}) and n is the total number of images. $Z = 1$ is an unconditional pass (clip present) and $Z = 0$ is an unconditional fail (clip missing).

4. Results

The six classifiers were trained using the original data set. The system was trained on 40% of the original data set and these results were then checked on 20% of the original data set. These images were chosen randomly. A final 40% was reserved for additional testing purposes. The results of this classification are shown in Table 2.

Ranking	Classifier	Clip	False Positives	False Negatives	% False Negatives	RMS Error E_{rms}
1st	Feature based with PCA and NN	Clip 1	0	0	0	0.0001
		Clip 2	0	0	0	0.0115
		Clip 3	0	0	0	0
		Clip 4	0	1	0.4	0.0547
		totals	0	1	(0.1)	0.066
2nd	Feature based with NN	Clip 1	0	0	0	0.0879
		Clip 2	0	2	0.8	0.0899
		Clip 3	0	0	0	0.0710
		Clip 4	0	2	0.8	0.0871
		totals	0	4	(0.4)	0.336
3rd	Eigenimage based with NN	Clip 1	0	7	2.6	0.1853
		Clip 2	0	4	1.5	0.1166
		Clip 3	0	3	1.1	0.1304
		Clip 4	0	1	0.4	0.1309
		totals	0	15	(1.4)	0.563
4th	Feature based with PCA and ANFIS	Clip 1	0	6	2.3	0.0721
		Clip 2	0	0	0	0.1102
		Clip 3	0	11	4.2	0.1748
		Clip 4	0	9	3.4	0.1509
		totals	0	26	(2.5)	0.508
5th	Eigenimage based with ANFIS	Clip 1	2	76	29	0.3932
		Clip 2	6	40	15	0.3319
		Clip 3	0	6	2.3	0.1512
		Clip 4	0	4	1.5	0.1908
		totals	8	126	(12)	1.067
6th	Feature based with ANFIS	Clip 1	1	29	11	0.9587
		Clip 2	2	9	3.4	0.3230
		Clip 3	10	12	4.5	0.5226
		Clip 4	6	6	2.3	0.2805
		totals	19	56	(5.3)	2.085

Table 2. Results with original data set (264 images per clip).

To improve performance a new set of features was used including the PCA colour technique and Rectangular Hough transform technique. These are shown in Table 3. The results of the Eigenimage backpropagation were not included in this table because they are same as those found in Table 2.

The final set of images that was examined was the new orientation set. There are four additional modes of failure and the lighting has changed. These additional modes of failure can be caught by other methods and the industrial partner does not need the vision system to inspect for these defects. However it does present an excellent set of data to test the robustness of the algorithm. As in the previous cases features were defined on the images. However these were not found relative to the centre of the beam because of difficulty finding the beam with the new lighting. The feature-based results are presented in Table 4.

Ranking	Classifier	Clip	False Positives	False Negatives	% False Negatives	RMS Error E_{rms}
1st	Feature based with PCA and NN	Clip 1	0	0	0	0.0047
		Clip 2	0	0	0	0
		Clip 3	0	2	0.8	0.0798
		Clip 4	0	0	0	0.0005
		totals	0	2	(0.2)	0.085
2nd	Feature based with NN	Clip 1	0	0	0	0.0488
		Clip 2	0	1	0.4	0.0647
		Clip 3	0	2	0.8	0.1043
		Clip 4	0	1	0.4	0.0908
		totals	0	4	(0.4)	0.309
3rd	Feature based with PCA and ANFIS	Clip 1	0	2	0.8	0.0621
		Clip 2	0	0	0	0.0688
		Clip 3	0	12	4.5	0.1416
		Clip 4	0	4	1.5	0.1284
		totals	0	18	(1.7)	0.401
4th	Feature based with ANFIS	Clip 1	1	6	2.3	0.5047
		Clip 2	1	3	1.1	0.1874
		Clip 3	0	5	1.9	0.1458
		Clip 4	1	26	10	0.3767
		totals	3	40	(3.8)	1.215

Table 3. Results with original data set and additional features (264 images per clip).

For the feature based results from Table 2 features were extracted from an image. These were a combination of holes, lines, circles and colours. These features are illustrated for Clip 1 in Figure 12.

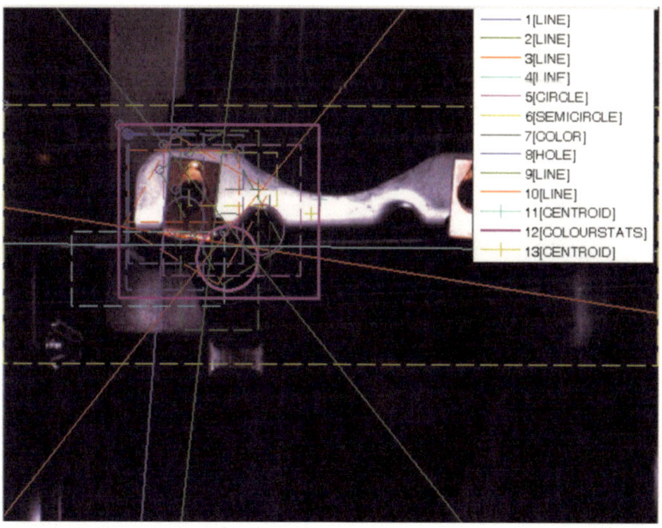

Fig. 12. Features defined on Clip 1 for original data set.

Ranking	Classifier	Clip	False Positives	False Negatives	% False Negatives	RMS Error E_{rms}
1st	Feature based with PCA and NN	Clip 1	0	0	0	0.0047
		Clip 2	0	0	0	0
		Clip 3	0	2	0.7	0.0798
		Clip 4	0	0	0	0.0005
		totals	0	2	(0.15)	0.085
2nd	Feature based with NN	Clip 1	0	0	0	0.0488
		Clip 2	0	1	0.4	0.0647
		Clip 3	0	2	0.7	0.1043
		Clip 4	0	1	0.4	0.0908
		totals	0	4	(0.3)	0.317
3rd	Feature based with PCA and ANFIS	Clip 1	0	2	0.8	0.0621
		Clip 2	0	0	0	0.0688
		Clip 3	0	12	4	0.1416
		Clip 4	0	4	1.5	0.1284
		totals	0	18	(1.6)	0.401
4th	Feature based with ANFIS	Clip 1	1	6	2	0.5047
		Clip 2	1	3	1	0.1874
		Clip 3	0	5	2	0.1458
		Clip 4	1	26	10	0.3767
		totals	3	40	(3.5)	0.710
5th	Eigenimage based with NN	Clip 1	9	0	0	0.1835
		Clip 2	0	0	0	0.0662
		Clip 3	0	0	0	0.0646
		Clip 4	0	0	0	0.0548
		totals	9	0	(0)	0.369
6th	Eigenimage based with ANFIS	Clip 1	12	2	0.9	0.2253
		Clip 2	2	5	2.29	0.1626
		Clip 3	1	1	0.5	0.1228
		Clip 4	1	2	0.9	0.0963
		totals	16	10	(0.9)	0.607

Table 4. Results with orientation data set (286 images per clip).

Again the best results were found with the PCA Neural Network, followed by the Neural Network. The feature-based results did better than the Eigenimage results and the Neural Network results were better than the ANFIS results.

One interesting result is the Eigenimage backpropagation results trained with a Neural Network. If Clip 1 is ignored the results are perfect.

5. Discussion

Six different classifiers were compared to the classification of clips on an automotive assembly. Tests were done on two images sets and with two different feature sets. Consistently it was seen that a Neural Network classifier whether used on feature data, on feature data with PCA applied or on Eigenimage coefficients performed better than the ANFIS system. The NN results always ranked higher on all the tests. It is reasonable to say that the Neural Network performs better than the ANFIS system. It has also been consistently seen that applying PCA on input data improves the results of classification. The results with feature extraction and PCA ranked higher that the results with feature extraction and no PCA on the majority of the tests.

For the original data set the performance of feature based techniques were better than region of interest techniques. In the case of the orientation data set in general the feature based techniques had similar perform to the Eigenimage techniques. However ignoring Clip 1 the Eigenimage Neural Network technique worked better than any other technique on this set of data. This shows that the Eigenimage technique is better able to distinguish multiple types of faults with brighter lighting than a feature based technique.

Applying PCA to a dataset eliminates the need to perform feature selection improving the result in a systematic way. The Eigenimage technique has the benefit of not needing to extract features. A region of interest is selected and the calculations can then proceed. The greatest benefit of these techniques is their speed of training, which makes the system more flexible.

6. Conclusion

The performance of six different classifiers has been compared as applied to the detection of missing fasteners. Traditional feature based classifiers were first used to train Neural Networks (NN) and Neuro-Fuzzy (ANFIS) systems, with and without Principle Component Analysis (PCA). As an alternate, a non-feature based Eigenimage classifier was used to generate the inputs for the classifiers. It was found that when there was only one type of defect, both the NN and Eigenimage based classifiers, but not the ANFIS based classifier, could achieve the required performance. On the other hand, when there was more than one type of defect, only the NN and ANFIS based classifiers could maintain the required level of performance. Finally, given that the Eigenimage based classifier takes much less time to set up and train, it is considered superior to the NN based classifier for practical applications.

7. Acknowledgment

The Authors would like to thank the generous support of AUTO21 and OCE, which has allowed this research to be carried out. They would also like to thank Queen's University for their support in this research.

8. References

Davies, E. R. (2005) *Machine Vision: Theory Algorithms, Practicalities,* 3rd edition, Morgan Kaufmann, New York, NY.

Fawcett, T. (2005) "An introduction to ROC analysis," *Pattern Recognition Letters,* Vol 27, pp. 861-874. Garcia H. C., Villalobos J. R., and Runger, G. C., (2006) "An Automated Feature Selection Method for Visual Inspection Systems", IEEE Transactions on Automation Science and Engineering, Vol. 3, No. 4, pp. 394-406.

Garcia, H. C., Villalobos J. R., and Runger, G. C. (2006) "An automated feature selection method for visual inspection systems", *IEEE Transactions on Automation Science and Engineering,* Vol. 3, No. 4, pp. 394-406.

Gayubo, F., Gonzalez J.L., del la Fuente E., Miguel F. and Peran J. R. (2006) "On-line machine vision systems to detect split defects in sheet-metal forming processes," *Int. Conf. Pattern Recognition (ICPR '06),* Hong Kong, August 20 to 24.

Hunter, J.J., Graham, J., and Taylor, C. J. (1995). "User programmable visual inspection", *Image and Vision Computing,* Vol. 13, No. 8, pp. 623-628.

Jackman, P., Sun D-W., Du, C-J., and Allen, P. (2009) "Prediction of beef eating qualities from colour marbling and wavelet surface texture features using homogenous carcass treatment," *Pattern Recognition,* Vol. 42, pp. 751 - 763.

Killing, J., Surgenor, B.W., and Mechefske, C.K. (2009) "A machine vision system for the detection of missing fasteners on steel stampings", *Int. Jrnl. of Advanced Manufacturing Technology,* Vol. 41, No. 7-8, pp. 808-819.

Kumar, A. (2003) "Neural network based detection of local textile defects", *Pattern Recognition,* Vol. 36, pp. 1645-1659.

Kwak, C., Ventura, A., and Tofang-Szai, K. (2000) "A neural network approach for defect identification and classification on leather fabric", *Jrnl. of Intelligent Manufacturing,* Vol. 11, pp. 485- 499.

Lee, K-M., Li, Q., Daley, W. (2007) "Effects of classification methods on color-based feature detection with food processing applications," *IEEE Transactions on Automation Science and Engineering,* Vol. 4, No. 1, pp. 40-51.

Miles, B.C., and Surgenor, B.W. (2009) "Industrial Experience with a Machine Vision System for the Detection of Missing Clip," Changeable, Agile, Reconfigurable and Virtual Production (CARV 2009), Munich, Germany, October 5-7.

Ohba, K., and Ilkeuchi, K. (1997) "Detectability, Uniqueness, and Reliability of Eigen Windows for Stable Verification of Partially Occluded Object," *IEEE Transactions on Pattern Analysis and Machine Intelligence,* Vol. 19, No. 9, pp. 1043 - 1048.

Reynolds, M.R., Campana, C., and Shetty, D. (2004) "Design of Machine Vision Systems For Improving Solder Paste Inspection", *ASME International Mechanical Engineering Congress and Exposition,* ASME Paper IMECE2004-62133 Anaheim, California, USA, November 13-20.

Roger Jang, J-S. (1993) "ANFIS: Adaptive-Network-Based Fuzzy Inference System", *IEEE Transaction on Systems, Man and Cybernetics,* Vol. 23, No. 3, pp. 665-685.

Sun, J., Sun, Q., Surgenor, B.W. (2007) "Adaptive Visual Inspection for Assembly Line Parts Verification," *International Conference on Intelligent Automation and Robotics (ICIAR),* San Francisco, California, USA, October 24-26.